Arpad Geyze Gerster

**The rules of aseptic and antiseptic surgery; a practical treatise for the use of students and the general practitioner**

Third Edition

Arpad Geyze Gerster

**The rules of aseptic and antiseptic surgery; a practical treatise for the use of students and the general practitioner**
*Third Edition*

ISBN/EAN: 9783337157166

Printed in Europe, USA, Canada, Australia, Japan

Cover: Foto ©berggeist007 / pixelio.de

More available books at **www.hansebooks.com**

# THE RULES

OF

# ASEPTIC AND ANTISEPTIC SURGERY

A PRACTICAL TREATISE FOR THE USE OF STUDENTS
AND THE GENERAL PRACTITIONER

BY

ARPAD G. GERSTER, M. D.

PROFESSOR OF SURGERY AT THE NEW YORK POLYCLINIC; VISITING SURGEON TO MOUNT SINAI HOSPITAL
AND THE GERMAN HOSPITAL, NEW YORK

ILLUSTRATED WITH TWO HUNDRED AND FIFTY-TWO ENGRAVINGS
AND THREE CHROMO-LITHOGRAPHIC PLATES

THIRD EDITION.

NEW YORK
D. APPLETON AND COMPANY
1893

# PREFACE TO THE THIRD EDITION.

The necessity for issuing within two years a third edition of this work may be safely assumed as a sign of the spread of antiseptic doctrine and practice among the members of the medical profession in this country.

The general outlines and scope of the book, being based upon the course of lectures yearly delivered by the author to a body consisting of practicing physicians, have retained their practical character. Principal accentuation was placed upon the points showing important divergence from older methods.

Additional new matter was introduced in the chapters on Herniotomy, Hæmorrhoids, Appendicitis, and the Surgery of the Kidney. As in the former editions, statistical material was brought in only when the typical and uniform character of the operations pertaining to one subject permitted its safe use as a gauge of the value of aseptic or antiseptic methods. The additional experience of two years' work was utilized in widening the basis of the conclusions thus drawn from the computation of numbers.

The casuistic material, all original and carefully recorded, forms, in the opinion of the author, the most valuable part of the work. Its quantity will be found materially increased in the chapters that have been rewritten, or newly introduced.

56 EAST TWENTY-FIFTH STREET,
    NEW YORK, *September 8, 1890.*

# PREFACE TO THE FIRST EDITION.

The object of this volume is a systematic yet practical presentation of the Listerian principle that has revolutionized surgery within the last fifteen years. Its adoption has wrought so many incisive changes in practice, has shifted the surgeon's standpoint regarding all the important disciplines of the art in such a radical manner, that most English text-books of surgery, even those recently published, have become partly or entirely inadequate to the wants of the modern physician.

To a large number of medical men the aseptic and antiseptic methods present an incongruous chaos of seemingly contradictory and often incomprehensible detail, arbitrary and varying, according to the predilections or whims of this or that teacher.

Yet the principle involved is based on the correct observation of a common biological process—namely, that of the decomposition of organic substances. The well-known methods employed since the earliest dawn of civilization for the preservation of organic, especially animal, substances, are based upon the empirical yet correct appreciation of the causes of putrefaction, and the practical adaptation of these methods to the healing of operative or accidental wounds contains the whole essence of the new surgery.

Evils that former generations of surgeons deplored, but could not effectually combat, such as septicæmia, pyæmia, hospital gangrene, and erysipelas, have been much abated, as a direct consequence of a clear understanding of their essential nature and causation.

*Prevention* has become the watchword of modern practice, and it can be said that, by the successful employment of the preventive methods of the present day, surgery has become a *conservative* branch of the healing art.

The elimination of the accidental disturbances of repair caused by wound infection has depressed the percentage of mortality following amputation of the extremities from an average of thirty-five per cent to about fifteen per cent.

The dread of undertaking and submitting to a surgical operation has greatly diminished, and timely—that is, early—surgical interference has become more and more frequent, to the great advantage of both patient and physician.

As a direct consequence of the implied obligation of rendering timely aid where possible, a laudable eagerness for an early diagnosis is developed, and, there being so much to be gained by diagnostic knowledge, thorough and practical study of the morbid processes requiring surgical aid has been greatly stimulated.

The fear of suppuration with its dreadful consequences does not stay now the hand of the surgeon as of old, when an operation was always considered a forlorn hope and a last resort. Strangulated herniæ, for instance, are not allowed to gangrene as often as formerly, and herniotomy is readily resorted to, as it is well known that the dangers of an aseptic herniotomy done on a healthy gut are diminutive in comparison to the certain and enormous danger of strangulation itself.

By the conviction that a fault of omission may be followed by irremediable mischief, the sense of responsibility is stirred up to vigilance, which again breeds self-reliance and firmness of purpose in advising and carrying out incisive measures, made clearly necessary by a well-recognized danger to life or limb. And an additional degree of responsibility is imposed by the very safety of aseptic operations.

It can not now be successfully denied that *the surgeon's acts determine the fate of a fresh wound, and that its infection and suppuration are due to his technical faults of omission or commission.*

The principle underlying antiseptic surgery has ceased to be the subject of serious controversy. The author does not undertake to prove each of his statements to the satisfaction of those who look but see not. His object is instruction rather than controversy. Every one will have to pass his period of apprenticeship with its blunders and lessons. But he who becomes a master, to whom the primary healing of a fresh wound remains not a curiosity but becomes a matter of course, will not doubt the great change that has come over surgery.

The purely practical tendency of the work made a rather free arrangement of the several parts of the subject-matter a necessity, or at least a convenience; yet a sufficiency of systematic order was preserved to give the collection of papers the character of a well-rounded, organic whole.

The author begs to state explicitly that completeness—that is, the inclusion of all the disciplines of surgery—was not aimed at, else a complete text-book of surgery would have resulted. The leading idea, traceable through all the matter contained in the book, is to illustrate the incisive practical changes that the adoption of aseptic and antiseptic methods has wrought in surgical therapy. Hereby the changes in wound treatment are meant, as well as the notable extension of active surgery into fields formerly considered a *noli me tangere*.

As a consequence of the stupendous growth of operative surgery within the last decade, a fruitful development of operative technique is to be noted also. In accordance with the desire of the author to present to the profession a vivid and true picture of contemporaneous methods, the terms used as the title of this work should be accepted in their widest significance.

Confinement to the meager details of those manipulations which, strictly speaking, constitute aseptic and antiseptic measures, would have yielded an inadequate and tedious compilation. On the other hand, it is hoped that the pathological and technical diversions, introduced for the sake of laying a rational foundation to the principles composing the essence of *antiparasitic* surgery, may be admitted as germane to the subject.

The methods of wound treatment herein explained are to a certain extent still undergoing changes, hence should not be accepted as final. Yet it is undeniable that, as the clearness of the comprehension of the simple *principle* of asepticism applied to wound treatment has advanced, so the frequent changes and bewildering vacillation characteristic of the experimental stage of the new discipline have naturally given way to steadier methods. At present, changes are not so frequent as formerly, yet progress, especially the conquest of new fields for the legitimate practice of active surgery, is not at a standstill.

The author is well aware that the practical directions recommended by him are not the only ones that lead to success. Yet, in the main, he has refrained from quoting other authorities. As reasons for this may be adduced, first, the disinclination to write a bulky text-book, and, further, the knowledge that the interest of the reader is proportionate to the directness and immediate character of the facts and thoughts contained in the work under perusal.

As far as possible, all important statements will be found borne out by illustrative examples taken from the author's personal experience.

The author is much indebted to the gentlemen composing the house staffs of the German and Mount Sinai Hospitals for the ready kindness and courtesy with which their help was proffered in tracing and extracting histories of cases, and in making the very numerous photographic plates that form the bulk of the illustrations.

Great technical difficulties, inherent to the unfavorable season, the small space and inadequate lighting of the operating-rooms of the mentioned hospitals, had to be overcome in exposing the sensitive plates. The matter was rendered still more difficult by the circumstance that operating and photographing were done by one and the same set of persons, and that the welfare and interests of the patients themselves had constantly to be sedulously considered.

In view of the defective character of many of the author's negatives, the greatest praise belongs to Mr. William Kurtz, to whose artistic taste, skill, and versatility is due their excellent reproduction by phototypographic process.

Proper credit is given for the lithographic plates copied from Rosenbach, for the excellent microphotographs reproduced from Koch's classical reports, and for a few other illustrations borrowed from Esmarch, Henke, and Bumm.

In conclusion, the author may be permitted to express the hope that, by publishing his share of experience gathered from a modest public and private practice, he may succeed to somewhat propagate and popularize the principles and practice of antiparasitic surgery.

NEW YORK, *September 3, 1887.*

# CONTENTS.

## Part I.—ASEPSIS.

### CHAPTER I.

| | PAGE |
|---|---|
| What are Sepsis and Asepsis? | 3 |

### CHAPTER II.

| | PAGE |
|---|---|
| Aseptic Wounds—Aseptic Treatment | 5 |
| I. General remarks | 5 |
| II. Rules of surgical cleanliness | 7 |
|     1. Hands | 7 |
|     2. The instruments | 7 |
|     3. Wound irrigation | 7 |
|     4. Sponges | 8 |
|     5. Materials for ligatures and sutures | 8 |
|     6. Drainage-tubes and elastic ligatures | 9 |
|     7. Disinfecting lotions | 10 |
|     8. Dressings | 11 |
|         (1) Types of dressings | 11 |
|             *a*. Simple exsiccation. Bismuth, iodoform | 11 |
|             *b*. Chemical sterilization combined with exsiccation. Dry dressings | 12 |
|             *c*. Schede's modification of the **dry** dressing, favoring the organization of the moist blood-clot | 12 |
|             *d*. Simple chemical sterilization. **Moist dressings** | 13 |
|         (2) Preparation of dressings | 14 |
|             *a*. Gauze | 14 |
|                 (*a*) Corrosive-sublimate **gauze** | 15 |
|                 (*b*) Iodoformized gauze | 15 |
|             *b*. Absorbent cotton, **or common cotton batting** | 15 |
|             *c*. Sawdust | 16 |
|             *d*. Moss | 17 |
| III. Practical application of rules | 17 |
|     1. In operating | 17 |
|     2. Change of dressings | 20 |
| IV. Aseptic measures in emergencies | 23 |
|     Operating bag and kit | 25 |

### CHAPTER III.

| | PAGE |
|---|---|
| Soiled Wounds.—Antiseptic Treatment.—Difference between Aseptic and Antiseptic Methods.—Illustration of Antiseptic Method | 27 |

## CHAPTER IV.

| | PAGE |
|---|---|
| SPECIAL RULES REGARDING THE TREATMENT OF ACCIDENTAL WOUNDS | 29 |
|    I. Temporary measures | 29 |
|   II. Definitive relief | 31 |
|       1. Contaminated wounds | 31 |
|       2. Aseptic wounds | 34 |
|       3. Gunshot wounds | 35 |

## CHAPTER V.

| | |
|---|---|
| SPECIAL APPLICATION OF THE ASEPTIC METHOD | 36 |
| *A*. General principles | 36 |
|    I. Technique of surgical dissection | 36 |
|   II. Sutures | 44 |
|  III. Drainage | 47 |
| *B*. Application of aseptic method to diverse organs and regions | 48 |
|    I. Ligatures of arteries in their continuity | 48 |
|   II. Extirpation of tumors | 52 |
|       Preservation of asepsis | 52 |
|       Safe removal | 52 |
|       Complete removal | 52 |
|  III. Amputation of limbs | 61 |
|      1. Aseptics and antiseptics of amputation | 61 |
|        *a*. Clean cases | 64 |
|        *b*. Mildly septic cases | 66 |
|        *c*. Septic cases of greater intensity | 67 |
|      2. Hæmorrhage | 69 |
|        *a*. Artificial anæmia | 69 |
|        *b*. Ligatures and final hæmostasis | 72 |
|      3. Securing of a good stump | 74 |
|  IV. Operations about non-suppurating joints | 76 |
|      1. Puncture and irrigation | 76 |
|      2. Arthrotomy | 78 |
|        *a*. Hydrops genu | 78 |
|        *b*. Vegetations | 79 |
|        *c*. Floating bodies of the knee-joint | 80 |
|        *d*. Suturing of the fractured patella | 80 |
|      3. Arthrotomy for irreducible or habitual dislocation, and for deformity due to fracture | 82 |
|   V. Operations for deformities | 86 |
|      1. Knock-knee and bow-leg | 86 |
|      2. Bony anchylosis in a vicious position | 87 |
|      3. Deformed callus | 88 |
|      4. Club-foot and pes valgus | 88 |
|  VI. Plastic operations | 91 |
| VII. Aseptics of the oral cavity | 96 |
| VIII. Laryngeal operations | 100 |
|      1. Tracheotomy | 100 |
|        *a*. Superior tracheotomy | 102 |
|        *b*. Inferior tracheotomy | 103 |
|      2. Laryngofissure | 107 |
|      3. Extirpation of the larynx | 108 |

## CONTENTS.

|   |   | PAGE |
|---|---|---|
| IX. Goitre | | 111 |
| X. Amputation of the breast | | 115 |
| XI. Abdominal operations | | 119 |
|   1. General remarks | | 119 |
|   2. Herniotomy | | 121 |
|     *a.* Herniotomy for strangulation | | 123 |
|     *b.* Radical operation for hernia | | 133 |
|   3. Laparotomy | | 139 |
|     *a.* Exploratory incision | | 139 |
|     *b.* Abdominal tumors | | |
|       (*a*) General remarks | | 140 |
|       (*b*) Special observations | | 147 |
|         ($\alpha$) Ovarian tumors | | 147 |
|         ($\beta$) Removal of uterine appendages | | 150 |
|         ($\gamma$) Supra-vaginal hysterectomy | | 151 |
|         ($\delta$) Nephrectomy | | 153 |
|     *c.* Gastrostomy | | 154 |
|     *d.* Colotomy | | 155 |
|       (*a*) Lumbar colotomy | | 156 |
|       (*b*) Inguinal colotomy | | 156 |
|       (*c*) Excision and suture of gut (enterorrhaphy) | | 158 |
| XII. Hydrocele, varicocele, and castration | | 163 |
|   1. Hydrops of the tunica vaginalis | | 163 |
|   2. Varicocele | | 164 |
|   3. Castration | | 165 |
| XIII. Aseptic operations on the rectum | | 167 |
|   1. General observations | | 167 |
|   2. Hæmorrhoids | | 167 |
|   3. Rectal tumors | | 171 |
| XIV. Aseptics of the bladder | | 173 |
|   1. Catheterism | | 173 |
|   2. Litholapaxy | | 175 |
|   3. Cystotomy | | 177 |
|     *a.* Perineal section | | 177 |
|     *b.* Suprapubic section | | 177 |

## Part II.—ANTISEPSIS.

### CHAPTER VI.

| | |
|---|---|
| NATURAL HISTORY OF IDIOPATHIC SUPPURATION.—TREATMENT OF SUPPURATION | 183 |
|   I. The cause of suppuration, or phlegmon | 183 |
|   II. Portals of infection | 185 |
|     1. Infection through lesions of the skin | 185 |
|     2. Infection through lesions of the mucous membranes | 186 |
|   III. Entrance, progress, and localization of the infection | 187 |
|     Mechanical irritation | 189 |
|     Chemical and caloric irritation | 190 |
|   IV. Development of phlegmon | 191 |

## CONTENTS.

|   | PAGE |
|---|---|
| V. Spread of suppuration | 193 |
| VI. Diagnosis and treatment of phlegmon | 198 |
|    1. General principles | 198 |
|       *a.* Superficial suppuration, or septic ulcer | 199 |
|       *b.* Cutaneous and subcutaneous phlegmon | 199 |
|       *c.* Deep-seated or subfascial phlegmon. Lymph-gland abscess | 203 |
|       *d.* Acute infectious osteomyelitis | 205 |
|       *e.* Chronic suppuration due to bone necrosis. Necrotomy | 208 |
|    2. Phlegmonous affections of some special regions | 222 |
|       *a.* Face. Floor of the mouth. Neck. Temporal and mastoid regions | 222 |
|          (*a*) Face | 223 |
|          (*b*) Neck | 225 |
|             (α) Fauces and pharynx | 225 |
|             (β) Submaxillary and parotid cynanche | 231 |
|             (γ) Acute glandular abscesses of the anterior and lateral cervical regions | 234 |
|             (δ) Glandular abscesses of the temporal, mastoid, and occipital regions | 235 |
|       *b.* Mammary and retro-mammary abscess | 237 |
|       *c.* Empyema | 240 |
|       *d.* Phlegmon of the palmar aspect of the hand, of the arm, and axilla | 244 |
|       *e.* Suppurative affections of the lower extremity | 253 |
|          (*a*) Ingrown toe-nail | 253 |
|          (*b*) Chronic ulcers of the leg | 255 |
|          (*c*) Acute suppuration of the prepatellary bursa | 256 |
|          (*d*) Acute suppuration of the knee-joint | 256 |
|          (*e*) Suppuration of the inguinal glands | 259 |
|       *f.* Perityphlitic abscess | 260 |
|          *a.* Acute appendicitis (without tumor) | 263 |
|             (*a*) Simple appendicitis (no tumor) | 263 |
|             (*b*) Perforative appendicitis (no tumor) | 264 |
|          *b.* Acute appendicitis with tumor; perityphlitic abscess | 265 |
|             Types of acute perityphlitic abscess | 266 |
|               1. Ilio-inguinal type (Willard Parker's abscess) | 267 |
|               2. Anterior parietal type | 268 |
|               3. Posterior parietal type | 270 |
|               4. Rectal type | 271 |
|               5. Mesocœliac type | 271 |
|          *c.* Chronic or relapsing appendicitis and perityphlitic abscess | 272 |
|       *g.* Abscess of the liver | 276 |
|       *h.* Lumbar abscesses | 276 |
|       *i.* Pyonephrosis, renal abscess, and calculous kidney | 279 |
|          (*a*) Nephrotomy | 279 |
|          (*b*) Nephrectomy | 281 |
|       *k.* Anal abscess. Fistula in ano | 284 |

### CHAPTER VII.

ERYSIPELAS AND PSEUDO-ERYSIPELAS . . . . . . . 289

## CONTENTS.

### Part III.—TUBERCULOSIS:

### ITS ASEPTIC AND ANTISEPTIC TREATMENT.

#### CHAPTER VIII.

|  | PAGE |
|---|---|
| NATURAL HISTORY AND TREATMENT OF TUBERCULOSIS | 293 |
| I. Etiology of tuberculosis. Tubercle bacillus | 293 |
| II. Complication of tuberculosis with pyogenic or suppurative infection | 297 |
| III. Treatment of tuberculosis | 297 |
|     General principles | 297 |
|     Local treatment of tuberculosis | 298 |
|     1. Cutaneous tuberculosis. Lupus | 298 |
|     2. Tuberculosis of the mucous membranes | 299 |
|     3. Tuberculosis of the lymphatic glands, or scrofula | 299 |
|     4. Tuberculosis of tendinous sheaths | 301 |
|     5. Tuberculosis of bone. Caries. Cold abscess | 303 |
|     6. Tuberculosis of joints. White swelling | 305 |
|         General part | 305 |
|             a. Technique of joint exsection | 305 |
|                 (a) Septic injection from without | 305 |
|                 (b) Complete removal of tuberculous tissues | 306 |
|                 (c) Control of hæmorrhage | 306 |
|                 (d) Preservation of function | 306 |
|            b. After-treatment | 307 |
|         Special part | 308 |
|             a. Shoulder-joint | 308 |
|             b. Elbow | 310 |
|             c. Wrist and hand | 314 |
|             d. Hip-joint | 315 |
|             e. Knee-joint | 319 |
|             f. Ankle and foot | 325 |

### Part IV.—GONORRHŒA:

### ITS ANTISEPTIC TREATMENT.

#### CHAPTER IX.

|  | |
|---|---|
| NATURAL HISTORY AND TREATMENT OF GONORRHŒA | 331 |
| I. Etiology of gonorrhœa. Gonococcus | 331 |
| II. Treatment of gonorrhœa | 333 |
|     1. Acute gonorrhœa. Clap | 333 |
|         a. Anterior gonorrhœal urethritis | 334 |
|         b. Deep-seated gonorrhœal urethritis | 336 |
|     2. Chronic gonorrhœa. Gleet | 339 |
|         a. Inflammatory stenosis (incipient stricture) and permanent or **cicatricial** stricture of the urethra | 339 |
|             (a) Anterior urethra | 339 |
|             (b) Deep urethral **strictures** | 343 |

## CONTENTS.

|  | PAGE |
|---|---|
| *b.* Vegetations of the urethra | 348 |
| *c.* Granular urethritis | 348 |
| *d.* Chronic catarrh of the posterior part of the urethra, and chronic cystitis | 348 |

## PART V.—SYPHILIS:

### ASEPTIC AND ANTISEPTIC TREATMENT OF ITS EXTERNAL LESIONS.

#### CHAPTER X.

| | |
|---|---|
| ASEPTICS AND ANTISEPTICS APPLIED TO EXTERNAL SYPHILITIC LESIONS | 353 |
| 1. Aseptic treatment of primary induration | 353 |
| 2. Antiseptic treatment of the primary syphilitic ulcer | 356 |
|    *a.* Chemical sterilization and surface-drainage by medicated moist dressings | 356 |
|    *b.* Chemical sterilization by strong caustics | 357 |
|    *c.* Sterilization by the actual cautery | 358 |

# PART I.

# ASEPSIS.

## CHAPTER I.

### *WHAT ARE SEPSIS AND ASEPSIS?*

It is not intended here to enter into an exhaustive exposition of the essence of suppuration and the whole complex of conditions known under the name of sepsis. It may suffice for the present to give a rough outline of the views that prevail regarding the causation of the conditions in question.

Albuminoid substances, such, for instance, as blood or blood-serum—in fact, all the tissues of the dead animal body—will become putrid under certain well-known conditions. These are, first, *moisture;* secondly, a certain temperature called *warmth*, for short; and, thirdly, the presence of living organisms, or fungi, named schizomycetes, better known under the name of *bacteria* and *micrococci*. If all these factors are present, the animal substance in question will ferment or putrefy. Absence of any one of these factors will be sufficient to prevent decomposition. To illustrate this proposition, we shall mention common facts. Fresh meat or fish, well dried, can be indefinitely preserved ; freezing and, to a certain extent, roasting will also prevent its spoiling ; and, lastly, exclusion of micro-organisms by air-tight packing or sealing, after boiling, will insure preservation for an indefinite length of time.

The active agents of decomposition are the micro-organisms, which will develop at once their disintegrating activity as the conditions favorable to their development (moisture and a certain temperature) are present.

We then either thoroughly dry the substance to be preserved or produce and preserve a very low or very high temperature in it, all of which will prevent the development of fungi. Exclusion of the fungi is herein unnecessary. The third mode of preservation is that employed in canning meats. They are first boiled thoroughly, then the vessel wherein this boiling was done is hermetically sealed while the substance is still very hot. Here we have a combination of first destroying the vitality of such fungi as are contained in the meat before boiling, and, secondly, exclusion of access of new micro-organisms to the sterilized substance.

NOTE.—The most effective sterilizer is the actual cautery. It not only destroys all the noxious germs contained within the tissues, but at the same time provides these with an often dry and always hermetic seal against further infection. If the eschar and its vicinity be well dusted with iodoform powder, it will often happen that complete cicatrization will take place beneath its protection, even before the detachment of the eschar.

An accidental or surgical wound presents conditions that are eminently favorable for the development of the fungi in question. The oozing blood and lymph, the bruised and dead cells of the various exposed tissues, furnish, severed from their natural connections, the moist pabulum of a proper temperature. The myriads of particles of filth or dust, filling the air in all inhabited localities, contain, according to indubitable evidence, a very large proportion of spores or seeds that, on falling upon the wound and its secretions, promptly develop into fungi, and at once set up a fermentative process known as decomposition.

The products of this fermentation are more or less highly poisonous substances—Bergmann's sepsin, or the ptomaïnes of the French authors. They promptly set up local changes in the shape of inflammation, and cause systemic trouble—that is, septic fever.

It is further necessary for us to know that in septic processes of a wound not only the ptomaïnes are absorbed by the lymphatics, but that often an actual invasion of the living tissues by the fungi will take place, and that the lymphatics and veins will also serve as channels for the importation of dangerous quantities of fungi into the circulation. Secondary deposits, *metastases*, will then easily occur.

Clinical observers properly distinguish between different, *more or less intense forms of septic infection*, in which bacteriology, however, does not always demonstrate correspondingly different forms of fungi. On the other hand, it is known that impoverished nutrition, but especially a certain morbid state, namely, diabetes mellitus, presents an extremely favorable condition for the development of bacterial sepsis.

Regarding syphilis and tuberculosis, this can not be said, as it is not difficult in these states to prevent suppuration of accidental or surgical wounds.

CASE.—In 1879 the author removed from the lumbar region of a young brewer a good-sized lipoma. His skin was covered at the time with a recent syphilitic roseola following a chancre. Under ordinary antiseptic precautions prompt union by the first intention followed, although the treatment was altogether ambulatory, the patient having been operated on and treated throughout at the German Dispensary.

Prompt primary healing of the wounds caused by the extirpation of syphilitic buboes is a rather common experience in the syphilitic ward of the German Hospital.

The excellent results obtained after exsections of tuberculous joints are also proof positive of the assertion that tuberculosis in itself does not dispose to suppuration and sepsis, and that prevention of septic processes in the wounds of the victims of tuberculosis is not difficult.

Diabetes mellitus, however, does undoubtedly heighten the disposition to septic conditions. Ordinary antiseptic precautions often fail to prevent suppuration; hence, an injury, or the necessity of a bloody operation in a diabetic, should never be treated lightly.

It is the immortal achievement of Lister to have first attributed to fermentative influences the disturbances of repair, and to have led wound-treatment into a rational, hence successful, direction.

*Modern wound-treatment is based entirely on the old and well-known principles of the preservation of organic substances.* Of the several modes of preservation, freezing is the only one that is inapplicable in human surgery. Exsiccation, however, and burning with the actual cautery (roasting); then chemical sterilization by germicides, and the combination of chemical sterilization with exsiccation, contain the essence of aseptic surgery. They insure wounds against decomposition, and are a secure preventive of suppuration.

## CHAPTER II.

### ASEPTIC WOUNDS—ASEPTIC TREATMENT.

#### I. GENERAL REMARKS.

SUPPOSING that the skin in the region to be operated on be shaved, then energetically scrubbed in hot water with soap and a clean brush for five minutes, then the surgeon's hands be scrubbed, likewise his knife, and now an incision be made through the skin; supposing that this happen in an atmosphere free from particles of dry filth called dust: such a wound could be safely termed a clean or aseptic one. All particles of filth adhering to skin, hands, and instrument were removed by this simple process of scrubbing, and no new particles could settle down out of the atmosphere, which we assumed to be free from dust.

Experience has taught that such a wound, however large, will heal without suppuration, first, if its edges be approximated by sutures made with a clean needle and clean wire, silk, or gut; and, secondly, if the immunity from an invasion of filth be maintained until the bloody serum marking the line of union become dry.

But we can vary our experiment, and show that a wound can heal without suppuration even if contact of the walls of the same be imperfect or none.

CASE.—Mrs. J. B., aged forty-nine; branchial cyst of the submaxillary region of the size of an orange. Had been punctured a number of times. *Oct. 7, 1882.*—Incision of six inches in length; difficult extirpation. The large vessels of the neck were freely exposed, a considerable affluent of the deep jugular vein was deligated. Catgut used was rather brittle. Suture and drainage of the large wound. Antiseptic dressings. Immediately after the operation patient had a severe coughing spell. *Oct. 12.*—On changing the dressings it was found that the interior of the wound was distended by a massive blood-clot, giving an appearance as though the tumor had not been removed at all.

Sanguinolent serum was discharging from the drainage-tube. Dressings renewed. *Oct. 16.*—Tumor much diminished in size. Drainage-tube removed. *Oct. 20.*— Wound firmly healed; outline of neck normal. Throughout, normal temperatures.

Here we see that undoubtedly secondary venous hæmorrhage had taken place into the large cavity of the wound. The distention did not reach a sufficient degree to produce a rupture of the line of sutures. The enormous clot was rapidly absorbed, and the wound healed without suppuration, though not by primary adhesion. If the wound had not been aseptic, putrefaction of the clot and dangerous septic processes would have inevitably followed.

Still more curious is the course of an aseptic wound that is not united at all, but is left gaping, provided that suitable means are employed to preserve its aseptic character.

CASE.—Mrs. C. T., aged forty-three, came from Ohio to have a syphilitic defect of the nose repaired. Total rhinoplasty, Sept. 18, 1883, at Mount Sinai Hospital. A suitable flap containing the periosteum was raised from the forehead. The edges of the frontal wound could not be drawn together, therefore a properly shaped, well-disinfected piece of rubber tissue was laid on it, and this was covered with an iodoform dressing. *Sept. 23.*—Stitches removed from nasal sutures. Dressing on forehead dry, therefore it was left undisturbed. *Oct. 1.*—Dressing of frontal wound being removed, the rubber-tissue covering became visible; after this was taken away the edges of the wound were found to be cicatrized to the width of half an inch on both sides. A moist, fresh-looking remnant of the blood-clot was still covering a strip of the middle of the wound. No suppuration whatever. Dressings renewed. *Oct. 6.*—Entire wound cicatrized with the exception of a spot as large as a penny at the upper end. *Oct. 10.*—Discharged cured.

Here, then, is an example of the now commonly observed fact that a gaping defect will cicatrize over without suppuration if putrefactive changes be excluded from the clot filling up the gap. This observation involves a radical difference from the old tenet that whatever wound does not heal by primary adhesion must heal by suppuration. A third possibility has become demonstrable, for which older pathology had no explanation.

It is necessary to state that in both of the latter examples the condition of a dustless atmosphere during the time of the operation was not present; the operations were done in ordinary rooms, openly communicating with the dusty streets of New York, yet the behavior of the wounds was perfectly correct.

The extreme difficulty of preparing and maintaining a dustless atmosphere in a room of an inhabited locality is well known to everybody, and, as a matter of fact, the general practitioner must and will have to do his surgery in more or less dusty rooms. Since the procurement of this condition is practically unattainable, frequent irrigation or rinsing of the wound becomes a necessity. But even a constant and powerful stream of fluids will not be able to dislodge all the particles of dust that may have settled down upon and insinuated themselves into the nooks and crevices

## ASEPTIC WOUNDS—ASEPTIC TREATMENT.

of a wound. Hence it is desirable to employ a liquid that, aside from its non-irritant quality, will have the property of extinguishing the noxious effects of those particles of dust that can not be washed away by the irrigation, but remain imbedded in the tissues. *This is chemical sterilization.*

Different disinfecting solutions are used for this purpose to answer various requirements. Their composition and uses will be mentioned hereafter.

NOTE.—Kümmel, of Hamburg, has shown that a dustless operating-room can be had in a well-appointed hospital, and Neuber, of Kiel, has excellent results from operations done in such a dustless room, with well-cleansed hands, apparatus, and instruments, *without the employment of antiseptic fluids.* Even the dressings used are not impregnated with any antiseptic chemical, but are merely "sterilized" by being exposed to dry heat. No sponges are used, all blood being removed with a sterilized solution of common salt (6 : 1,000), which is absolutely unirritating, and certainly forms the most gentle manner of cleansing a wound.

### II. RULES OF SURGICAL CLEANLINESS.

1. Hands.—The hands and forearms, *especially the finger-nails,* of the surgeon and his assistants should be well scrubbed in hot water with soap and brush for five minutes; likewise the region of the body of the patient to be operated on after carefully shaving off the hair. After this follows an immersion of the hands in alcohol, and then in corrosive sublimate lotion (1 : 1,000) for one minute.

NOTE 1.—Kümmel's recommendation of green soap (potash or soft soap) is excellent, on account of its great solvent properties.

NOTE 2.—Rings, especially those having stone settings, should never be worn by the surgeon or his aids in an operation. Bangles and bracelets of female nurses should not be tolerated. Every one's arms should be bared and scrubbed to the elbows.

2. The instruments should be subjected to a careful and minute cleansing with soap and brush, especial care being taken to remove dry particles of blood, pus, etc., from the grooves and behind the clasps of the more composite instruments, which ought to be taken apart each time for cleansing. Hollow instruments (trocars), or those that can not be taken apart, should be boiled in water for thirty minutes. They should be immersed for ten minutes in a three-per-cent solution of carbolic acid before use.

NOTE.—The surgeon should learn to get along with as few instruments as possible. In selecting instruments, preference should be given to the most simple. The best instruments are those having smooth and well-polished surfaces; grooved or roughened handles are hard to clean, and unnecessary.

3. Wound Irrigation.—*During the operation* the wound should be frequently irrigated with the proper kind of a disinfecting fluid; the hands of the surgeon and his assistants should be also washed at not too long intervals in a disinfecting fluid (corrosive sublimate, 1 : 1,000); the instruments should be kept immersed in a three-per-cent solution of carbolic acid (which is the least injurious to them). The most convenient form of irrigator is the well-known "fountain syringe" of vulcanized rubber.

NOTE.—Whenever any one of those engaged at an operation touches a not disinfected object —hands a chair, opens the window or door, helps the anæsthetizer during a vomiting spell of the patient, scratches his face, or wipes his nose—*it is absolutely necessary* that his hands be *scrubbed and disinfected anew*. Instruments that are accidentally dropped should be left untouched. Raw assistants, and *especially nurses*, male and female, trained or untrained, should be earnestly instructed beforehand, and *constantly watched* afterward, regarding this all-important discipline.

4. **Sponges** should be beaten free from calcareous particles, then immersed for fifteen minutes in dilute muriatic acid to dissolve the remnant of lime, washed in cold water, then thoroughly kneaded by hand with green soap in hot water for five minutes, rinsed, and then immersed in a five-per-cent solution of carbolic acid, in which they remain until required for use. Sponges used once in an aseptic operation can be used again. Careful washing out with green soap and hot water of all the remnants of fibrin and blood, then immersion in a five-per-cent solution of carbolic acid, are sufficient. It is not good to use too many sponges at an operation. When saturated with blood at an operation, they should be washed free from it in *tepid water*, then thrown into a basin filled with carbolic solution, and hence handed to the surgeon. Carbolic acid is preferable for preservation of sponges until use, because it does not become decomposed and inert, as, for instance, corrosive sublimate.

NOTE.—Selected Florida sponges are cheap and good. In New York a pound can be bought for about two dollars, each sponge costing on an average two cents.

5. **Materials for Ligatures and Sutures.**—Well-prepared *catgut* of different thicknesses will answer every purpose for ligatures and sutures. The finest suture work on the intestines can be neatly and reliably done with catgut No. 0. The most massive pedicle can be safely tied with catgut No. 4. For ordinary ligatures and sutures No. 1 will be most convenient, and should constitute the bulk of the surgeon's supply.

The simplest way of preparing catgut is Kocher's: Wash in ether, then immerse catgut for twenty-four hours in good oil of juniper (ol. juniperi baccarum, *oil of the berry*, not the oil gained from the wood); transfer into and preserve in absolute alcohol 1,000, corrosive sublimate 1, until use. Alcohol keeps catgut hard and firm, yet flexible. Where it is desirable to prevent too early absorption, as, for instance, in intestinal sutures, a hardening process should be added to the disinfection. After disinfection the article should be washed in alcohol, then placed into a quart of a five-per-cent solution of carbolic acid containing thirty grains of bichromate of potash. Forty-eight hours' immersion will produce catgut that will resist the action of the living tissues for a week or longer. Large-sized catgut needs a longer immersion. Wind up on bobbins.

NOTE 1.—Good catgut can be procured from L. H. Keller & Co., 64 Nassau Street, New York, for a moderate price. Dry preservation makes catgut more suitable for transportation: Immerse the prepared article for five minutes in ether, 100; iodoform, 5. Take out and place in a well-corked, wide-mouthed bottle. A film of iodoform will cover each thread.

NOTE 2.—The author observed once unmistakable *wound infection by improperly kept catgut*. CASE.—Jenny Marks, servant-girl, aged twenty, admitted November 10, 1883, to Mount Sinai Hospital with habitual subcoracoid dislocation of the right shoulder-joint. "Sprain" had been diagnosticated by a physician, seven weeks previous to her admission, who ordered a liniment. On admission, reduction was easily effected by manipulation, but the weight of the limb was sufficient to reproduce the dislocation. A plaster-of-Paris jacket, inclosing the reduced arm, was applied and worn for four weeks without any effect. *Dec. 11th.*—The joint was freely opened by an anterior longitudinal incision, when it became evident that the tendency to dislocation was due to laxity or redundancy of the anterior part of the capsular ligament. By two semi-elliptical incisions, a piece of the capsule one inch long and half an inch in width was removed. The capsular as well as the muscular and the skin wound were united by three tiers of interrupted catgut sutures, a drainage-tube having previously been carried just within the capsule. The next day moderate fever (101° Fahr.), but great dejection, headache, and vomiting were observed, the patient complaining of much pain in the joint. *Dec. 13th.*—The thermometer indicated 103° Fahr., with a corresponding increase of the general disturbance. The patient was anæsthetized, and the wound was exposed. No redness, only slight œdema was visible. The wound was reopened. Firm agglutination was present everywhere except in four places, where swollen, discolored ligatures applied to the circumflex artery and some smaller vessels were seen surrounded by a halo of yellowish, semi-fluid, broken-down tissue, evidently representing small abscesses that were forming about the catgut ligatures. They were removed, the wound was irrigated with carbolic lotion, and packed with gauze. The fever fell off at once, and no further complication interrupted the course of healing. The habitual luxation was also cured.

**Silk** and **common cotton or** linen thread **can** be rendered unirritant **either by** boiling **it for an hour in a** five-per-cent solution of carbolic acid (Czerny) or by immersion **during** twenty-four **hours in a** solution of alcohol 100, corrosive sublimate 1, **then** preserving **in alcohol.**

*Silk-worm gut* is excellent material for suturing. **It is prepared like** silk, and before use should be soaked awhile **in carbolic** lotion **to** make **it** supple. Its advantage : it is easy **to** thread.

6. **Drainage-tubes and elastic ligatures** are **cut** into proper lengths—that is, a little shorter than the height of the wide-mouthed bottle in which they are kept. This is filled with a five-per-cent solution **of** carbolic acid, that should be renewed from time to time. The **tubes will** always occupy an upright position in the bottle, and can be taken **out easily.**

NOTE.—Rubber tubing **of** black material is preferable to the **coarser and unyielding white** stuff, on account of its softness and pliability.

Theoretically speaking, **a** *perfectly* aseptic wound does not require any drainage. If the secretions **following an** operation or injury do not contain anything that is capable **of inducing** putrid changes, **they** will be absorbed, and will not **cause any disturbance** in the wound or the general health. The large blood-clot around **a fractured bone** is harmlessly absorbed ; a large blood-clot in an aseptic **operation wound** will be also absorbed without local or general disturbance, as Mrs. **B.'s** case (see page 5) has shown. The experienced surgeon who has mastered **the** technique of asepticism will not hesitate to close up without drainage **a** small wound, **as,** for instance, after deligating the subclavian or iliac **arteries. But,** in operations where large surfaces were long exposed, and where the **wound is** very irregular, the possibility of a however slight and unavoidable contamination should always be kept in view. Vents should therefore **be provided in the** shape of properly placed drainage-tubes for the **easy egress of secretions,** possibly containing elements of future decomposition. **If** the healing be prompt, the tubes can be withdrawn on the third, fourth, or sixth day. In case of suppuration, bland or destructive, they will be in place and very opportune.

7. **Disinfecting Lotions.**—With a few exceptions (very large wounds requiring prolonged irrigation, and in operations involving the peritoneum), two lotions will be found sufficient. For the immersion of the instruments, a three-per-cent solution of carbolic acid, and for the irrigation and disinfection of hands and skin, a solution of corrosive sublimate of 1 : 1,000—1,500.

NOTE.—The almost exclusive use by the author of carbolic acid and corrosive sublimate as germicides is intentional. It was determined by the fact that these substances are, *first*, thoroughly reliable and highly effective; *secondly*, procurable almost everywhere, in the country store as well as in the city; *thirdly*, because adherence to certain carefully selected substances results in a thorough knowledge of their proper use under varying conditions.

*Boiled* water is preferable as a solvent. It alone would be sufficient if we were absolutely sure against the introduction of filth into the wound.

NOTE.—A ready and handy way of mixing the lotions is the following one:
*Carbolic Acid.*—One tablespoonful or four teaspoonfuls to a "quart bottle" of hot water will make a lotion of the strength of about three per cent, reckoning 650 grammes to the ordinary wine-bottle, erroneously called a "quart bottle."
*Corrosive Sublimate.*—Keep on hand a few ounces of an alcoholic solution of the salt of 1 : 10 in a glass-stoppered bottle (in boxwood case for transportation). One teaspoonful of this added to a quart bottle of hot water will make about a 1 : 1,500 solution, which can be weakened by dilution. The addition of one teaspoonful of cooking salt will prevent disintegration of the mercuric preparation.

*Boro-Salicylic Lotion.*—In cases where carbolic or mercurial poisoning could be produced by the use of mercuric or carbolic irrigation, *Thiersch's solution* is commendable as a substitute. It consists of salicylic acid 2, boric acid 12, and hot water 1,000 parts. It is non-poisonous, very bland, and the peritoneum can be washed with it with impunity. External wounds of large size should be also irrigated with this lotion. A final thorough irrigation with corrosive sublimate should sterilize the wound before closing it.

*Creoline Emulsion.*—Somewhat more convenient than Thiersch's solution is a mixture of creoline with water in the proportion of from one half to two per cent. It is also non-poisonous, and does not irritate the skin or corrode instruments.

NOTE.—The selection of different lotions should be governed by the following experiences: *Carbolic lotions are dangerous to small children*, even in great dilution, and should never be used on them. Corrosive sublimate is also poisonous, causing salivation and occasionally fatal diphtheritic inflammation of the ileum and the thick gut, if its use is immoderate. Wherever superficial ulcers or inflammations of the cutis require the antiphlogistic action of the very diffusible carbolic lotion, it should be employed in the strength of two or three per cent. The continued use of higher concentrations will corrode the tissues, and is otherwise dangerous.

Where a direct application of the lotion to the wounded or diseased surface is desirable, as, for instance, in all bloody operations, mercuric bichloride deserves the preference over carbolic acid. Even weak solutions (as 1 : 5,000) have a decided germicidal power, and can be used on very extensive wounds for hours without serious danger of intoxication. The final irrigation of an operation wound should always be done with a stronger (1 : 1,000) solution. Abscess cavities will always require the stronger solutions.

The greatest advantage of corrosive sublimate over carbolic acid is, however, to be sought in its different effect upon the fresh blood-clot and the tissues exposed to its action in a fresh wound. It will be seen that irrigating an amputation wound, for instance, with carbolic lotion, will each time provoke very profuse oozing. Vessels that had stopped bleeding by the formation of a clot within their cut orifices begin to bleed anew **after** carbolic irrigation. **This** is caused **by the** peculiar macerating effect of carbolic acid upon the fresh **blood-clot.** Its color turns from dark red to a light brick-red, its toughness and cohesion **are** lost, and **the** slightest touch of a sponge will suffice to detach it from the orifice of cut vessels, thus renewing the hæmorrhage. Another disagreeable effect of carbolic lotions upon wounds is the profuse discharge **of** bloody serum continuing for one or two days **after** the operation, rendering one or more changes of dressings necessary within a day or **two**, and thus depriving the wound of needed rest at the most critical period of repair.

Corrosive sublimate does not dissolve clots, hence oozing stops by natural means during its use. It does not irritate the **vaso-motor** nerves as carbolic acid seems to do, hence the oozing subsequent upon an operation done with its aid is very scanty. Drainage is easier, can often be altogether spared; no early change of dressings is required, and cure under one dressing is possible, and, in fact, is the rule after its proper use.

8. **Dressings.**—We have mentioned that there are two ways of preserving the aseptic character of a wound, viz., by exsiccation or by sterilization of the secretions. These two methods can also be advantageously combined.

### (1) *Types of Dressings.*

*a.* SIMPLE EXSICCATION.—Small, or comparatively small wounds, admitting **of an exact coaptation of the deeper** as well **as** their superficial parts by suture, are **exquisitely fit for this method** of treatment. Plastic operations about the face may serve as **a fair** type.

*Bismuth and Iodoform.*—Certain finely powdered **substances, as** iodoform or subnitrate of bismuth, have the quality of rapidly inspissating blood and serum to **a dry crust.** Accordingly, after the hæmorrhage has been controlled **and** the wound closed by suture, a quantity **of the** substance chosen is dusted over the sutures. No further dressings are applied. The escaping bloody **serum** forms a paste with the powder, which by its sterilizing property prevents decomposition, while **the** paste remains moist. Free access **of** air will hasten exsiccation, and **the** dry, hard crust once formed will securely prevent further ingress of **dust** into the wound. In cases where the powder is washed away by profuse oozing, the dusting has to be repeated every half-hour after the operation, until the object—the formation of a dry crust—is accomplished.

NOTE.—Elderly subjects are prone to iodoform poisoning if the agent is too freely used. In these cases a mixture of equal parts of iodoform and bismuth is safer.

Small cuts, **abrasions, and burns** can also be similarly treated, care being taken to first **render the injuries** aseptic by ablution with corrosive sublimate lotion.

NOTE.—*Acetic Acid.*—An excellent way of **treating small injuries** is to wash them as soon as possible—after staunching the hæmorrhage—with **pure acetic** acid; or, if this can not be procured, with ordinary vinegar. The intense smarting is soon controlled by the application of cold water. After this the part is dried with a towel. The dry but flexible eschar produced by the union of the acid with the exposed tissues gives excellent protection, **under** which the wound heals without reaction or suppuration. The great advantage of this form of treatment will be especially appreciated by physicians, as the eschar is insoluble, and the injured or chapped hands treated in this manner can be washed repeatedly without compromising repair or risking new infection by contact with pus.

More extensive burns or denudations are, within reasonable limits, also adapted to the exsiccative treatment. However, to prevent injury of the granulations at change of dressings, due to their matting into the meshes of the gauze, protecting the burned surface by a layer of rubber tissue will be found very useful and commendable. But the larger the absorbing surface, the more caution is needed in the use of iodoform.

*b.* CHEMICAL STERILIZATION COMBINED WITH EXSICCATION. DRY DRESSINGS.—In extensive injuries or large operation wounds the amount of oozing is generally so large that dusting alone will not suffice to control decomposition. Besides the patient's person, the bedding or splints will be uncomfortably soiled; hence it is necessary to provide a receptacle for the absorption of the secretions. For this purpose absorbent dressings are used that have been rendered aseptic by saturation with a chemical germicide: iodoform, corrosive sublimate, or carbolic acid. A small surplus of the chemical used will suffice to prevent decomposition of the absorbed serum or blood. No impervious covering (Mackintosh) should be used on the outside of the dressing, as the free admission of dustless air is desirable. It will hasten the exsiccation of the absorbed secretions, and thus insure the protective action of the dressings, even if the chemical employed become evaporated or inert. As evaporation of the deepest parts of the dressing— those nearest the skin and farthest from the surface—is the most difficult, and is made still more difficult by their greater saturation with serum, a few layers of iodoformized gauze placed immediately over the line of union will be of very great service in hastening exsiccation. These are covered with an ample mass of dressings impregnated with corrosive sublimate, which are held down with a roller bandage.

This is the method of dressing most commonly resorted to nowadays, and has been found the most simple and effective by the majority of modern surgeons.

*c.* SCHEDE'S MODIFICATION OF THE DRY DRESSING, FAVORING THE ORGANIZATION OF THE MOIST BLOOD-CLOT.—There is a considerable number of cases where extensive loss of substance consequent upon an injury or an operation precludes approximation of the walls of the wound, and renders healing by primary adhesion impossible. In these cases a blood-clot forms and fills up the defect soon after the injury or the operation. In an aseptic wound this blood-clot serves a highly useful purpose in protecting the raw surfaces, preserving their vitality, provided that the integrity of this blood-clot be again protected from exsiccation on one and from putrefaction on the other hand. If this condition is fulfilled, granulations will gradually consume, as it were, the blood-clot; and, by the time the clot disappears, cicatrization will be completed. When healing under the moist blood-clot is aimed at, the dressings will have to be arranged as follows: Immediately over the wound is laid a suitably trimmed piece of fine rubber tissue, previously well soaked in carbolic solution. It should just overlap the edges of the wound. This is covered with a **layer of** iodoformed gauze, and the whole is well enveloped **in an** ample covering of dry corrosive sublimate gauze. The outer dressings will absorb and render innocuous the surplus of blood and serum; **the** film **of rubber tissue will** preserve the underlying clot in a moist condition.

NOTE.—Tissues of low vascularity, as bone, fasciæ, and tendons, will certainly undergo superficial or deep-going necrosis if exposed to evaporation, even if asepsis be rigidly preserved.
*Case.*—George Braun, German Hospital, aged sixty-six. Rodent ulcer of the nose. *Feb. 19, 1886.*—Extirpation of diseased parts followed at once by partial rhinoplasty. Sutured parts dusted with iodoform. Large defect on forehead (the flap including periosteum) inadvertently covered with iodoform gauze, without interposition of rubber-tissue protective. **When** the dressings were removed ten days later, no suppuration was found, but **the** surface of the frontal bone was seen to be exposed (no blood-clot), and very dry. After four weeks the **first sparse** granulations were observed sprouting **out** of the denuded bone, which eventually **became cicatrized** over *in the fall of the same year.* Had the protective not been omitted, **rapid cicatrization** would have **been secured.**

*d.* SIMPLE CHEMICAL STERILIZATION. MOIST DRESSINGS.—A moderately moist condition of the **outer** dressings is **very favorable** to rapid absorption. This fact is **parallel with the** phenomenon seen if a thoroughly dry sponge is thrown on **water.** It will **not** absorb rapidly and sink, but, on the contrary, will float on the **surface** for a considerable period of time. But moisten this sponge first thoroughly, then squeeze it out completely, and then throw it into water, **and it** will **at** once become filled and sink. Where rapid **absorption is** desirable, as in the presence of septic or fetid discharges, and where clogging of the drainage-holes by inspissated secretions is **to be avoided,** dry dressings will be advantageously replaced by a *moist dressing.* By applying **a** piece of **impermeable** material to the outside **of the** well-moistened dressings, **evaporation and** exsiccation will **be** prevented. **The dressings will remain in a moist condition for an** indefinite period **of** time, **and will** act like a poultice.

*Rubber tissue* (not rubber sheeting) is an excellent and cheap substitute for Lister's "Mackintosh" and his "protective." It can be had **in** all **rubber stores.** A rather stout **quality** is the best article, as it is not apt **to tear, and can be repeatedly used as** the outer covering of moist dressings. *It always forms the outermost layer of what is called throughout this book a "moist dressing."* Oiled silk, well soaked in carbolized lotion, is a tolerable **substitute for rubber tissue.** Another substitute **is** waxed paper, **or "tracing paper."** A piece of **stout,** brown paper, such **as** is used by shopkeepers for **packing, well soaked in** grease, preferably tallow, will answer on a pinch. **If none of these** articles **can be** had, frequent moistenings of the dressings **will have to be employed in** order to prevent evaporation. One or more teaspoonfuls **of carbolic or** mercurial **lotion** instilled into the dressings every half-hour or so will have the desired **effect.** This form of moist wound-treatment was very extensively employed **by** the author in his seven-years' service at the German Dispensary, and has been found so satisfactory both to patients and surgeons that it is still the standard form of moist dressing **used at** that institution.

## (2) *Preparation of Dressings.*

*a.* GAUZE.—Gauze, called in the trade cheese-cloth, or tobacco-cloth, forms undoubtedly the most convenient material for wound-dressings. It is cheap, can be bought everywhere, absorbs well, is soft and pliable, and can be easily prepared for use by every practitioner. For hospital pur-

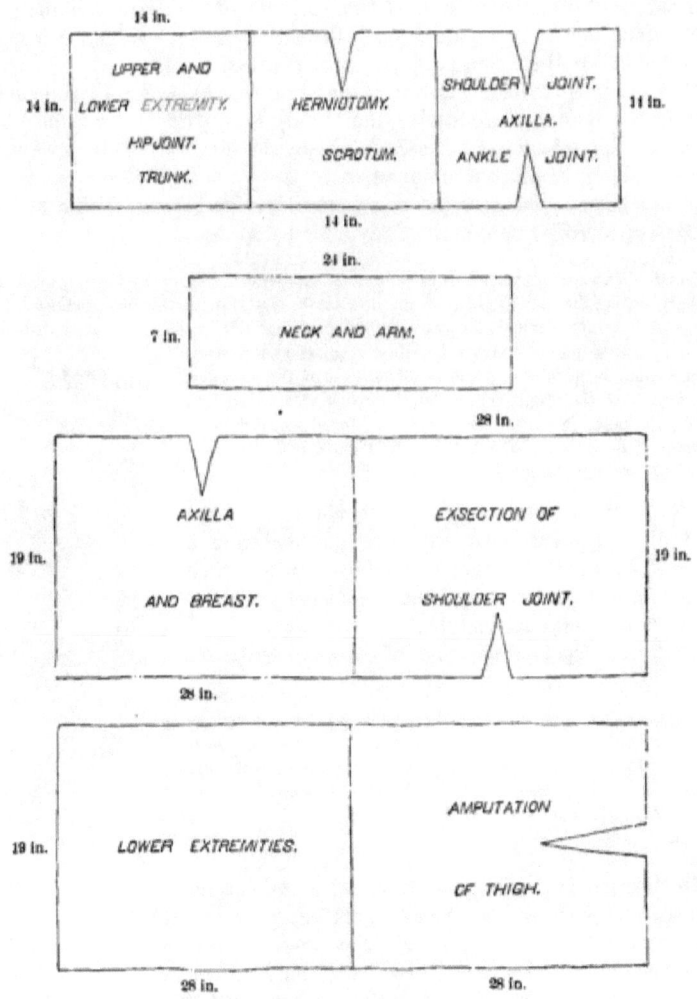

FIG. 1.—Patterns for various dressings, modified from Neuber.

poses, moss or peat dressings in the shape of cushions or bags are more convenient. In the practice of the country physician, however, they are out of the question.

(a) *Corrosive Sublimate Gauze.*—The raw gauze is treated as follows: To free it of its oily contents, and thus to make it more absorbent, twenty-four yards of the fabric are boiled for an hour in a wash-kettle filled with sufficient water to cover the material, to which should be added two pounds of washing-soda or a pint of strong lye. After this the stuff is washed out in cold water, passed through a clothes-wringer, and immersed in a sufficient quantity of a 1 : 1,000 solution of corrosive sublimate for twenty-four hours, then passed again through a clothes-wringer, dried, and put away in a well-covered glass jar until required for use.

The fabric is so folded by the manufacturer that each fold is just one yard long. It is best to divide the twenty-four yards into segments of six yards each, which can be again folded by the surgeon into large or small, square, oblong, or narrow compresses to suit each individual case. If a long time has elapsed since the preparation, reimpregnation with a 1 : 1,000 solution of corrosive sublimate is advisable before use.

NOTE.—In a small proportion of cases, contact with corrosive-sublimate dressings will cause an angry-looking dermatitis, which at the first blush very closely resembles erysipelas. The absence of fever and sickness, the exact limitation of the rash by the extent of the dressings, will soon disperse possible doubts. Profuse application of vaseline or some other bland ointment will readily dispose of the irritation. The strength of the impregnation should be then also reduced by washing the gauze in water. If it should be found that mercury is not borne at all, it should be substituted by carbolic-acid solution or Thiersch's boro-salicylic lotion, or creoline emulsion.

(b) *Iodoform Gauze.*—The moist, absorbent gauze is evenly sprinkled with iodoform powder from a pepper-box, or the author's iodoform duster, well rubbed into the meshes by hand, and then put away in a wide-mouthed bottle.

*Roller bandages* are made out of corrosive-sublimate gauze.

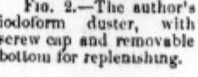

FIG. 2.—The author's iodoform duster, with screw cap and removable bottom for replenishing.

NOTE.—Roller bandages made of a starched fabric known as "crinoline," or "crown-lining," are very useful in completing every dressing. They are moistened in water, and applied over the dry roller-bandage. They soon become stiff again, and make a very compact and neat dressing, that will not shift easily. The stuff is the same that is used extensively for plaster-of-Paris bandages.

In emergencies various substances of absorbent qualities can be utilized as dressings ; such are, for instance, cotton, moss, and sawdust.

b. ABSORBENT COTTON, or COMMON COTTON BATTING, well soaked in corrosive-sublimate solution, then wrung out, will make a tolerable dressing. Its drawbacks are that it packs and gets hard and lumpy, but, properly used, it will answer every practical purpose. Care should be taken not to tear the cotton into irregular masses. After unrolling it, suitably large, square pieces should be cut off with the scissors ; these pieces should be folded, then soaked in the lotion, squeezed out hard, and unfolded again, thus preserving their shape and uniform thickness. Two or more of these pieces laid one over another will make a very passable dressing.

CASE.—Michael B., aged sixty-three, sustained, early in the morning of November 13, 1883, a compound fracture of the left elbow-joint. He was put to bed, and, under the advice of the family attendant, applications of cold water were made to the injured part. Twelve hours after the injury, the author found a Y-shaped fracture of the lower end of the humerus, the conical sharp point of the upper fragment protruding through a small wound above the olecranon. The joint was filled with a large clot, and some oozing from the perforation was noticed. The edges of the perforation wound were snugly fitting around the protruding bone, and during the subsequent manipulations good care was taken not to allow the bone to slip back. Not having been informed of the nature of the injury, the author arrived unprepared at the patient's bedside. The case, however, did not brook delay, hence everything had to be extemporized. Several ounces of a ten-per-cent alcoholic solution of corrosive sublimate and a little iodoform were ordered from the nearest druggist, and at the same time several bundles of common cotton batting were procured. Soon plenty of a 1 : 1,000 corrosive-sublimate solution was ready, in which square pieces of cotton were soaked as described. The patient's poverty compelled an economical management of affairs. An old but clean bed-sheet was ripped up into roller-bandages, which were likewise impregnated. This done, soap and hot water were applied to the elbow, and the skin was shaved clean all around, but especially near the perforation. This was followed by a vigorous rubbing off of the skin and protruding bone with the mercuric lotion, which at the same time was copiously poured over the region of the elbow from a pitcher. After this, reduction of the protruding bone and adjustment of the fragments by extension of the arm was effected. The size of the perforation-hole at once became much smaller. In order to provide some drainage, a small fillet of cotton, well dusted with iodoform, was inserted into the cutaneous part of the outer wound, which was also liberally dusted. Over this were placed four layers of cotton pads, which were snugly bandaged to the limb. Two lateral splints, made of a pasteboard box, secured the extended position, in which the arm was suspended from a nail in the ceiling. The temperature never rose above 100° Fahr. *Nov. 19.*—The dressings were removed. The swelling, due to the effusion of blood, had disappeared to a great extent. Oozing had ceased; no suppuration. The fillet of cotton was withdrawn, and the arm was put up in a plaster-of-Paris splint flexed at a right angle. Passive motion was commenced on removal of the splint, four weeks after the injury. Ultimate result was ascertained in October, 1884: Flexion was normal; extension could not be carried beyond 140°.

*c.* SAWDUST.—With a view to the occasional impossibility of procuring any of the common dressing materials in times of war or some other public calamity, the author has tested the efficacy of sawdust as a dressing during his service at Mount Sinai Hospital, extending from August 1, 1883, till February 1, 1884. Clean pine, spruce, or hemlock sawdust was impregnated with a 1 : 1,000 solution of corrosive sublimate for twenty-four hours ; then it was spread on sheets of muslin to dry, and finally was inclosed in different-sized bags made of cheese-cloth gauze. To prevent the shifting of the sawdust, a thin layer of wood-shavings, called by the trade "excelsior," was first inserted into the open bag ; then a proportionate quantity of sawdust was evenly strewed into the meshes of the "excelsior," and then the bag was closed by stitches made with threads soaked in mercuric lotion.

## ASEPTIC WOUNDS—ASEPTIC TREATMENT.

The thickness of the bags varied, according to their size, from one to two inches. After the wound was drained and sewed, some iodoform gauze was placed next to it; then came one, two, or more smaller bags, and on top a large bag, the whole being snugly fastened with roller bandages.

Aside from the trouble of preparing the bags, they were found very convenient in applying and quite efficient in absorbing blood and serum, and preventing decomposition.

*d*. Moss.—The different species of sphagnum, coating the surface of peat-bogs and the trunks of dead trees in our northern forests, are excellent material for making dressing-bags. On account of its cheapness, small weight, elasticity, and great absorbing power, moss has displaced other dressings at almost all of the surgical clinics of Germany. Its preparation is very simple. It has to be gathered with some care—that is, with no admixture of the soil. After being dried, it is impregnated with corrosive sublimate, inclosed in gauze bags, and is ready for use. Moss-bags are in daily use at the German Hospital since 1884, and can not be praised enough both for their handiness and effectiveness. But, like other similar dressings, they are not adapted to the needs of the general practitioner, and will find their principal employment in hospital practice.

### III. PRACTICAL APPLICATION OF RULES.

1. In operating.—In order to gain a coherent idea of the practical workings of the aseptic apparatus, we shall now rehearse all the steps of a well-conducted operation.

Assuming that a cancerous breast is to be removed in the rooms of the patient, it is first necessary to select a suitable person to act as nurse. Her duty is to administer a laxative the day before the operation, and to carefully scrub with soap and brush the patient's breast, corresponding shoulder, and axillary space on the day preceding and on the day of the operation. A clean, well-lighted room is selected, out of which all unnecessary furniture, hangings, etc., should be removed. A bare, well-scrubbed floor is preferable to a carpet. One or two narrow kitchen-tables, covered with a quilt and provided with a straw pillow, will make a capital operating-table. A piece of rubber cloth (3×4 feet) is placed over the quilt, and a clean sheet is laid on top. The nurse provides soap, nail-brush, plenty of hot and cold water, and towels. The operator and his assistants arrive at least a half-hour before the appointed time of the operation. Everybody's hands are washed in hot water with soap and brush. The necessaries are now unpacked and arranged, and the solutions of carbolic acid and corrosive sublimate are mixed, for which purpose six or eight well-cleansed quart bottles should be held in readiness by the nurse. A fountain syringe is filled with sublimate solution, and suitably suspended from a nail or chandelier near the operating-table. A new pail or bucket is filled with hot water for rinsing the blood out of the sponges; alongside of it is placed a basin filled with a three-per-cent solution of carbolic acid for the reception of the cleaned sponges, from which they ought to be handed to the assistants by the nurse. Two more japanned tin or earthenware basins are filled with a corrosive-sublimate solution, and placed on chairs to the right and left of the operating-table for the occasional rinsing of the hands of the operator and assistants. The instruments are arranged on an adjacent table in a certain order, which, to prevent confusion, should be rigidly adhered to during the entire operation.

NOTE.—The author has found that it is very convenient to be independent of the patient's resources, as far as the necessary vessels for sponges and instruments are concerned. A nest of four good-sized, flat-bottomed block-tin wash-basins, six tin soup-basins (six inches diameter), and four tin bake-pans, will serve every purpose, and the small expense will be abundantly repaid by the cleanliness and sense of comfort that will result. Two coats of some reliable oil-paint or japan will prevent corrosion of the metal vessels. The asphalt is dissolved in gasoline or ether, then applied evenly. As soon as the solvent evaporates, the pans will be ready for use.

All vessels are wiped clean. The knives, sharp and blunt retractors, scissors, anatomical, mouse-tooth, and dressing forceps, probes, and grooved director should be put into one pan with carbolic lotion; all the artery forceps by themselves into another one. Between the two pans is placed a third one, filled with hot water, in which all the instruments not in actual use should be rinsed free from blood before being returned to the carbolic lotion. This will keep them and the carbolic lotion clean and bright all the while, and no time will be lost in hunting for them in the bottom of a turbid pool of soiled carbolic solution. In a smaller tin basin, ligatures, in another one needles, are arranged, threaded with fine (No. 0) and coarser (No. 1 or 2) catgut. A third small basin will hold the drainage-tubes and a number of safety-pins.

The dressings are now attended to. Eight or ten small (6 × 8 inches), and just as many large (19 × 28 inches), compresses of gauze are cut, care being taken not to make the dressings too scanty, as an ample first dressing may save the trouble of many subsequent dressings. The best rule is to let the outermost compresses overlap the wound on all sides by at least eight inches. To this should be added a sufficient number of strips of iodoformed gauze, three or four rather wide gauze roller-bandages, and the same number of starched or crinoline roller-bandages. All this should be wrapped in a clean towel and laid aside in a secure place until needed.

These having been attended to, anæsthesia may commence in an adjacent room. The anæsthetizer should be provided with ether and a cone, a tin basin for the reception of ejecta in case of vomiting, a towel, a hypodermic syringe, a wide-mouthed bottle with morphine solution for injections in case anæsthesia be imperfect, a similar bottle with whisky to be used in case of heart-failure; finally, with a dressing-forceps and gag for withdrawing the tongue if it should sink back on the epiglottis.

The anæsthetized patient is placed on the operating-table, and the parts, being exposed, are freely soaped and shaved. After this a piece of rubber cloth (3 × 4 feet) is so placed over the patient's body as to leave exposed only the field of operation. Now the parts are well rubbed off with a towel

## ASEPTIC WOUNDS—ASEPTIC TREATMENT.

dipped in corrosive-sublimate solution and freely irrigated, and a number of clean towels wrung out of the same solution are suitably spread around the field of operation, protecting the operator and assistants against contact with the clothing or body of the patient, and providing for a clean place where instruments or sponges may be laid down for a moment if necessary. The end of a wet towel is tucked under the breast and armpit of the side to be operated on, and is hung over the edge of the table in such a manner as to conduct the blood and irrigating fluid into a bucket placed on the floor underneath. It serves as a drip-cloth. Every assistant should strictly attend to the duty allotted to him, and not meddle. All unnecessary talk should cease, and the work proceed in an orderly manner. The first assistant should keep his eyes open, and know and aid the operator's intentions. He should be alert, but not over-zealous.

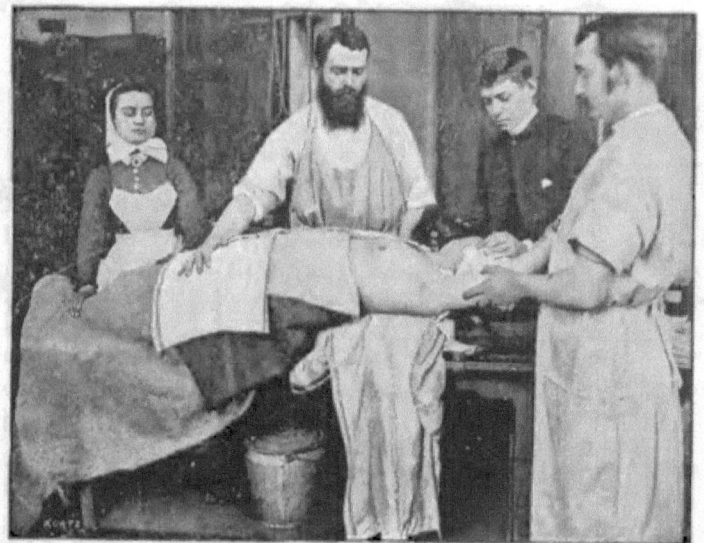

FIG. 3.—Patient made ready for amputation of mamma.

The anæsthetizer must take good care that, in case of vomiting, no ejecta are thrown on the wound or its vicinity. Towels soiled by vomit should be at once replaced by clean ones.

Now the parts are distributed. The trustiest man serves as first assistant over against the operator; a younger physician at the left of the operator is second assistant, and irrigates or helps as need may require; another physician takes charge of the instruments and ligatures, and the nurse attends to the sponges, and keeps in readiness "sublimated" and dry towels and a pitcherful of corrosive-sublimate solution.

Aprons are donned, everybody's hands are finally scrubbed with soap and brush, rinsed in mercuric solution, and the operation begins.

NOTE.—The employment of copious irrigation during operations requires measures for protecting the person and clothing of the surgeon against the influence of the chemicals commonly used. An ample apron, made of light rubber sheeting, and reaching from the chin to the toes, is most convenient, and can be easily cleaned. The surgeon's shoes may be protected by a pair of light rubbers. However, they are apt to sweat the feet. The author overcame this drawback by the use, at the hospital, of wooden pattens (French *sabots*) worn over the shoes. They are donned and doffed without the aid of the hands, and keep the feet warm and dry, and can be bought at 75 Essex Street, New York.

In removing the breast and contents of the axilla, hæmorrhage should be carefully attended to by ligaturing every bleeding vessel with catgut. Having removed the diseased parts, the wound is carefully irrigated, each recess being attended to in succession; drainage and sutures are applied. The projecting end of the drainage-tube cut off "flush" is transfixed with a safety-pin, the wound is once more irrigated through the tube so as to clear it of clots, and the clots and irrigating fluid are removed from the wound by gentle pressure exerted with a sponge or two. Iodoformed gauze strips are next placed along the suture and around the drainage-tube, passing under the safety-pin, and a few pads of gauze are held pressed against the wound while the patient is slightly raised to cleanse her back and face and the table from blood. The soiled towels are replaced by dry ones, and the dressing completed by applying as many gauze compresses as required. These are fastened rather tightly with gauze bandages, the other breast and arm-pits being first padded with absorbent cotton. A large, square piece of absorbent cotton, somewhat overlapping the dressings, is next applied, and snugly held down by crinoline roller-bandages; the corresponding arm is included by the bandage or is placed in a sling; the patient is brought to bed, and an opiate is administered.

2. **Change of Dressings.**—In most cases where the rules above given are conscientiously and intelligently observed, no fever will follow the operation. After the effects of the anæsthesia are over, the patients will be found cheerful and contented, feeling no pain or sickness, their only complaint being the tightness of the bandage, which they will soon learn to bear. The temperature will range during the first three days at about 100° Fahr.; after that it will sink to the normal standard. Sometimes, especially if the drainage is not properly placed, and some serum or a blood-clot is retained in the wound, the thermometer will indicate from 100° to 103° Fahr. As long, however, as the patient is cheerful, and does not feel sick with headache and general dejection, as there is no sharp, throbbing pain about the wound, or some other grave disturbance of the local or general comfort, no alarm need be felt. In these cases we have to deal with an elevation of temperature benign in character, and identical with the harmless fever observed after almost every simple fracture. It is due to the absorption of the extravasated blood or lymph, bland and harmless on account of the absence of putrefactive changes. This is Volkmann's "aseptic fever."

The temperature soon becomes lowered, appetite reappears, and the dressings need not be disturbed.

Should, on the other hand, the patient complain of chilliness, headache, sickness, general dejection, and drawing pains in the limbs, or persistent and increasing pain about the wound, the thermometer indicating at the same time a high or only a moderate elevation, the dressings should at once be removed, and a search instituted for the cause of the disturbance.

Previous to this a new dressing should be prepared similar to the one to be removed. This and a tin pan containing carbolic lotion, with a dressing-forceps, anatomical forceps, scissors, scalpel, grooved director, and a piece of drainage-tube, together with another vessel holding a few small pads of cotton wrung out of the same solution, should be placed on a small table near the bed. An irrigator filled with *warm* carbolic or mercuric lotion should be suspended from the bedpost or a nail, and a pail for the

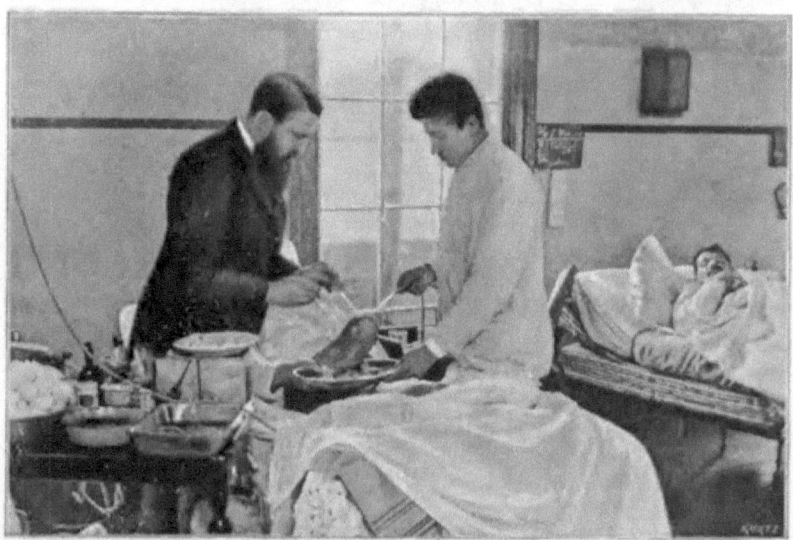

FIG. 4.—Change of dressings after amputation of the thigh.

reception of the soiled dressings should be at hand. A piece of rubber cloth covered with a draw-sheet and spread under the patient's back will protect the bed, and a pus-basin or square tin pan held alongside of the patient's thorax will receive the irrigating fluid.

After this the turns of the roller-bandage are cut through without jar, and the outer layers of the dressing are gradually removed. As the deeper parts are being raised, irrigation should commence, in order to moisten the gauze and aid in its gentle removal. Care should be taken not to disturb the drainage-tubes. *After the removal of the soiled dressings, the physician's hands should be carefully cleansed before touching any part of the wound.* While the irrigating stream is playing, the vicinity of the wound is gently wiped with a small pad of moistened cotton, in order to remove clots of blood or fibrin that can not be dislodged by irrigation.

If the edges and vicinity of the wound look normal, the skin pale, not swollen, and not painful to touch, it should be forthwith redressed. A careful physical examination of the internal organs will then certainly reveal, as the cause of the fever, some internal complication, as, for instance, pneumonia, or, at any rate, some newly developed or overlooked disorder independent of the wound.

If the aseptic measures employed were insufficient, the edges of the wound will be found swollen, reddened, and painful; the wound will have lost its aseptic character, and is the seat of a septic process ending in suppuration. Prompt action is required to limit the inevitable destruction of tissue, and to check the further poisoning of the system.

From this moment on, *aseptics* must give way to *antiseptics;* prevention having failed, curative measures must step in to eliminate the mischief that might have been prevented by the exhibition of more care, attention, or skill.

The therapy of septically infected or suppurating wounds will be treated in the following chapter.

In case that the course of the healing of the wound is correct, as indicated by the absence of local or general disturbance, the first dressing may remain unchanged for from seven to forty days. Flesh-wounds should be dressed on the fourth to the seventh day, as it is desirable to remove the drainage-tubes and sometimes the stitches. The finer catgut sutures will generally be absorbed by this time, and their exposed part can be simply wiped away. Where stout retention sutures were employed for the approach of the edges of a wide, gaping wound, they will be found cutting through the tissues by this time, and quite useless. They should be removed, and the stitch-holes dusted with iodoform. According to the completeness of the result, the dressings will have to be changed every third, fifth, or seventh day, their bulk decreasing with the diminution of the secretions. Finally, the few granulating spots need only a dressing consisting of a patch of some unirritant plaster, such as empl. cerussæ or empl. hydrarg., and an occasional touching with nitrate of silver, to aid final cicatrization. Where the operation has involved parts of the skeleton, as in amputations of extremities, exsections of joints, necrotomies, etc., the dressings have to be left undisturbed much longer. After exsections of the knee-joint, for instance, where bony ankylosis is aimed at, the first dressing is not removed without a clear indication before the thirtieth or fortieth day. No patient should be discharged "cured" before cicatrization is complete, as it has happened that such "cured" cases, left to their own care, contracted erysipelas the day after their discharge, and died of it.

NOTE.—All the manipulations about a freshly agglutinated wound should be very deliberate and gentle. In removing stitches, a forceps should gently raise the thread; then it should be cut as close to the stitch-hole as possible, and lightly withdrawn. Drainage-tubes are grasped at the projecting end, gently rotated to and fro till they are freely movable, then withdrawn. Sometimes it will be found that a painless fluctuating swelling occupies some deeper part of the wound. In these cases retention of serum is generally caused by clogging of the drainage-tube by a clot. On withdrawing the tube, a quantity of clear or turbid yellowish serum will escape. In these cases it is good to replace the cleared tubing to prevent further retention, and thus to bring about contact of the separated walls of the wound, which will at once become adherent. At the subsequent change of dressings, the tube can be definitively removed.

CASE.—Mrs. Clara G., aged forty-six. *Alveolar glandular cancer of an aberrant (detached) lobe of the right breast.* Tumor of the size of a small fist, situated in the axillary space close to the edge of the pectoralis major muscle. It was connected by a stout pedicle with the adjacent part of the breast-gland proper. *Jan. 16, 1885.—* Amputation of mamma; total evacuation of axillary fat and glands. Drainage by counter opening made through the latissimus dorsi muscle. Suture of the entire wound except a part of axilla, where the skin had been extensively removed. Course of healing feverless. Change of dressings on the tenth day. Primary union of all the sutured parts. Axillary wound granulating. Under the lower flap of the breast-wound a painless, soft, fluctuating swelling discernible. By gently inserting a probe between the corresponding edges of the united wound, entrance into this sac was effected, whereupon about two ounces of a yellow, slightly turbid, and very viscid serum escaped. A small drainage-tube was inserted, and the wound was redressed. *Jan. 30th.—*Walls of the cavity were found firmly adherent. Tube removed. No suppuration.

The interior of freshly healed wounds of normal appearance should never be syringed ; the injection of a strong jet of fluid is unnecessary and often injurious, as it tends to separate tender adhesions.

## IV. ASEPTIC MEASURES IN EMERGENCIES.

Unremitting attention to, and a severe self-discipline in always carrying out the measures of strict cleanliness known to be necessary to uniform success in the management of wounds will gradually become, however irksome in the beginning, a mere matter of accustomed routine. As the mind and senses learn to exercise vigilance without special effort, the surgeon's results will become more and more gratifying. His attention, freed from the severe strain unavoidable in acquiring command of the detail of a difficult business, will concentrate itself upon higher objects, and the smooth routine resulting from long and severe training will not divert attention from the finer detail of his special work.

It is a great mistake, paid for by the loss of limbs and lives, to believe that the mastery of practical cleanliness or asepticism can be acquired without a clear comprehension of the principle, and without earnest and severe training in the *handicraft of asepticism*. The wholesome truth, that failure of achieving primary union in fresh wounds is mainly and almost always due to one's own lack of knowledge and skill, and that these attributes can be secured only by the exercise of great diligence and many, often unsuccessful trials, should be constantly present in our mind. Failures are bitter lessons, but their honest study will inevitably bring to light the causative deficiencies, and will teach us to avoid them.

The school for learning to employ the principles of asepticism is open to every general practitioner in the treatment of the many affections and injuries pertaining to minor surgery. Mistakes made in the removal of a wen or the treatment of an incised wound of the hand are easily found out and easily corrected. They carry much and sometimes more instruction than a large operation. It is wicked to attempt to learn the first lessons of aseptic surgery in capital operations, when, possibly, the surgeon's experience is bought with the life of his trusting patient. The attempt of removing an ovarian tumor, for instance, should be permitted only to those who have learned to invariably heal a fresh wound by primary adhesion, as this is the first and sole test of the possession of the ability justifying such a grave undertaking.

Emergencies will necessarily involve varying modifications of the means, *never a deviation from the principle of asepticism.*

A hasty tracheotomy for the removal of a foreign body, a herniotomy to be done in the dead of night amid the squalid surroundings of a tenement, or the first care of a compound fracture or a gunshot-wound, will present special and varying difficulties, to be overcome only by good training, circumspection, and versatility. They can be overcome, as many examples in the experience of every successful surgeon testify.

In addition to the case of compound fracture of the elbow-joint quoted on page 16, another instructive case may be told from the author's experience.

Case.—Herman John, laborer, aged sixty-one. Right, irreducible, strangulated femoral hernia. Rupture of long standing, strangulated since the evening of April 1, 1882. Symptoms of great acuity necessitated prompt action. Dr. H. Wettengel, the family attendant, administered the anæsthetic in the middle of the afternoon of the following day, while author was making the necessary preparations for the presumably inevitable operation. The place was a narrow, dark, rear room of a rear house of a squalid tenement, and a lamp had to be procured. The divested patient's pubic and inguinal region was shaved, while anæsthesia progressed. A flat bake-pan was covered with one of the few clean towels to be had; on this were spread the instruments, and over them was poured a quantity of a five-per-cent carbolic lotion. No sponges were on hand, as the summons had been very hasty, and no time was afforded for preparations. Therefore, a part of a clean bed-sheet was torn into a number of small pads, which were well soaked in the same lotion to serve as sponges. A remnant of the lotion was saved in a pitcher for purposes of irrigation. After an unsuccessful attempt at reposition, the inguinal region and the surgeon's hands were once more well soaped and washed off with the carbolic lotion. The epigastric artery had to be tied, and external herniotomy was performed. A small knuckle of gut slipped back easily into the abdominal cavity, but evidently did not represent all the contents of the sac, within which an additional soft body could be felt that resisted every gentle effort at reposition. The sac being opened, a slender portion of omentum was found to be adherent to it. This, being dissected away, was replaced into the abdominal cavity. The outer wound was well irrigated, and united by a number of catgut sutures. A few strands of catgut were inserted into the lower angle of the wound for drainage. In the absence of other dressings, a clean sheet was used for the manufacture of a number of compresses and roller-bandages. These, being well soaked in carbolic lotion, were applied to the wound in the shape of a spica bandage. Vomiting ceased. Oozing being very scanty, the dressings soon became dry, and, the patient's condition being excellent in every respect, they were not disturbed until a fortnight after the operation, when the wound was found healed throughout by the first intention.

# ASEPTIC WOUNDS—ASEPTIC TREATMENT.

Yet it must be said that such conditions render operating very risky, and in every way uncomfortable. If unavoidable, the additional risk must be shouldered by the patient as well as the surgeon.

**Operating Bag and Kit.**—Timely preparation made in the shape of procuring a well-arranged hand-bag, containing the most necessary articles for operating in an emergency, will well repay the small expense and trouble.

A leather hand-bag, about sixteen inches long, will be sufficiently large.

Have a sufficiently long, rather stout strap sewed to one side of the interior of the

FIG. 5.—Author's operating bag, with tin pans and rubber cloths strapped to it.

bag, so as to provide loops for five or six bottles, which will be held safely in the upright position. The first loop will be occupied by a half-pound tin can of ether; the second is allotted to a two-ounce bottle of corrosive-sublimate solution (ten per cent alcoholic); the third to a four-ounce bottle of pure carbolic acid; the fourth to a wide-mouthed bottle containing catgut and silk of different sizes on spools; the fifth to a wide-mouthed bot-

FIG. 6.—Interior of operating bag.

tle filled with drainage-tubes of different sizes in carbolic lotion; the sixth to a wide-mouthed fruit-jar with tight cap, containing two or three dozen sponges in carbolic lotion. A stout pair of scissors for cutting the dressings, a dressing-forceps for the anæsthetizer, and a razor can be conveniently stuck in behind the bottles. On the other side of the bag two more spaces are reserved for a dusting-box filled with iodoform-powder and a wide-mouthed vial for an assortment of surgeon's needles. The bottles containing pure carbolic-acid and corrosive-sublimate solution should be inclosed

FIG. 7.—German instrument-pouch.    FIG. 8.—Interior of German instrument-pouch.

in boxwood or tin cases for safety. A side-flap will hold nail-brush, safety-pins, and one complete dressing rolled up in a clean towel. The body of the bag is reserved for the instruments, which are rolled up in another clean towel, and for three or four small tin basins, together with a fountain syringe and ether cone, each kept in a separate rubber sponge-bag.

To the bottom of the hand-bag is strapped on the outside a nest of four oblong tin pans of fitting size.

Such a bag contains all the necessaries for an emergency, and has been used by the author many years with much satisfaction.

NOTE.—*Surgical pocket-cases*, as generally sold by surgical cutlers, are mostly incomplete and unsatisfactory. Their main objection is the small size and frailty of the instruments contained in them. The instrument-pouch depicted in Figs. 7 and 8 is very complete, and is worn strapped to the waist underneath the coat. It contains, besides the instruments held by a complete pocket-case, a sharp spoon, a key-hole saw, a flat oblong *iodoform dusting-box* of hard rubber, and a set of diverse detachable knife-blades, that can be fitted to smooth hard-rubber handles, all very easy to clean. In an emergency, the hip-pouch will be found large enough for the reception of one complete dressing to a moderate-sized wound.

## CHAPTER III.

*SOILED WOUNDS.—ANTISEPTIC TREATMENT.—DIFFERENCE BETWEEN ASEPTIC AND ANTISEPTIC METHODS.—ILLUSTRATION OF ANTISEPTIC METHOD.*

In the preceding chapter the treatment of freshly made, clean, or uncontaminated wounds was discussed; its subject was the *aseptic form of treatment*—that is, the manner in which a fresh or clean wound has to be managed in order to prevent its septic infection.

The aseptic discipline is a purely preventive one.

*Antiseptic treatment*, on the other hand, refers to such wounds as have become the seat of infection, causing inflammation, suppuration, or the higher forms of sepsis—phlegmon and gangrene. The object of the antiseptic treatment is the limiting and elimination of *established* septic processes by drainage and disinfection. It is also preventive, but in a narrower sense than the aseptic method. There all mischief is prevented from the outset; here further extension of present mischief is sought to be checked. The aseptic method will generally preserve all the parts involved; the antiseptic method can not restore the integrity of parts destroyed by ulceration, suppuration, or gangrene.

*Illustration of Antiseptic Method.*—For the sake of illustration, let us go back now to our former example of breast-amputation.

Some gross fault having been committed, such as, for instance, the use of unclean instruments, or a sponge that, having fallen to the floor, was picked up by the nurse and was handed for use in the wound. The mild course of the case is compromised, and trouble will follow.

In such cases the patient's general condition is deeply disturbed, more or less high fever is present, with headache, sickness, general dejection, and drawing pains in the limbs. The tongue is foul, much thirst and loss of appetite are complained of. The wound is painful and throbbing, and the patient dreads any movement lest the sore parts be hurt.

Under these circumstances an immediate examination of the wound is imperative. The preparation mentioned in the preceding chapter being made, the wound is exposed. Its edges and the vicinity will be found angry-looking, swollen, hot, and tender.

The stitches should be all removed. The point of the grooved director should be inserted between the edges of the wound, which are gradually separated till the index-finger can be insinuated. Exerting gentle pressure, the wound is thus opened throughout its entire extent. One or more small foci containing pus will be laid open and discharged. The wound should be carefully irrigated with warm mercuric lotion till the slight hæmorrhage ceases, and lightly filled with sublimated gauze. After this the outer dressings, with the addition of an externally placed piece of rubber tissue to prevent evaporation, should be renewed, and the timely interference will be soon rewarded by a decided improvement in the patient's condition. In these cases the dressings must be changed as often as they become soiled through. If the fever should continue, renewed search must be instituted for overlooked points of retention.

In some cases examination of the wound will reveal only partial or quite circumscribed inflammation. In locating the exact point of retention, the sensations of an intelligent patient will greatly aid the surgeon. If the retention be near the edges of the wound, the grooved director will easily separate them and find its way into the focus. A dressing-forceps should be then insinuated along the director, and withdrawn with its branches partly opened. Pus escaping, a slender drainage-tube should be inserted into the track.

If the point of retention be remote from the edges of the wound, and its locality well marked by redness and pain, an incision will best answer the purpose, and often may prevent suppuration of the rest of the wound.

Let us assume that for one reason or another nothing efficient was done to relieve the patient on the second or third day after the operation. Finally, the increasing severity of the symptoms will compel some action, and, the wound being laid bare, the following state will be generally met with: The wound will be more or less gaping, ichor or pus escaping everywhere; the skin will appear flushed, swollen, and painful; the edges of the wound will be marked by a grayish-yellow, closely adherent coating, that extends through its whole interior. This coating represents molecular, often deep-going necrosis of the wound surface. Independent abscesses will often be found established along the connective-tissue planes contiguous with the wound, and should be forthwith incised and drained. The wound should be well irrigated and loosely filled with sublimated gauze. Over this should be applied a *moist dressing* of ample proportions, covered with an overlapping piece of rubber tissue to prevent evaporation and inspissation. The secretions will thus be readily and continuously drained away and disinfected, and the warm moisture of the dressings will at the same time exert a very soothing influence upon the inflamed parts. Frequent, at least daily, change of dressings is proper, accompanied by copious irrigation. Detached shreds of necrosed tissue should be removed with thumb-forceps and scissors. If new abscesses form, they must be found and opened promptly. The fever will soon abate, and the wound will gradually assume a clean granulating appearance. As the amount of secretion diminishes, the dressings should be changed less frequently.

Essentially, the so-called "*idiopathic*" *phlegmon*, or *spontaneous suppuration* (abscess) is a form of local septic infection which can be traced back to an infection extending from a lesion of the skin or the mucous membranes.

Even the suppurative or infectious form of osteomyelitis must be classed under this heading.

## THE TREATMENT OF ACCIDENTAL WOUNDS.

But, on account of the great practical importance of the subject, requiring special consideration of several anatomical regions involving important modifications of the antiseptic procedure, it is deemed expedient to treat of this theme in a special chapter (page 183).

# CHAPTER IV.

## SPECIAL RULES REGARDING THE TREATMENT OF ACCIDENTAL WOUNDS.

### I. TEMPORARY MEASURES.

TAKING charge of a fresh case of accidental wounding, the surgeon should bear in mind that, on the one hand, by the avoidance of suppuration, a complete or almost complete restitution of normal conditions can be accomplished in a great majority of cases ; on the other hand, suppuration will enormously increase the gravity of a given injury. A compound fracture of the leg, or an incised wound of the wrist, with opening of joints and severing of arteries, veins, and tendons, may serve as examples.

In approaching a fresh case of bloody injury, we should always consider the possibility that the wound may be surgically clean, or may still be aseptic, and that our first ministrations should not carry septic contamination into the wound, and thus harm the patient instead of aiding him. As a matter of fact, a large proportion of incised and lacerated wounds, of compound fractures by blunt force or gunshot, *are aseptic*. They need no disinfection. The surgeon's first object should be in these cases *not to spoil matters by hasty action and ill-considered zeal*. With the comparatively rare exception of injuries to large vessels accompanied by dangerous hæmorrhage, where immediate action is unavoidable, conditions should be created by the surgeon, under which safe—that is, aseptic—approach to the wound is made possible. Temporary protection of the wound in the shape of a simple dressing is meant thereby. Iodoform-powder dusted profusely over the wound and its vicinity, a compress made of a clean towel dipped in hot water or carbolic lotion, also well dusted with iodoform and tied on to the wound, will be sufficient. The addition of a temporary splint in cases of compound or gunshot fracture will make transportation to the patient's home or to a hospital possible, and will thus afford time for the absolutely necessary preparations. Extensive or even superficial examination of an accidental wound by probing or digital exploration in the street, on a train, or in a railroad-station or drug-shop, is strongly to be condemned, as it almost necessarily exposes the wound to unavoidable infection. Meddlesome and untimely surgery of this kind smacks of ostentation, is unnecessary, and in many cases positively more dangerous than the injury itself.

Bergmann's experience during the Russo-Turkish war has shown that most gunshot wounds are aseptic, and that, with the exception of those cases where shreds of soiled clothing or gun-wads were carried along by the projectile into the bottom of the wound, healing without suppuration can be confidently expected if the wound is not infected by meddlesome and uncleanly surgery. These experiences refer principally to gunshot fractures of the knee-joint.

As a matter of fact, it may be safely assumed that an examination by probing or digital exploration, performed on the filthy floor of a public place or on the street pavement, even by the most experienced surgeon, can not be, and is not cleanly or aseptic. It is extremely dangerous, unnecessary, hence culpable. Even in most cases of profuse arterial hæmorrhage, mesial constriction with an extemporized tourniquet, as, for instance, the "Spanish windlass," or digital compression of the afferent arterial trunk, can be successfully employed, while the patient is transferred into a suitable locality, where permanent relief can be safely afforded by deligation.

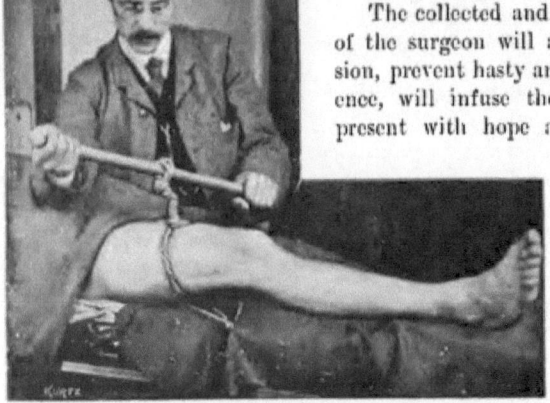

Fig. 9.—Extemporized tourniquet—"Spanish windlass."

The collected and businesslike manner of the surgeon will at once allay confusion, prevent hasty and injurious interference, will infuse the patient and those present with hope and confidence, and will facilitate well-considered and rational action.

As a rule, the fate of a fresh wound is determined by the views and training of the physician who first attends to it. If the patient be so fortunate as to fall in with a man fully imbued with the spirit, and familiar with the practice of aseptic surgery, he is truly to be congratulated, because his chances of avoiding suppuration are excellent. If his first attendant be one of those to whom wound infection by dust or filth adherent to hands or a probe be a myth, he is to be pitied. Without previous cleansing, immediate probing of the gunshot wound of a vertebra, for instance, accompanied by digital exploration, will be performed on the patient extended on a mattress laid on the dirty floor of a railroad station.

Of course, the bullet will not be found, and nothing beyond the infection of the wound will be accomplished. A dressing will be applied anyway, and the patient will be taken home. Suppuration, that otherwise might have been avoided, will surely set in, and the patient is doomed. No amount of consulting can devise a way, for no surgical skill can establish efficient drainage of the inaccessible parts of the wound. The chances for recovery were thrown away here from the outset.

On taking charge of a fresh wound, the fearful and often irremediable consequences of a first false step should be always present to the mind of the surgeon, and his attention should be directed chiefly to the avoidance of septic infection. A temporary aseptic dressing having been applied, the general condition and comfort of the patient should be looked to by the administration of stimulants or sedatives. After transfer home or to a hospital, the necessary measures for permanent relief should be carried out as soon as the patient's general condition will permit.

## II. DEFINITIVE RELIEF.

Preparations, comprehensive and thorough, as required for an aseptic operation, should now be made in the manner described in Chapter II.

The patient is well stimulated if necessary, is anæsthetized if the case require it, and, his clothing being removed by cutting or in some other proper manner, he is placed on the operating table.

After this should come a careful cleansing and sterilization of the surgeon's and his assistant's hands by scrubbing with soap and brush and immersion in a germicide lotion, followed by a likewise thorough cleansing of the integument in the vicinity of the wound. Plenty of soap-lather, with the use of a razor, scrubbing with soap and brush, rubbing and washing off with a solution of corrosive sublimate, will soon accomplish this.

1. **Contaminated Wounds.**—The character of further procedures will have to be decided by the answer to the question: *Is the wound clean or is it contaminated?* Gross evidence of contamination, such as, for instance, street-dirt imbedded in the wound or the clots, or the knowledge that the wounding was done with a filthy instrument, as, for instance, a foul and fetid butcher's cleaver, will answer the question in the affirmative. In these cases the leading object should be thorough cleansing and disinfection of the wound, followed by very comprehensive measures at drainage. If the external wound be small, it has to be well enlarged, so as to afford a good insight. Every nook and recess of the wound should be systematically gone through, cleansed of clots and dirt, thoroughly irrigated, and well drained. Great care must be taken not to overlook recesses, as one particle of filth left behind unawares, may cause very grave trouble.

Drainage of the more remote recesses should be made as direct as possible; that is, a rubber tube carried to the surface from a distant corner of the wound through a properly placed counter-incision, will be more direct, therefore better, than a long tube bent or twisted and brought out through a distant opening.

Hæmorrhage must also be, of course, well stanched by ligature or otherwise.

Divided tendons, nerves, muscles, or fractured bones are next united by suture, and, if the edges of the wound be viable, they are also approximated by sutures. Where extensive loss of substance precludes uniting of the edges, or where uncontrollable oozing prevails, the wound should be packed. This is best done by first lining the entire wound with one layer of iodoformized gauze, within which are packed a suitable number of loose balls of sublimated gauze. After a final irrigation and clearing of the drainage-tubes, the wound and its vicinity are enveloped in a moist dressing that should be protected from evaporation by a large piece of rubber tissue or Mackintosh. In case of fracture, the limb is supported by a splint. On account of their frequency, and their gravity in case of suppuration, injuries to the cranium and their treatment may receive special mention.

*Scalp-wounds* have been held undeservedly in bad repute on account of their alleged tendency to suppurate. They heal as kindly as, and in fact, on account of their great vascular supply, heal better than, many other wounds, provided that they be first carefully cleansed, well drained before suturing, and sufficiently protected by a suitable dressing from subsequent contamination.

In case of a greater denudation of the cranium, the loose scalp should be raised (after shaving and thorough cleansing of the skin), blood-clots should be turned out, and the wound well irrigated and rubbed out with corrosive-sublimate lotion. A bistoury is inserted into the deepest part of the recess formed by the flap, and thrust out through it. Into this opening a short piece of slender tubing is placed, after which the edges of the wound are brought together by an exact line of sutures. A dry dressing will be proper in these cases.

If a compound fracture of the cranium be present, the first care of the surgeon should be to ascertain that no septic material remain imbedded in the recesses of the wound. The external wound must be adequately enlarged to permit of thorough inspection, cleansing, and disinfection. After this the nature of the fracture should receive due attention. Often foreign matter, such as street dirt or the hair, will be found impacted between the depressed fragments. In this case the edges of the fractured area are to be sufficiently removed by the aid of the chisel and mallet to permit of an easy elevation or extraction of the fragments and foreign matter. This is followed by a thorough cleansing of the exposed dura mater, especially of the recesses formed by the stripping off of this membrane from the internal surface of the skull. If the foreign body or fragments of bone have injured the dura mater, the rent must be widened, in order to permit of careful extraction and cleansing. A slender drainage-tube having been inserted in the dural rent, its edges are approximated by a few catgut stitches. The chips or button of bone, removed either by the chisel and mallet or by the trephine, should be saved and preserved in a boro-salicylic solution till the operation be completed, when they are replaced in the cranial defect, over which the skin is united by an external suture, leaving sufficient space for the emergence of the drainage-tube. A moist dressing is appropriate in these cases.

CASE.—Regino Libertello, aged thirty, an Italian cobbler, was admitted to Mount Sinai Hospital, on November 30, 1889, with a fresh scalp-wound, two inches long, over the left half of the occipital bone, and parallel to the sagittal suture. On exposure of the bottom of the wound a deeply depressed fracture was noticed, and within one of the fissures a small bundle of the patient's hair was found to be imprisoned, where it had been evidently driven by the brickbat that had caused the fracture. The general condition was good, no cerebral symptoms of any gravity being present, with the exception of a marked dilatation of the right pupil. The presence of hair within the cleft of the fracture was considered an ample indication for thorough disinfection; hence the man being anæsthetized, the scalp shaved and disinfected, the wound was well enlarged, and the periosteum raised from the skull. The edges of the oblong depression were carefully removed by the chisel and mallet, the loose fragments of the outer and inner table extracted, when the uninjured dura mater came in view. A portion of the fractured inner table was left *in situ*, and, after thorough wiping and irrigation, a fillet of iodoform gauze was applied to the dura, and carried out of the wound by its lower angle. A number of external catgut stitches and a moist dressing completed the procedure. The depressed area was ovoid, measuring one by three quarters of an inch. Immediately after the operation the pupils were equal in size. The drainage was removed on December 4th. The patient was discharged cured on December 14th.

If the steps described above are adequately taken, as a rule no septic fever and no destructive suppuration will follow an accidental injury; though aseptic fever, due to absorption of non-decomposed secretions, may often enough be observed.

Tissues or bone whose vitality was compromised by the crushing force causing the injury will be gradually detached. This will be accompanied by a rather scanty secretion of thinnish sero-pus, and very little fever, if any.

CASE.—P. S., agent, aged forty-six, was, January 26, 1886, while in a state of deep alcoholic intoxication, run over by a heavily laden truck, and was at once brought to the German Hospital, where he was anæsthetized about two hours after the accident. There was hardly any shock noticeable. The soiled and torn garments were cut away from the extremity, which was then carefully cleansed from adherent street-dirt and blood-clot by the application of soap, hot water, the scrubbing-brush, and a razor. There was very little external hæmorrhage present, but the appearance of the member gave unmistakable evidence of extensive and serious injury. A laceration of the integument in front of and corresponding to the middle of the left leg, four inches long, was found. Also compound comminuted fracture of the tibia and fibula. The tibia was broken into four, the fibula into at least three fragments. Severe hæmorrhage from the torn tibialis antica artery had caused an enormous infiltration of the leg, which had attained double the size of its fellow, and was quite cold. The integument was much discolored in the vicinity of the external wound, and very tense. Elsewhere the skin appeared abnormally pale and glossy. Esmarch's bandage was applied, the external wound was enlarged to about eight inches, the massive clots, some containing particles of street dirt, were turned out of the muscular interstices, and from between the fragments one perfectly detached piece of the tibia was extracted. From the middle of the main cavity into which the fragments protruded, a counter-incision was made backward through the calf of the leg, into which a large-sized drainage-tube was placed. Three more counter-incisions, corresponding to as many recesses, were made. The torn artery could not be found. A large moist dressing was applied, and the limb fixed between two well-padded lateral board splints, held together by a pure gum bandage. Moderate oozing soiled the dressings somewhat during the following night, wherefore the elastic bandage was removed in the morning, and the soiled parts of the underlying dressing were well dusted with iodoform. Another envelope of gauze was laid on top of the old dressings and the splints were replaced and fastened with muslin bandages. *Jan. 31st.*—The patient's temperature had not risen above 100° Fahr., he complained of very little pain, no hæmorrhage had followed, the circulation of the limb was good, hence the dressings were not disturbed until this date. The wound was found to be in good condition; some blood-clots were still adherent to the drainage-tubes. Wound was re-dressed and limb put up in a solid plaster-of-Paris splint. In the beginning the dressings were changed about weekly; from February 15th, every fortnight. *March 3d.*—After the exuberant granulations surrounding it had been scraped away, the entire belly of the tibialis anticus muscle was found to be of a grayish-yellow color and necrosed. It was not putrid, although a good deal of secretion was present. The wound was enlarged and the necrosed muscle was removed. Thereafter the secretion diminished materially, although five sequestra were consecutively removed. Consolidation was rather slow, but finally complete, so that the patient was able to walk without support in October of the same year. Shortening about one inch. If left to themselves, deep-seated and extensive contaminated wounds, presenting a small external orifice, are, for obvious reasons, most dangerous. Free exposure, thorough-going cleansing and disinfection, together with good drainage, are then imperative.

2. **Aseptic Wounds.**—The nature of many wounds and their causation are such as to preclude the probability of contamination. Most gunshot wounds and many compound fractures belong to this class. In these cases interference should be very discreet. It should consist of thorough cleansing of the integument, ordinarily an aseptic *dry dressing*, or, in case of doubt, of superficial drainage and a *moist* dressing, together with reduction and support and retention by splint where a fracture requires it.

CASE.—John D., aged thirty-two, December 4, 1885, sustained a compound comminuted fracture of the upper half of the tibia by a horse-kick. Dr. W. T. Kudlich, of Hoboken, saw him immediately after the accident, cut off the clothing, disinfected the vicinity of the small wound, and dressed it amply with iodoform gauze. A temporary splint was also applied, and *probing or examination was thoughtfully refrained from.* The patient was brought to his home, where, the next day, he was anæsthetized. The temporary splint and dressings were removed, the vicinity of the wound was carefully cleansed and disinfected, and, with the observance of all necessary *cautelæ*, a thorough examination of the injury was instituted. A compound comminuted fracture was easily made out, and three loose fragments of bone were removed. The laceration of the soft parts and ecchymosis were found very moderate, and confined to the tissues anterior to the tibia. A couple of short drainage-tubes were inserted into two recesses, and, the wound being well irrigated, was enveloped in a moist dressing. The limb was put up in a solid plaster-of-Paris splint, with the knee bent at an obtuse angle, and was suspended from a frame.

The temperature remained normal or almost normal throughout.

*Dec. 18th.*—Appearance of wound normal. Moderate secretion due to limited necrosis of a loose fragment of bone. *Dec. 28th.*—Second change of dressings. Exuberant granulations have filled up the defect. *Jan. 18th.*—A fenestrated silicate-of-soda splint was applied. The secretion continued to be scanty. In May consolidation was perfect, but a small sinus remained until October, when, after the extraction of several small spicula of bone, definitive healing of the wound ensued. No appreciable shortening resulted.

NOTE.—In the more extensive injuries of the extremities caused by crushing force, the gravity of the case hinges more upon the extent of the injury to the soft parts than to the bones. A compound fracture by direct force—for instance, the blow of a hammer upon the tibia, where the crushing and laceration of the soft parts are comparatively limited—is by far not as dangerous as, for instance, the stripping off of the entire integument of the lower extremity, or the crushing and pulpification of the large muscles, vessels, and nerves situated on the anterior and internal aspect of the thigh, though these latter injuries be uncomplicated with fracture. The shock and the presence of extensive thrombosis, in addition to the fact that, with the large quantity of mortified tissues, preservation of the aseptic state is extremely uncertain and difficult, class these injuries among the most grave and dangerous.

3. **Gunshot Wounds.**—The fact that most fresh gunshot wounds are aseptic has been pointed out by **Esmarch**, and is now well established. Reyher and Bergmann's experiences in the Russo-Turkish war put the fact beyond controversy.

Wise precaution against infecting a fresh gunshot wound will be richly rewarded by excellent results. In most cases cleansing and disinfection of the skin in the vicinity of the points of entrance and exit, together with a *dry dressing*, will be sufficient. If the case is complicated by fracture, a suitable splint, preferably plaster of Paris (Bergmann), should be added.

If the course is free from septic fever and suppuration, this will be manifest within the first three or four days; in that case, the first dressing and the splint can be left undisturbed for the length of time required for the accomplishment of bony union.

Flesh-wounds will be healed within a fortnight or three weeks. Gunshot fractures will require a longer time for healing and consolidation, but are in no way different from ordinary compound fractures.

The projectile will cause very little or no irritation in aseptic—that is, non-suppurating—gunshot wounds. Generally it will become encysted. Search for the projectile in the bottom of the wound is rarely indicated. It can occur, however, that pressure of a projectile or its fragment, or a sharp spiculum of bone on a nerve-trunk, may necessitate search and extraction. This must be done under careful asepsis.

It is even not necessary to remove a projectile lodged under the skin. It will do no harm if left there until the channel which it cut by its passage through the tissues is obliterated, when its removal by incision can not lead to an infection of the bullet-track.

In cases of injury to large vessels or the intestines, immediate interference can not be delayed, but should be carried out under most rigid antiseptic precautions.

NOTE.—Recent successes (W. T. Bull) achieved by immediate laparotomy and suture of the wounded intestines justify the procedure.

Where the nature of the charge or the short distance from which the shot was delivered makes the entrance of a gun-wad probable, or where the examination of the superjacent clothing shows a large defect, rendering the probability great that shreds of soiled cloth have been carried to the bottom of the wound, dilatation, search, and extraction may be indicated. But it is better to wait in cases of doubt, as even these foreign substances may become encysted and harmless.

Should suppuration follow, the patient will not be worse off than if a fruitless search had been made at the outset, and the use of the suppurating track as a guide will materially facilitate the finding of the irritating body.

NOTE.—Reyher's observations (Volkmann's "Sammlung," Nos. 142, 143, 1878) may serve as a fair sample of the radical change that has taken place in the results of the treatment of gunshot fractures.

Gunshot fracture of the knee-joint was formerly considered an indication for immediate amputation. Reyher treated eighteen fresh cases aseptically—that is, by simply cleansing and disinfecting the skin about the wound, and occluding the same by an antiseptic dressing. Where the wound was gaping, or where there was ground to suspect the entrance of dirt or shreds of clothing into the bullet-track, dilatation, irrigation, and extraction of the foreign body, with subsequent drainage, was practiced before the wound was sealed up. Of these eighteen cases, fifteen recovered, with movable knee-joints—83·3 per cent of recoveries. One patient died of fatty embolism in twenty-four hours after the injury; another of hæmorrhage from the divided popliteal artery and vein on the fifth day; and the third one of pyæmia.

Of nineteen that came under his care several days after the reception of the injury, with well-established suppuration, eighteen died, and one recovered with a stiff joint. In spite of an energetic antiseptic treatment by incisions, drainage, and irrigation, a mortality of 85 per cent was noted.

Of twenty-three that were not subjected to any form of antiseptic treatment, twenty-two died, one survived, a mortality of 95·6 per cent—clearly justifying the practice of the older surgeons, who at once performed amputation in cases of gunshot fracture of the knee-joint.

Infected accidental wounds or gunshot injuries that become the seat of suppuration can be classed under the heading of phlegmonous processes, and their treatment will be dealt with in a subsequent chapter.

## CHAPTER V.

### SPECIAL APPLICATION OF THE ASEPTIC METHOD.

#### A. GENERAL PRINCIPLES.

#### I. TECHNIQUE OF SURGICAL DISSECTION.

MODERN surgery demands that the invasion of the uninflamed tissues of the human body by the surgeon's knife should be surrounded by all the safeguards that are known to be effective in preventing suppuration. The mortality following operations sanctioned by pre-antiseptic surgery has been remarkably depressed by a conscientious and intelligent adherence to the principles of surgical cleanliness. A large number of recently devised useful operations have become legitimate under the assumption that suppura-

# SPECIAL APPLICATION OF THE ASEPTIC METHOD.

tion can be excluded. The cranium, large joints, the tendinous sheaths, and the peritoneal cavity are now safely accessible for curative or even diagnostic purposes.

The statement that a real observance of asepticism offers a sure guarantee against suppuration, be the performance of a bloody operation however clumsy, rough, and unskillful, is true, but can not be pleaded as an excuse for the absence of that equipment of pathological and anatomical knowledge and technical skill which go toward forming a good surgeon. Although the general standard of safety and success in surgery has been considerably raised, excellence will be attained by those only who unite the qualities of a good diagnostician, pathologist, and anatomist with the tact, energy, and technical skill of the accomplished surgeon.

The technique of surgical dissection is based upon principles, the observance of which enables us to safely explore and manipulate any accessible part of the human body.

Aside from the ever-present desideratum of preventing infection, the avoidance of accidental injury of important organs and the control of hæmorrhage first deserve attention.

*The principle of doing every step of an operation under the guidance of the eye, is the most important discipline of dissection to be acquired.* It should never be sacrificed without the most stringent necessity. Its non-observance is the source of most that is embarrassing, appalling, and disastrous in operative work.

*Upon this principle is based the rule to always make an ample and adequate incision,* which should be gradually deepened layer by layer, until the part sought after is freely exposed.

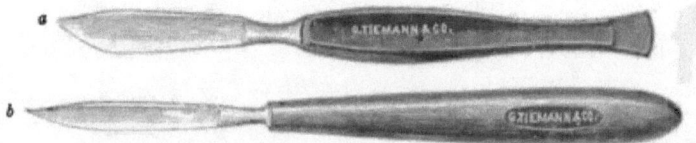

FIG. 10.—*a*, Bellied scalpel for cutaneous incision. *b*, Sharp-pointed scalpel for deeper dissection.

For the cutaneous incision a bellied scalpel, held like a fiddle-bow, is the most useful. A careful and clean incision will insure a lineal cicatrix. As soon as the skin is divided, the subcutaneous vessels will become visible. If they are crossing the line of incision, they should be grasped between two artery forceps, divided between, and safely tied off with catgut. In cutting through the fascia, the grooved director used to play an important part in former times. Its use has been supplanted by a safer mode

FIG. 11.—Manner of holding the knife for the cutaneous incision. (Esmarch.)

of preparation, known as *cutting between two thumb-forceps*. The author once observed that, in thrusting a grooved director underneath the fascial coverings of a hernia, the hernial sac was opened, and the adherent gut nearly torn through. As it was, only its serous covering was lacerated. In another instance, puncture of the deep jugular vein by the point of the grooved director happened, and led to very annoying hæmorrhage from the deepest parts of the wound, which made exposure and ligature of the injured vein very difficult. It may be said that, unless very thin layers are taken up by the grooved director, the surgeon never can tell beforehand what he is going to cut through while using it. Veins especially are easily injured, as, being put on the stretch, they become empty. Stretched, they lose their identity to the eye, and look exactly like ordinary connective tissue.

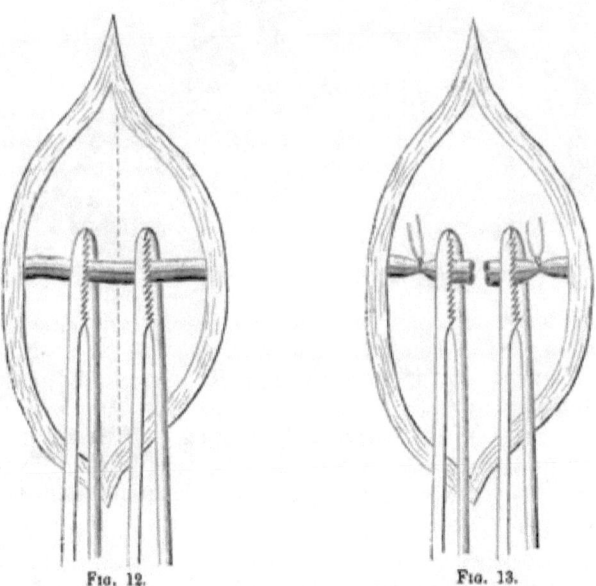

FIG. 12.   FIG. 13.
Securing and tying vessels traversing the line of incision.

Cutting between two forceps has the peculiarity that, a thin layer of tissue being raised before each cutting, air enters into and rarefies its meshes, rendering clearly visible the vessels, which can be easily isolated and secured before they are cut. From this result two very great advantages: First, the patient does not lose one drop of blood from a vessel secured previous to its division; and last, but not least, the wound remains dry and clean. No time is lost in hunting for a retracted vessel in a pool of blood, there is no occasion for hasty and rough sponging, and everybody preserves an even tenor of mind very essential to success.

The advice, so often met with in text-books, that the knife should be laid aside where the tissues are loose, and that tearing or scraping with forceps or the finger-nail is safer, is, to say the least, very questionable. This advice is born of the fear of unexpected hæmorrhage, which, however, can be always avoided by cutting between two forceps. The beginner, especially, is prone to carry this mode of blunt preparation to great lengths, and laceration of large veins, the peritoneum, or cysts is the result.

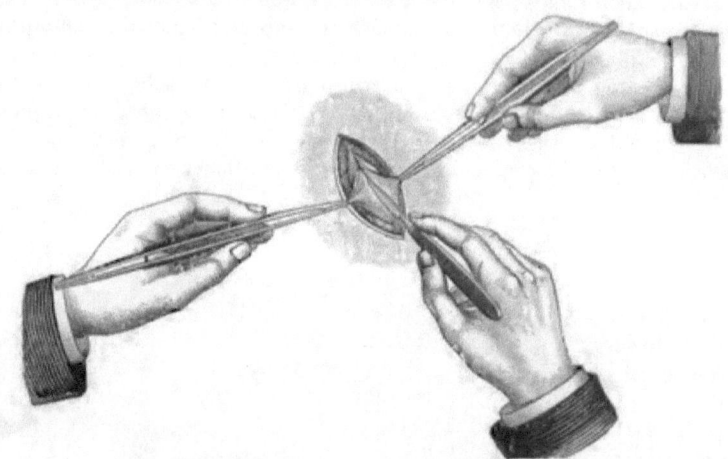

FIG. 14.—Cutting between two thumb-forceps. (Esmarch.)

A consideration of no small importance is the fact that a clean-cut wound will sometimes heal in spite of some local reaction and fever. This means, that the blood- and lymph-vessels of the parts concerned being not much bruised, sufficient nutriment is carried to the walls of the wound to overcome a moderate degree of micrococcal infection. Where the nutrition of the parts is seriously interfered with by tearing and bruising pertinent to blunt dissection, a much higher degree of asepticism is required to secure absence of suppuration.

NOTE.—The old surgical tenet, that torn and bruised operative wounds are not prone to heal kindly, is based upon the fact that devitalized tissues form an especially favorable pabulum to microbial development. The observation that very well nourished tissues, as, for instance, those of the face, will heal readily under almost all circumstances, and without the observance of antiseptic precautions, is explained by the fact that they are very well vascularized, and a rich supply of oxygenated blood is one of the strongest germicides. We often saw the parts become red, swollen, and painful, and were expecting suppuration, but in vain, as all the local symptoms and the fever receded, and good union followed.

As the wound is gradually deepened, sharp or blunt retractors should be employed to well expose to view its bottom, in which is centered the surgeon's interest. The skin, muscles, fasciæ, tendons, or the periosteum can be held back by sharp retractors; vessels and nerves, the peritoneum, and friable glands or cysts should never be hooked up by them, blunt retractors deserving the preference.

Most of the retractors commonly sold by the instrument-dealers are worthless. A useful retractor must have a good, ample curve, a proportionate and safe grasp, a smooth, solid handle, and a strong shank, so as to be able to sustain a good deal of pressure without bending or breaking.

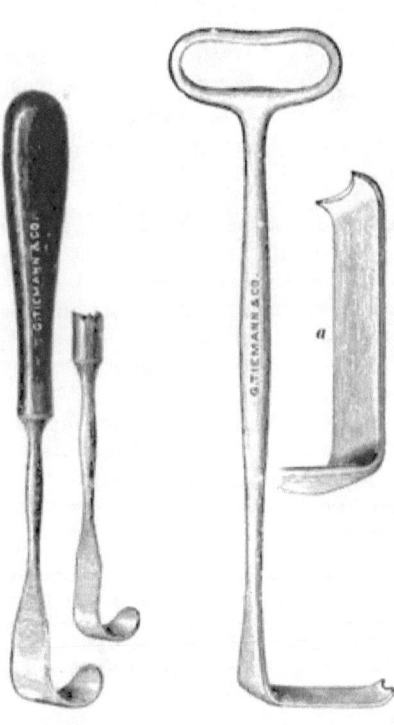
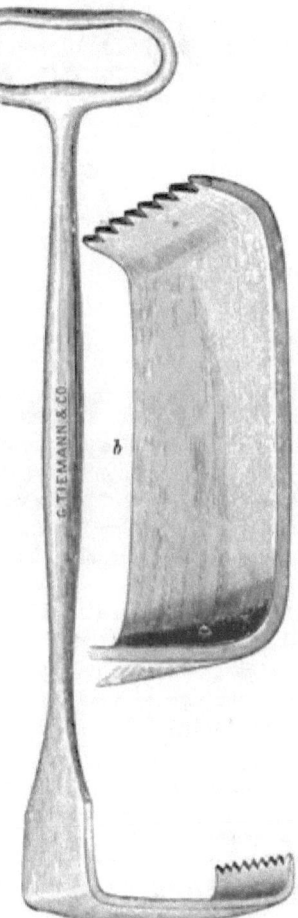

Fig. 15.—Small blunt retractors.

Fig. 16.—Medium-sized blunt retractor. *a*, Actual size.

Fig. 17.—Large-sized blunt retractor. *b*, Actual size.

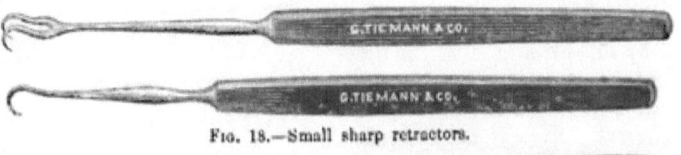

Fig. 18.—Small sharp retractors.

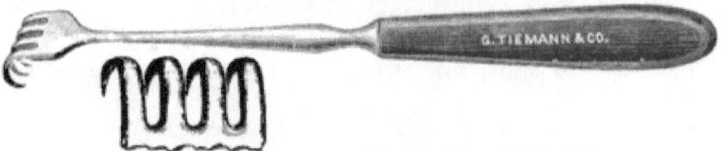

Fig. 19.—Large four-pronged sharp retractor (Volkmann).

The shapes and sizes most useful for general surgical work are depicted by Figs. 15, 16, 17, 18, and 19.

The deeper the knife penetrates, the nearer it approaches important organs, the shallower its strokes should become. A somewhat pointed scalpel should be used, and its strokes, especially where they sever dense tissues, should be made with the very point of the instrument, which should be held like a pen, but rather steeply.

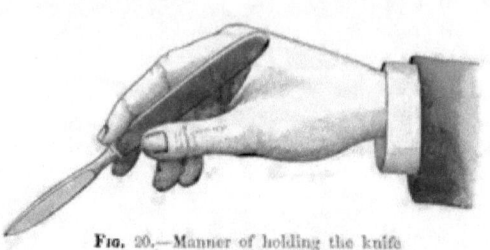

FIG. 20.—Manner of holding the knife for deep dissection. (Esmarch.)

Use of the grooved director, or the scissors, or the *sickle-shaped bistoury* in the bottom of a deep wound is always unsafe, as it may lead to unexpected hæmorrhage or something worse. Especially dangerous is the last-named instrument, as its very nature renders impossible the observance of the principle of *not cutting what we do not see*. It cuts from within outward, takes up unseen tissues, and may become the cause of unnecessary trouble and embarrassment.

Should it become evident, as the wound deepens, that the first incision is inadequate, and that, in order to afford access, its edges must be subjected to severe tension, and that work is thereby cramped, an extension of the first incision is in order. This should be done methodically from without inward until the wound is sufficiently enlarged.

NOTE.—The author once saw an ovariotomist make abdominal section with exaggerated minuteness, layer by layer, until the belly was opened, tying each small vessel as it was exposed. When a digital exploration had made evident the insufficiency of the incision, he enlarged it *by cutting through the entire thickness of the abdominal wall with a stout pair of scissors at one stroke*. Of course the incision was uneven, some layers being further cut than others, hæmorrhage was considerable, and finding and securing of the retracted vessels not easy.

The shape of every operation wound should be such, if possible, as to afford the best conditions of access, and, later on, for natural drainage. *The funnel shape* (Fig. 21, A) is meant by this—that is, that the first incision should be the longest, the next one a little shorter, the last one the shortest. Even if no drainage-tube is inserted in such a wound, as long as the closing stitches are not too tight and too many, the interstices of the suture will afford ample drainage.

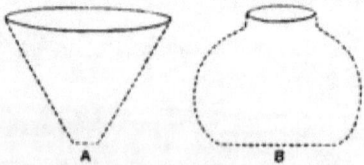

FIG. 21.—A, Funnel-shaped wound. B, Bottle-shaped wound.

*Bottle-shaped wounds* (Fig. 21, B) are disadvantageous in every way. They result from a too small cutaneous incision, are uncomfortable and unsafe during the operation, and after closure offer poor conditions for natural drainage. They always require a drainage-tube, and, even with a tube, if not absolutely aseptic, become a very hot-bed of suppuration, as the discharges of infected recesses may not find ready egress.

Where the incision must be carried through *condensed or inflamed tissues*, preparation between two forceps will be generally impossible. All the more stress should be laid upon the amplitude of the first cut, and *upon the adequate dilatation of the wound by serviceable and solid retractors*. As the wound deepens, the hooks should be alternately released and inserted deeper, so as to follow up closely the work of the knife.

On account of their hyperæmic state and density, hæmorrhage will be found a great deal more profuse in inflamed than in normal tissues. The presence of vessels will become manifest only by the hæmorrhage caused in cutting them. The smaller arteries can be easily controlled by increasing the tension exerted by the retractors on the edges of the wound. Larger vessels must be tied off. But the density and often the brittleness of the tissues prevent grasping of the bleeding points with artery-forceps, hence another expedient must be used.

An ordinary curved, or, better, a perfectly round hæmostatic needle, armed with catgut, is carried with a needle-holder through the tissues adjacent to the bleeding point in two or three

Fig. 22. Hæmostatic needle.

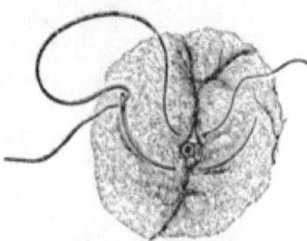

Fig. 23.—Manner of applying hæmostatic needle (Esmarch).

stitches, so as to surround it like a purse-string. Being tied, it closes the bleeding orifice.

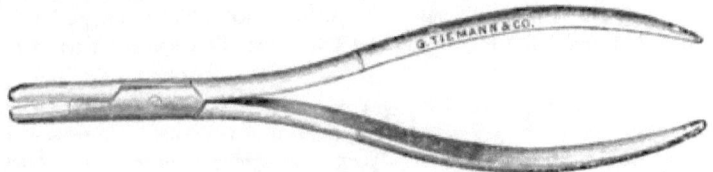

Fig. 24.—Dieffenbach's needle-holder.

When a plexus of considerable vessels, especially veins, is encountered in the bottom of a wound, or where, for some reasons, it is desirable to hasten operative work, the employment of mass ligatures will be found an expedient and safe way to rapid progress.

*Thiersch's spindle and forceps* is an invaluable apparatus for applying mass ligatures to dense tissues in difficult and deep situations. A blunt, probe-pointed, curved needle and a straight ivory spindle, armed with stout silk or catgut, and an appropriate forceps, make up the apparatus. The

# SPECIAL APPLICATION OF THE ASEPTIC METHOD. 43

probe-pointed needle is grasped by the beak of the forceps, and is cautiously insinuated under the plexus or mass to be tied off. Veins and arteries are not apt to be injured by the blunt point, as they are inclined to slide off from it. As soon as the ligature thread is drawn through under the mass, a knot is made, and, the spindles serving as solid handles, it can be tightened with a great deal of firmness and security. The mass can be safely divided between two of these ligatures.

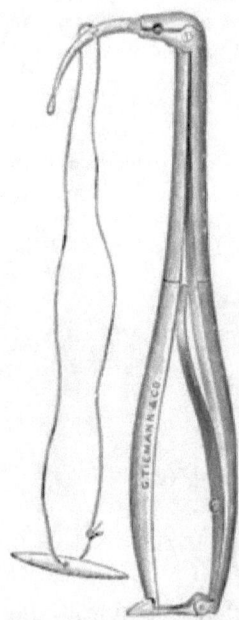

Fig. 25.—Thiersch's spindle apparatus.

The treatment of veins in operative wounds is similar to that applied to arteries. There are some points, however, that constitute an important difference, and deserve special attention. The tension exercised by retractors is very apt to obliterate the normal characteristics of veins. The dark blood they contain is driven out of them, and they can not be distinguished from ordinary connective tissue. Especially in blunt preparation, lacerations of veins are apt to occur and cause serious difficulty. To find a bleeding vein is not as easy as to locate an injured artery, readily marked by its jet of blood. And, even if the bleeding point is recognized, it is not always easy to stop a torn vein, as the laceration may be, and in fact frequently is, an irregular and extensive slit. On the other hand, venous hæmorrhage can often be effectively checked by simple pressure or plugging. If the finding of a torn and retracted vein should be difficult and involve too much time, it will be found a good expedient to plug up the place from which the hæmorrhage issues with a strip of iodoformed gauze, held in place by light finger-pressure until coagulation occurs. Formerly the author used a bit of sponge for this purpose, but the following experience has shown that sponge is not a safe material:

CASE.—Theresa Kops, housewife, aged forty-eight. *February 10, 1883.*—Amputation of left breast, with evacuation of the contents of the axilla for scirrhus of the mammary gland. Wound sutured throughout; drainage by counter-incision through latissimus dorsi. Aseptic dressing. After feverless course, first change of dressings on February 21st, when the wound was found united. Drainage-tube was withdrawn. *Feb. 22d.*—Severe chill, phlegmonous infiltration of axillary region. *Feb. 23d.*—Incision through cicatrix, and evacuation of a large quantity of pus, followed by a small fragment of sponge; drainage. Uninterrupted healing of the axillary abscess by granulation.

In removing the axillary glands a small vein was put on the stretch, and, being ruptured, retracted so far that it could not be found. A good-sized sponge was stuffed temporarily into the recess from which the hæmorrhage issued, and the operation was finished. When the sponge was extracted, it came away, as usual, with some resistance, due to the matting of the blood-clot into its meshes. The sponge was a very soft and brittle one, and its own cohesion was apparently less than the cohesion of its surface to the tissues matted to it. A small portion of the sponge tore off and was left behind in the wound. It caused no trouble for eleven days, and only after the disturbance of its relations by the removal of the drainage-tube did its decomposition set in. Since that time a strip of iodoformed gauze was used for the mentioned purpose by the author, which would not tear, and could not be overlooked, as its end is carried out of the wound for a mark.

Close attention to the details enumerated above will secure a dry and easily accessible wound. No sudden and uncontrollable hæmorrhage will occur to create flurry or alarm; no embarrassment will cause undue haste or an ill-considered move; the patient will fare well, as, even with the seeming deliberation, the operation will be speedily accomplished, and, what is the main thing, no unnecessary loss of blood will be sustained.

## II. SUTURES.

Primary union with a linear cicatrix is the ideal of the healing of an aseptic wound. As it depends to a great measure upon an exact coaptation of its edges in such a manner, that circulation of the integument should not be interfered with, and as exact coaptation under varying circumstances requires a variation of the procedure, a discussion of the important differences in the technique of suturing may receive some consideration.

Exact coaptation of the corresponding points of the edges of the wound by finger-pressure or otherwise, *before* and *while passing the stitch*, is the first condition of a true suture. Where there is no considerable loss of integument, and where the edges of the wound are equally thick and have sufficient body, this can be done easily by compressing the edges between the index and thumb until they touch on the same level. A good-sized *curved needle* is then passed through both edges of the wound, which will be retained in their correct relation by simply tying the catgut thread.

Where one of the edges is thick and the other rather thin, coaptation is more difficult, as the thinner edge is apt to slip back, leaving a portion of raw surface exposed. Or where both edges of the wound are thin, as, for instance, on the neck, the scrotum, and the dorsum of the hand or foot, they have the tendency to curl under, raw being in contact with epidermidal surface. Both of these relations will produce an uneven line of suture, and will frustrate exact primary union. Partial healing by granulation is then unavoidable.

Under these circumstances the best result will be achieved by the following plan: The edges of the wound are brought together and pinched up by index and thumb in such a way as to form a continuous ridge, on

SPECIAL APPLICATION OF THE ASEPTIC METHOD. 45

the top of which should appear the line of incision. A *straight needle* is thrust transversely through the base of this ridge, and the suture is tied while the fingers still retain their position. The appearance of the completed suture is rather grotesque; but, when the stitches are absorbed or removed, the peculiar-looking ridge will flatten out spontaneously, and the result will be a beautiful fine cicatrix. See Figs. 26 and 27.

In tying a surgical knot, a certain little knack will be found extremely useful, especially where good assistance can not be had. It consists in jamming down the first or double cast into the angle of the suture nearest to

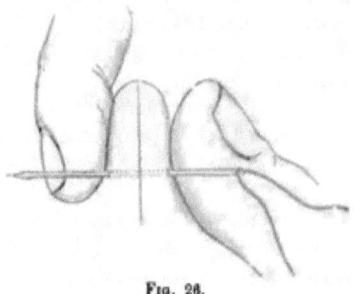

Fig. 26.

the operator by a slight jerk, made upon the distal end of the thread, while the mesial one is held steadily on the stretch. This jamming of the catgut will be just sufficient to hold the edges of the wound together, until with the second cast the knot is tied. It will even hold together edges approximated with some degree of force.

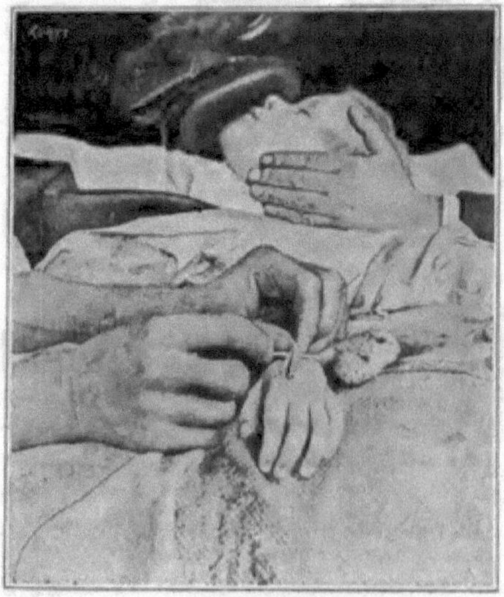

Where there is much loss of integument, as in many cases of breast amputation, or where the sutures may have to stand a good deal of strain, as, for instance, the abdominal stitches after ovariotomy, aside from the sutures of coaptation above mentioned, *supporting or retentive sutures* are necessary.

They have to embrace a good deal more integument than the finer stitches, and should be inserted from one half to two inches away from the edges of the wound. Lateral concentric pressure by the hands of an assistant will very much facilitate the proper placing of these sutures.

Fig. 27.

They can be made in several ways. The simplest one is to pass three or four or more interrupted catgut sutures of wider scope, and then to tie them while the edges of the wound are firmly supported by an assistant (Fig. 28). The required number of finer stitches is passed afterward. Another good way is the application of a mattress suture, illustrated in Fig.

29, combined with a continuous coaptation suture, all done with one piece of catgut.

Where silver wire or silkworm-gut are available, the quill suture or Lister's button suture will give much satisfaction. Both of these forms of

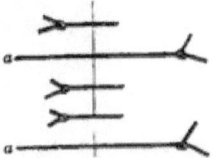

Fig. 28.—*a*. Interrupted retentive suture.

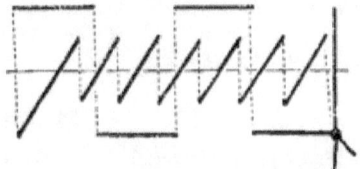

Fig. 29.—Combined mattress suture and Glover's stitch.

retentive suture will be very proper after abdominal operations. For the quilled suture, small cylindrical pieces of well-disinfected wood will answer. Buttons for Lister's retentive suture (Fig. 30) are cut out of stout sheet lead with a pair of scissors. It is sold by dental-supply traders under the name of "suction lead." The wire or gut is armed with a perforated shot,

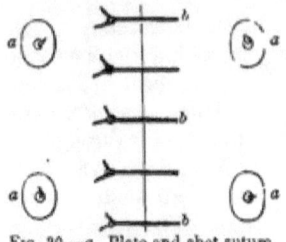

Fig. 30.—*a*. Plate and shot suture. *b*. Interrupted suture.

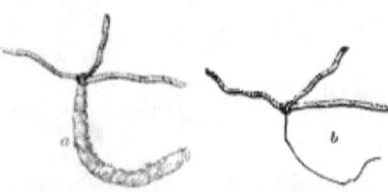

Fig. 31.—*a*. Catgut suture from suppurating stitch-hole. *b*. Catgut from sweet stitch-hole, nearly absorbed.

which is clamped to its end; over this is slipped a button. The suture is passed, and the needle is unthreaded. Over the second end a button and shot are slipped, the stitch is tightened, and the shot is clamped.

In uniting more extensive wounds, it is better to commence at the middle and not at the angle, as the latter way may result in uneven distribution and puckering.

After abundant trial and comparison, the conclusion was arrived at by the author that, as a rule, the interrupted suture is in every way preferable to the continuous one. The exceptions are mentioned at the proper place.

The chief advantage claimed for the continuous suture—namely, the saving of time—is illusory. As regards safety in holding and exactitude of adaptation, the interrupted suture has no peer.

*Secondary Suture.*—Kocher and Bergmann have taught us to combine the advantages of the open treatment with those of the suture of wounds. Where it is deemed unsafe, for various reasons, to close a wound at once by suture, the wound is packed from the bottom with iodoform gauze. A suitable number of silk-worm gut or silver wire stitches are then passed through the edges of the wound. They are not closed, but their ends are fastened together and arranged alongside of the wound, which is dressed as usual. Thus the escape of the serous discharges is absolutely unimpeded, and no retention can take place. On the fourth day the packing is removed, and, by the aid of the stitches left *in situ*, the wound is closed, provided that its condition is sweet. Wounds treated thus behave like fresh ones, and usually heal by agglutination.

Secondary sutures are used with great advantage, also, for hastening the closure of widely gaping wounds.

## III. DRAINAGE.

Small aseptic wounds of a favorable, that is funnel shape, do not require drainage by rubber tubing. As few stitches should be taken, however, as possible, to permit the escape of the oozing between them. Small wounds of bottle shape will do very well with a narrow strip of iodoform gauze placed in one angle for capillary drainage, which should be renewed on the third day. Larger wounds, especially those with a sinuous cavity, require drainage by rubber tubing.

Before using the tube, a number of oval holes should be clipped out of its side.

"*Through drainage*," with a view to subsequent irrigation, is best effected by placing the mesial end of the tube just within the cavity to be drained. Drawing a long piece of tubing transversely through the cavity does not afford the best conditions for thorough irrigation, as

FIG. 32.—Perforated rubber drainage-tube.

the bulk of the irrigating stream will pass directly through the tube without entering the cavity at all. Where two or more short pieces of tubing are placed just within the cavity, the entire mass of the irrigating stream is thrown into the cavity, to escape through the opposite opening only after having washed the entire extent of its interior.

Aseptic rubber tubes never cause "irritation." Increased discharge or irritation of any kind is due to infection introduced into the wound by means of the tube at change of dressings. If the withdrawn tube is touched by unclean hands and is then reintroduced, it is apt to cause irritation. But it is not the tube but the dirt adhering to it that is the cause of the trouble.

The persistence of sinuses after certain operations, notably exsections, was also attributed to the use of drainage-tubes. This mistake is now explained by the knowledge, that the sinuses in question do not heal on account of reinfection by tubercle bacilli, extending along the tubes with the discharges from an incompletely evacuated tubercular focus.

In aseptic wounds, the office of the drainage-tube is performed by about the end of twenty-four hours after the operation. But other considerations, notably the unwillingness of disturbing the rest of the wound and of the patient, make it inexpedient to reopen the dressings so soon for the purpose of withdrawing the tube. It is generally left *in situ* until the first change of dressings. If there is no purulent discharge visible in the dressings removed from the third to the fifth day, the tubes can be safely withdrawn. If the healing was not entirely faultless, as seen from the presence of more or less pus in the dressings, it will be safer to reintroduce a short piece of tubing for the purpose of keeping patent the external end of the tube-track until the discharges shall have become scanty and serous.

When a wound is in good condition and no pyogenic or tubercular infection be present, the surgeon will find it a very difficult matter to keep a tube in place for a long time, should he desire to do so. The cicatrization of the deeper parts of the drainage-hole will irresistibly expel the tube, or granulations will invade the lumen of the tube through its lateral fenestræ, and will simply fill it up completely.

The tube should be always extracted for inspection at the first change of dressings. If it is found to be filled up with a more or less solid clot of sweet blood or fibrin, the interior of the wound can be assumed to be in good condition. Should the clots be foul and semi-fluid, the tube must be shortened and replaced after thorough cleansing.

The decalcified bone drainage-tubes, devised by Neuber, have been abandoned by the author on account of their many inconveniences not overbalanced by the advantage of their absorbability.

Glass drainage-tubes, provided with a number of lateral holes, are used extensively in abdominal surgery. By placing within their hollow a wick of iodoform gauze, tubular and capillary drainage can be combined to great advantage.

It may be said, on the whole, that rubber tubing has so far not been supplanted by anything better for purposes of ordinary wound drainage.

## B. APPLICATION OF ASEPTIC METHOD TO DIVERSE ORGANS AND REGIONS.

### I. LIGATURES OF ARTERIES IN THEIR CONTINUITY.

With due observance of the rules of surgical dissection and of the landmarks pointed out by anatomy, the exposure and deligation of the larger arteries will present no serious difficulty.

The treatment of the vascular sheath deserves some special remark.

Free incision of the sheath will be found to facilitate very much the isolation of the vessel. No fear need be entertained of causing thereby necrosis or suppuration in an aseptic wound.

The sheath should be grasped and raised with a pair of mouse-tooth forceps, and the cone thus formed should be incised with the knife held horizontally. The incision can be extended to half an inch in length. (See Fig. 33.)

Isolation of the vessel is best accomplished by gently insinuating into the slit the point of a bent silver probe, while the edge of the cut is held up

FIG. 33.—Incising the vascular sheath. (Esmarch.)

by the mouse-tooth forceps. As soon as the point of the probe emerges on the opposite side of the artery, it is followed up by an aneurism-needle armed with a catgut thread, which is tied in a square knot.

Encircling a vessel with an aneurism-needle having a sharp or even a too slender point may lead to piercing of the artery wall by the instrument.

CASE I.—Carl Tompert, carpenter, aged forty, noticed in October, 1881, a pulsating swelling on the left side of his neck. By February, 1882, it had attained the size of a goose's egg. *March 2d.*—Ligature of left common carotid between the heads of the sterno-mastoid muscle at the German Hospital. In passing aneurism-needle under the artery without the exertion of unusual force, suddenly a jet of arterial blood was seen to spurt up from the wound. Traction on the aneurism-needle controlled the hæmorrhage. A catgut ligature was passed around the artery above and another below the aneurism-needle, and both were tied. The artery was divided between the ligatures, and then it was ascertained that the aneurism-needle had made a longitudinal slit into the artery wall. No drainage-tube was used, and the wound was closed by a few catgut sutures. Pulsation of the tumor had ceased, and subsequently it shrunk away to a stout cord-like structure. The wound healed by the first intention and no fever occurred, but the first two days following the operation very profuse general perspiration was observed. Patient was discharged cured, March 20.

In this and the subsequent cases, as well as in all other operations done by the author since 1877, catgut was used exclusively as ligaturing material with the greatest satisfaction. Only one case of suppuration occurred in which the infection could be traced to the use of impure catgut (page 8). Secondary hæmorrhage by slipping of the ligature was observed twice (page 72). Even in suppurating wounds, catgut has been found to be a safe ligaturing material. It is in every way preferable to silk, and in no case was its use ever regretted. Those who have been accustomed to tie vessels with silk, usually employ too much force in tightening catgut ligatures. They overtax the strength of the animal thread, and to their great annoyance constantly break it. A small amount of traction is sufficient to safely tighten the knot, as it is not necessary nor desirable to sever the inner coat of the artery. The many cuts, so common on the ulnar side of surgeons' fingers at the time, when silk was generally employed for tying vessels, are very rarely seen nowadays. To preserve its strength, catgut should never be immersed in any kind of a watery solution, as it is apt to become swollen and soft when brought in contact with water. The dish holding the ligatures at an operation should be dry, or should contain absolute alcohol.

In all the cases here reported, no drainage-tube was used, reliance being placed on natural drainage. The catgut sutures employed were few and loose, and permitted a free escape of the oozing during the first twenty-four hours.

Primary union of the wounds occurred in every case.

CASE II.—Herrmann Stinze, fishmonger, aged forty-six, admitted to German Hospital January 3, 1880, with aneurism of the femoral artery, situated just underneath Poupart's ligament, displacing it forward and upward. Syphilis admitted. Causation, severe effort at rowing fifteen months before admission to hospital. Direct compression of swelling was unsuccessfully employed for eighty hours. *Jan. 17th.*—Deligation of external iliac artery. No drainage-tube. Catgut suture. Prompt establishment of collateral circulation. Primary union. Discharged cured February 28th. Patient examined March 28th, when at the site of the aneurism a cord of the size of the middle finger could be felt.

CASE III.—Henry Greenwald, clerk, aged fifteen. End of June, 1882, sustained stab-wound of left palm, followed by copious hæmorrhage, which ceased spontaneously. Development of pulsating swelling of palm, which, by the direction of the family physician, was kept tightly compressed with a leaden bullet. *Aug. 17th.*—In the Catskills severe arterial hæmorrhage from pressure-sore over swelling, when bullet was removed and another compressory bandage was applied. *Aug. 20th.*—Renewed hæmorrhage. Esmarch's band being applied, the clot was turned out of the open sore, the sac of the size of a hazel-nut was split and excised, and both afferent vessels were tied. Suture. Primary union followed.

CASE IV.—August M., agent, aged forty-one, suffering from progressed ataxia, cut his ulnar artery August 20, 1881, in a suicidal attempt. Hæmorrhage was arrested by pressure made by a physician who attended to the patient immediately after the attempt. *Aug. 23d.*—Secondary hæmorrhage. Esmarch's band being applied, the wound was dilated, and, the partially cut artery being exposed, was doubly tied and cut through between. Suture. Primary union.

CASE V.—Alexander Goerlitz, engraver, aged thirty-four. Had chancre eleven years ago, and had been in the habit of folding his legs while at work. *June, 1883.*—Noticed pulsating swelling in right popliteal space. *Sept. 15th.*—Circumference of left knee, thirteen, of right knee, sixteen and a quarter inches. Knee semi-flexed. Skin over aneurism dusky and hot. Esmarch's constrictor applied above and below swelling for an hour under ether without success, circumference increasing to seventeen and a quarter inches. *Sept. 19th.*—Ligature of right superficial femoral artery in middle of thigh. *Sept. 21st.*—Swelling hard, non-pulsating. Paralysis of dorsal flexors of foot and of extensors of toes. No necroses. Primary union. *May 17, 1884.*—Knee can be fully extended, paralysis disappeared, muscles of leg have regained their normal bulk, tumor shrunken to a small, hard mass.

CASE VI.—August Bente, cigar-maker, aged fifty-one. No syphilis. In the summer of 1883 felt neuralgic pains in right arm, followed by wasting of the brachial muscles, cyanosis, formication, and hyperidrosis of the extremity. In December severe dyspnœa supervened, and a pulsatile swelling under the right sterno-clavicular junction and in the lower cervical triangle was made out by Dr. John Schmidt, who directed the patient to the author, then on duty at the German Hospital. Aneurism of the innominate and subclavian arteries at their junction was diagnosticated, and simultaneous ligature of the right common carotid and the axillary arteries was performed January 16, 1884. The latter vessel was tied in Mohrenheim's triangle, just below the outer third of the clavicle. No drainage-tubes; suture. Immediately after the operation the pulsation of the swelling became more pronounced, and for the next four weeks the shooting pains in the arm were much complained of. Both wounds healed by primary intention. Toward the end of February decrease of the swelling and moderation of the subjective symptoms became manifest. In March and April thirty hypodermic injections of Bonjean's ergotine were made in the abdominal region, and seemed to hasten the shrinking of the tumor. By May, the cyanosis, sweating, glossy skin, and formication, as well as the neuralgic symptoms, had very much abated, and the patient had gained ten pounds of flesh. Under massage, the application of faradism, and active exercise, the atrophy of the muscles had also materially improved, and in June the patient could resume his occupation. *Nov. 11, 1884.*—Patient was presented to the Surgical Society. Pulsation had almost entirely disappeared, and what there was of it seemed to be transmitted. Bruit was not noticeable. A well-perceptible fullness and resistance could still be made out in the right supraclavicular fossa. Occasionally short and mild attacks of shooting pains were felt in the arm and nape of the neck. A claw-like deformity of the nails of the right hand remained unaltered. In August, pulsation and other signs of relapse were noted, with increasing pain, radiating toward the occiput. Renewed injections of ergot were without avail. In October, during the author's absence from town, Dr. Adler incised an abscess pointing in the supraclavicular space, and a few days later performed tracheotomy for threatening asphyxia. A sharp pneumonia followed, from which the patient recovered only to succumb in November to sudden suffocation. No autopsy was permitted.

CASE VII.—John H. Nittinger, grocer, aged forty-five. No syphilis; had had articular rheumatism seven years before. Pulsating swelling of left popliteal space of the size of a man's fist. Leg had been œdematous for three months; marked emaciation. *Jan. 20, 1885.*—Ligature of left femoral artery in Scarpa's triangle. Primary union of wound. Recovery retarded by circumscribed necrosis of integument over tuberosity of calcaneum (due to pressure?). Discharged cured, March 30, 1885.

CASE VIII.—Emmanuel Luecke (see history on page 186).

CASE IX.—Robert Klaile, school-boy, aged fourteen. Congenital arterio-phlebectasia of anterior part of left foot; pulsating, dusky swelling, of doughy feel, of dorsum and planta pedis. Along the course of saphenous nerve were seen a series of flat, hard, dark-blue, rough nodes, some of them as large as a silver quarter, their size tapering off toward ankle. Two of them were ulcerated and covered by a dry scab. Left foot on the whole larger than its mate. Pulsation of femoral arteries abnormally strong. Heart hypertrophied. Ablation of diseased parts was declined. *July 7, 1885.*—Ligature of superficial femoral artery. Short stoppage, and return of pulsation. Immediate ligature of external iliac of same side. Wounds sutured; no drainage. Primary union. Necrosis of terminal phalanges of first and second toes, of the integument of the external side of leg, and of peroneus longus muscle. Scanty aseptic suppuration, and very slow detachment under antiseptic dressing. Tardy cure. The cicatrices on the toes became ulcerated in the winter, and the pulsation of the tumor, which had not diminished in size, had returned. *Jan. 29, 1886.*—Pirogoff's amputation. Unusual number of ligatures required on account of many abnormally large arteries. Cap of calcaneum was fixed to tibia by steel nail driven through from below. Catgut suture.

Drainage through counter-incision alongside of tendo Achillis. No fever followed. First change of dressings was done February 19th. Primary union was observed throughout, except where a narrow strip of the integument had necrosed along the anterior part of the incision. Dry dressing. *February 24th.*—All firmly healed. Patient walks well without support.

Case X.—Louis Wiersch, aged forty-two, diffuse cirsoid aneurism of right temporal region. *July 24, 1888.*—Deligation of external carotid and of four large collateral vessels. Primary union. Marked diminution of pulsation. Discharged improved, August 4th.

Case XI.—Carlo Somma, laborer, aged fifty-three, fusiform aneurism of right axillary artery. *March 2, 1888.*—Ligature of subclavian artery, third division. Immediate cessation of pulsation, and subsequent rapid shrinkage of tumor. Discharged cured, March 16th.

## II. EXTIRPATION OF TUMORS.

In removing tumors, three requirements have to be commonly held in view:

*Firstly,* the avoidance of septic infection from without or from within.
*Secondly,* the complete removal of the neoplasm.
*Thirdly,* its safe removal.

How to avoid infection from without was seen in previous chapters of this book. By infection from within, two kinds of infection are meant.

One is the contamination by septic contents of the tumor that may escape into the wound through an accidental cut or a laceration of the tumor, caused by rough handling or the careless use of sharp retractors, as, for instance, in extirpating suppurating glands.

Case.—Sarah Barn, servant, aged sixteen; old Pott's disease of the cervical vertebræ; large glandular swelling of right submaxillary region, with several sinuses leading down toward the spine. It was pretty certain that no serious degree of the affection of the vertebræ could be present, as the function of the cervical spine was nearly normal. *November 4, 1886.*—Flap incision and exsection of the large mass of tubercular glands at Mount Sinai Hospital. Though the utmost care was exercised in not grasping the

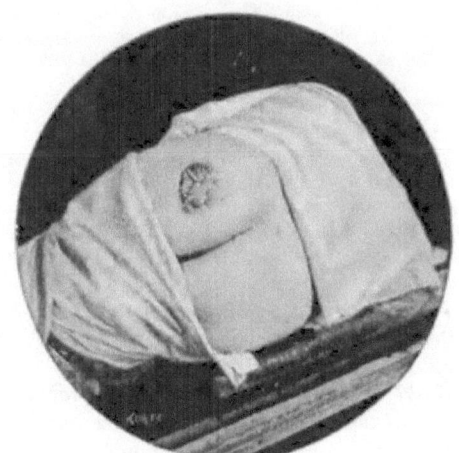

Fig. 34.—Gluteal tumor before extirpation.

glands with sharp-pointed instruments, one of them broke down, and poured out its contents into the large wound. As subsequent events demonstrated, seemingly thorough irrigation with a strong solution of corrosive sublimate did not disinfect all the parts of the wound. The dissection mainly extended into the intermuscular space —namely, the slit between the scaleni and the posterior border of the sterno-mastoid.

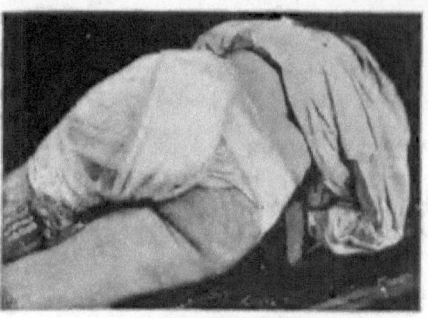

Fig. 35.—Gluteal dressing.

After the removal of the mass, the finger was easily inserted into a track leading toward the second vertebra, the anterior surface of which was found rough and bare of periosteum. It was thoroughly scraped and irrigated (the instrument could be felt *in situ* from the oral cavity); the outer wound was drained, sutured, and dressed. *Nov. 5th.*—High fever, with much dejection. Skin below ear red, painful, and swollen. The flap was reopened, and a small abscess was detected just under the base of the flap, where probably irrigation had been insufficient. Open treatment. Temperature fell off to normal at once. The patient was discharged cured December 1st.

The other kind of infection is the dissemination through the lymphatics of cancerous or sarcomatous cell-elements into the body caused by pressure due to rough manipulation of the tumor.

NOTE.—It is a well-known fact that, in some cases of malignant tumor of slow growth, after operation, a large number of secondary nodes will spring up and develop with great rapidity in the neighborhood of the cicatrix. Two causes, either singly or combined, may be at the bottom of this phenomenon.

Either the operation was incomplete—that is, the surgeon's dissection hugged the tumor too closely, leaving behind a number of outstanding microscopical foci,—or the forcible manipulations of the tumor during the operation have disseminated along the lymphatics and veins embryonal cell-elements of malignant character into the vicinity of the wound or throughout the body. This is commonly called "change of the character of a malignant neoplasm, due to mechanical irritation."

Undoubtedly there are many cases where an incomplete operation leads to wide dissemination of the elements of the neoplasm. In these cases relapse in the unhealed wound or in the fresh cicatrix is observed, together with the simultaneous appearance of regional and more distant nodes of new formation.

Thus an incomplete or rough operation may, by generalization of the disease, hasten instead of retarding the patient's death.

Reasonable hope of the complete removal of a malignant new-growth is the main justification for operative interference. There is, to be sure, a considerable class of cases where complete removal is from the outset out of the question. Great discomfort from putrescence of a sloughing tumor or frequent hæmorrhages do sometimes indicate partial removal. But, wherever possible, complete removal is to be aimed at by all permissible means, as the non-return of the disease depends solely upon the fulfilment of this condition.

Our third object must be to remove the tumor with the least possible amount of immediate danger to the patient's life. Careful and deliberate dissection, guided by anatomical knowledge, limiting of the hæmorrhage to a minimum, and avoidance of accidental injury to important organs, is meant hereby.

The most important condition to be fulfilled in eschewing these dangers is *an adequate incision.*

A too large incision never can do any harm, its worst consequence being the necessity for a few more suture-points. An insufficient incision, on the other hand, may be the source of great danger to the patient, and of much embarrassment to the surgeon.

When the incision is ample, the new-growth and its connections can be readily exposed without the use of much traction from sharp or blunt hooks, and forcible grasping and dragging to and fro of the tumor itself will be unnecessary. Most of the vessels that are to be divided will be noticed, and can be cut between two artery forceps without loss of blood. Accidentally injured vessels can be easily secured and tied off.

The wretched expedient of digging a malignant tumor out of its capsule, and leaving behind the latter, should never be resorted to, as a speedy relapse is certain to follow.

Dissection should be done altogether with the knife, and exclusively in healthy tissues. Blunt methods of preparation are not to be used at all, since they are unnecessary, and involve a certain amount of rough force.

In removing infiltrating or ill-defined malignant new-growths, the surgeon's knife should give the tumor a wide berth, and all cosmetic or functional considerations not involving present danger should be disregarded, the first object being the complete eradication of the disease.

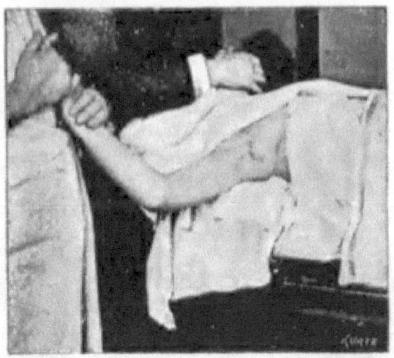

Fig. 36.—Axillary tumor before extirpation.

NOTE.—In extirpating malignant new-growths, which are known to cause an infection of the contiguous lymphatic glands at an early stage of their development, the rule of removing these involved lymphatic glands together with the fat wherein they are imbedded, should never be disregarded without a very cogent reason of expediency. The absence of a gross lymphatic tumor is no evidence of the freedom from infection of the pertinent lymphatic glands. The additional traumatism caused by this complementary step is richly repaid by the vast improvement of the patient's chances against a speedy relapse.

In an ample wound the tumor can be handled with the necessary gentleness, and the main attack can be directed upon its adhesions to the surrounding tissues. With rare exceptions, sharp retractors are never to be plunged into the tumor. They should be used on the edges of the wound for dilatation, the tumor itself being held by hand throughout.

The softer the mass of the tumor, the greater care must be exercised not to injure it. Cysts especially require very tender treatment. Lipomata and fibromata will stand a good deal of rough handling without harm.

NOTE.—In former days lipomata used to have a bad reputation. It was said that their extirpation was often followed by erysipelas and phlegmon. One of the first operations ever witnessed by the author was done upon a healthy young man in 1868 in Prof. D.'s clinic, at Vienna, for a lipoma of the shoulder. It caused the patient's death from septicæmia. This peculiarity, noted by surgeons in times gone by, was undoubtedly due to the readiness with which a phlegmonous process will spread in loose and ill-nourished adipose tissue. Of course, the infection always came from the hands and apparatus of the surgeons themselves.

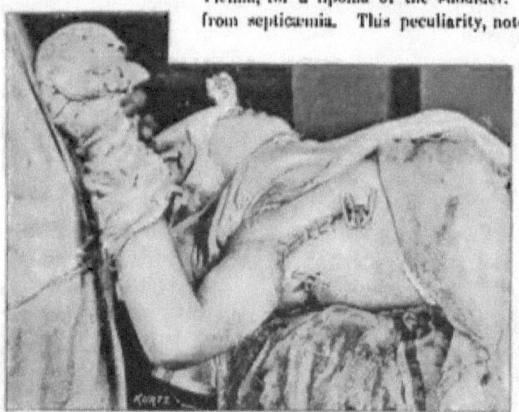

FIG. 37.—Axillary wound, united, after extirpation of tumor.

*Where should dissection first be directed to,* is a question that puzzles every beginner, and it is not indifferent from which side we approach a tumor. Surgery owes to Langenbeck a clear exposition of the principle which should guide us in this matter.

*In excising tumors holding close relations to large vessels,* as, for instance, those in the neck, axilla, and in Scarpa's triangle, *the greatest safety lies in first exposing these vessels above and below the tumor, so as to have full control* of them during the subsequent steps of the operation. This precaution offers great security against injury of those vessels, and at the same time reduces to a minimum the otherwise formidable dangers of such accidental injury, should it occur. If it become evident that the tumor has involved the walls of the adjacent large vessels, a ligature above, another below the growth, will permit of a safe and complete exsection in one mass of the tumor and the diseased parts of the vessel.

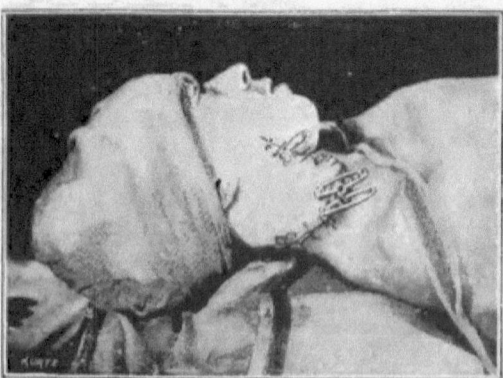

FIG. 38.—Flap incision for removal of tumor of neck. Wound drained and sutured.

NOTE.—It is the common tendency of young surgeons to carry too far the dissection of a vessel adhering to a tumor. This is actuated by the desire of preserving the integrity of the vessel in question, and by the natural disinclination of complicating the operation by double ligature, which again involves extra dissection. The consequence of this tendency may be twofold: either portions of the tumor adhering to the vessel wall are left behind to cause speedy relapse, or the vein is cut or torn.

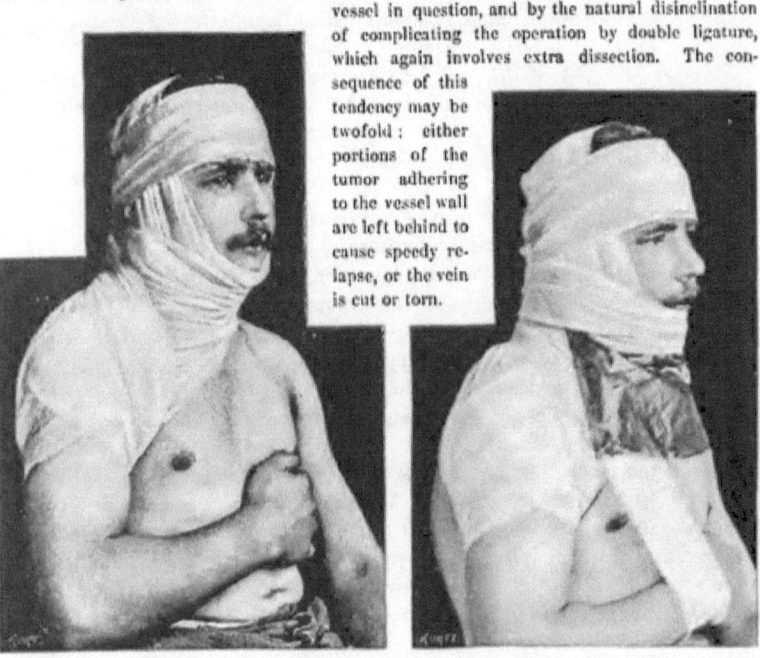

FIG. 39.—Dressing for neck wounds.   FIG. 40.—Dressing of neck wound completed by rubber-tissue bib and arm-sling.

*Whenever the surgeon has succeeded in forming a pedicle* to a tumor situated in the vicinity of large vessels, *cutting of such a pedicle without first tying it off is a very risky step.* Traction upon the tumor will obliterate any vessels included in the pedicle, and, when cut, the innocent-looking mass, closely resembling ordinary connective tissue, may open up into unexpected and overwhelming springs of welling blood. The stump will at once retract, and finding and securing the retracted vessel in an inexhaustible pool of blood is a terribly difficult, sometimes impossible, thing. Should it be an artery, the tips of two or three fingers must be thrust at once into the place from which the hæmorrhage is issuing. The blood must be mopped up by rapid sponging, to enable the surgeon to find the vessel, in order to secure it with an artery forceps, or to surround it by a suture passed through the adjacent tissues. His mettle will be put to the severest test, and it will be a lucky day if his patient do not succumb on the table.

In trying to secure the stump of a large vein accidentally cut across, the wide extent of its circumference will offer much difficulty, as an ordinary artery forceps is too small to take in the entire lumen of the vessel. One or more great leaks will remain, even if the vessel be fortunately grasped by one forceps. Two, three, or more additional instruments have to be brought

## SPECIAL APPLICATION OF THE ASEPTIC METHOD. 57

into requisition till the end is accomplished. The haste, natural and almost unavoidable on such occasions, will easily lead to further tearing of the soft walls of the vessel, and, finally, salvation will have to be sought in plugging with iodoform gauze. Here prevention is much easier than cure.

*Lateral tearing or slitting of a large vein* is another accident to which disregard of Langenbeck's rule may lead. There are two ways out of this contingency. One is to expose and deligate the vein above and below the laceration, while the fingers of an assistant compress the injured part of the vessel. The other one is the application of a lateral ligature or a continuous suture of fine catgut occluding the rent.

Both of these latter methods, however, are difficult and not very reliable, though they have succeeded in the hands of several surgeons, including the author's.*

Fig. 41.—Lateral ligature and continuous suture of injured vein.

They were bred of the fear of tying large veins, formerly so prevalent on account of the dangers of phlebitis and, in the extremities, of gangrene. In cases where a large portion of the vein wall is lost by sloughing or cutting, and the resulting aperture is very large, lateral ligature and suture are impossible. Whenever feasible, a double ligature should be applied, whether it concerns the deep jugular or axillary and femoral veins. Langenbeck's advice to tie the accompanying large artery has been much impugned lately, as it was found that gangrene of the extremity followed its adoption. On the other hand, a growing number of cases are on record, where deligation of the femoral or axillary vein led only to temporary disturbance of no great import.

CASE.—Henry Rickriegel, carpenter, aged twenty-three, admitted to German Hospital, March 2, 1887. Two days later the house-surgeon extirpated a mass of suppurating glands from Scarpa's triangle of the right side. The saphenous vein, which passed into the tumor from below, was tied and cut across. Likewise were treated a number of larger veins entering the tumor from above. The femoral vessels were not exposed, but the pulsation of the artery could be distinctly felt, and it was carefully held aside. Finally, the

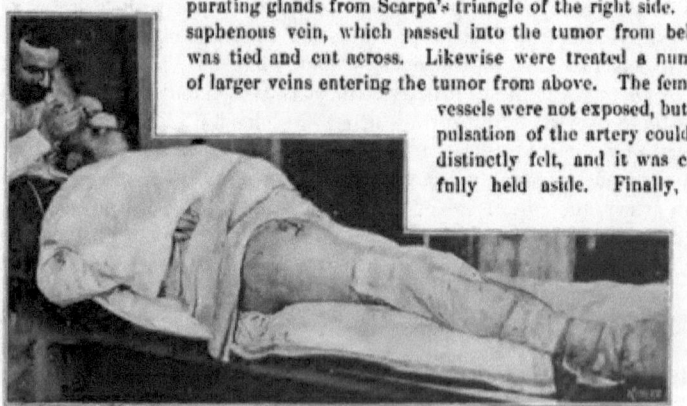

Fig. 42.—Periosteal myxosarcoma of thigh before removal.

* In a case of exsection of lymphomata of the neck, done in 1880 in the German Hospital, where the deep jugular was injured. The patient recovered.

mass was freed all around, until a stout pedicle was formed, which was seen entering the oval foramen of the fascia lata. This pedicle was tied with catgut and was cut through. In the mean time the patient had become semi-conscious and began to struggle, whereupon, suddenly, an enormous jet of venous blood was seen to well up from the bottom of the wound. The operator plunged his fist into the pool of blood, and thus succeeded in checking the hæmorrhage until Dr. Bachmann, the chief of the house-staff, appeared, who luckily succeeded, with the aid of Thiersch's spindles, in passing two ligatures, one below, the other above the bleeding point, effectually stopping the formidable loss of blood. Immediately, deep

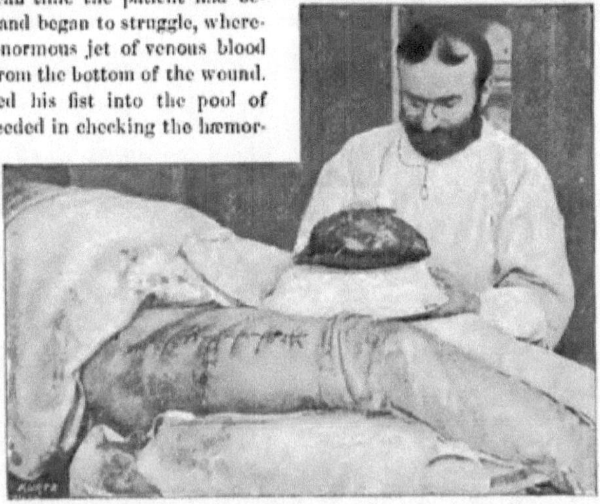

Fig. 43.—United wound after removal of myxosarcoma of thigh.

cyanosis and œdema of the lower extremity developed, and the author, who saw the patient directly after the operation, ordered elevation of the limb, which was brought about by its vertical suspension in a wire cradle. *March 5th.*—Cyanosis disappeared, œdema much diminished. Temperature, 101·5°. Circulation of limb good. The wound did well, but, March 18th, temperature rose to 103° Fahr., and signs of phlebitis of the femoral vein in the middle of the thigh appeared in the shape of a cylindrical, painful, and hard infiltration. This and a number of similar attacks were subdued by the application of an ice-bag. The persistent œdema was combated by elastic compression with Martin's bandage, supplemented later on by massage. *May 15th.*—The patient was discharged cured, very little of the œdema being still noticeable.

In this case, apparently, a portion of the trunk of the femoral vein was drawn into the cone of the pedicle containing the root of the saphenous vein, and was excised along with the tumor.

The ligature slipped off, and a wide gap was opened in the side of the femoral vein corresponding to the place of entrance of the saphena. The peculiarity of the walls of large veins to yield to lateral traction is well known to surgeons, and is a just source of anxiety, as the extended vein becoming empty can not be recognized.

Double ligature of the vein will be insufficient to check the hæmorrhage when a large branch inosculates between the two ligatures. Such branch must be separately exposed and tied.

CASE I.—*March 27, 1880.*—The surgeon in charge of the ward for syphilis and skin diseases at the German Hospital excised a large glandular tumor from Scarpa's triangle on John Te Gempt, aged twenty-four. The operation was finished without accident, and, according to the then prevailing custom, the wound was mopped with an eight-per-cent solution of chloride of zinc. *April 11th.*—A large slough of the vein wall was detached, and fearful hæmorrhage ensued, which Dr. Loewenthal, the house-surgeon, could not check completely by local pressure. When the author saw the patient, he was nearly exsanguinated, though conscious. No pulse could be felt. Without anæsthesia the femoral vein was exposed below the opening in its wall, while pressure by three finger-tips completely controlled the hæmorrhage.

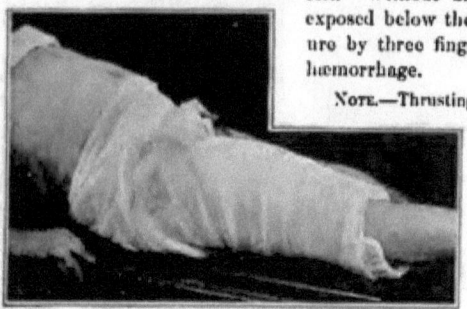

Fig. 44.—Dressing after removal of myxosarcoma of thigh.

NOTE.—Thrusting of the fist or of a sponge into the wound will not check hæmorrhage effectually in these cases. The tips of the fingers pressed exactly upon the bleeding orifice, and without much force, will always succeed in controlling the vessel.

As the vein bled from above, too, Poupart's ligament was cut across, and the external iliac vein was tied. After this the loss of blood became very much diminished, but a considerable vein inosculating just opposite the defect in the wall of the femoral vessel required separate exposure and deligation, whereupon the hæmorrhage ceased completely. Unfortunately, the total loss of blood had been so considerable that the patient survived the operation only a short time, and died in collapse from acute anæmia.

In a similar case the difficulty caused by the presence of an inosculating branch, situated between the two ligatures, was overcome by plugging.

CASE II.—Ferd. Brenner, aged forty-nine. *November 22, 1889.*—The removal of a very large relapsed sarcoma, located deeply in Scarpa's triangle, was attempted at Mount Sinai Hospital. The femoral vein was found imbedded in the tumor mass. In essaying to ascertain whether the attachment was loose or more intimate, a piece of the vein wall measuring three fourths of an inch, and involving the entire width of the vessel, came away with the tumor when it was raised. The very profuse hæmorrhage was promptly checked by exact finger-pressure upon the aperture in the vein. Two ligatures, one above the other below the tumor, were thrown about the vein, but hæmorrhage continuing, it was concluded that a large branch, inosculating between the two ligatures, probably the profuda vein, remained unoccluded. As the excision of the tumor was out of question on account of its diffuse character, search and deligation of the inosculatory branch was deemed inexpedient. Hence the entire lumen of the vein included by the ligatures was tightly packed with strips of iodoform gauze, the ends of which were brought out through the defect in the vein wall. The wound was also packed and a compressory bandage applied. Slight cyanosis persisted for a few hours, but had disappeared by the next day. Packings removed November 26th. No hæmorrhage. *December 25th.*—Patient discharged, with granulating surface.

*Deligation and partial exsection of the axillary vein* for ingrowing cancer of the axillary glands has been often performed by various surgeons with entire success, and *can be undertaken without hesitation* whenever unavoidable.

CASE.—Betty Lowy, aged forty-two. *April 26, 1889.*—Amputation of right breast for extensive carcinoma. The axillary glands are found to be very much involved and enlarged, the axillary vein passing through the middle of the tumor mass. Excision of two inches of the axillary vein between two catgut ligatures. Drainage, suture. No cyanosis or œdema of arm followed. Primary union. Discharged cured May 30th.

*In deligating the deep jugular vein, avoidance of the pneumogastric nerve will require close attention.* When there is enough space to expose and liberate the vein freely, this will not be found very difficult. Low down at the root of the neck however, the decision of the question whether the ligature encompasses the nerve or not may occasionally be impossible.

CASE.—Mrs. Catharine Plunkett, aged sixty-four. Extirpation of recurrent lymphosarcoma of neck, December 22, 1886, at Mt. Sinai Hospital. A tumor of the size of a hen's egg was located low down in the supra-clavicular fossa. Though it was freely movable, its close relation to the large cervical vessels was anticipated. A flap incision and careful dissection laid bare the jugular vein above and below the tumor, when it became evident that it would be impossible to remove it without excising a corresponding portion of the vein. The lower ligature had to be applied somewhat behind the sterno-clavicular rim, and on account of the lack of space this was very difficult. Isolation of the vein had to be done with the greatest caution to avoid its injury. Finally a silver probe wormed its way around the vein, and the question arose, Was or was not the pneumogastric nerve included in the ligature? To test this the thread was firmly tied in a single knot. No change whatever of the respiration or pulse being noted, it was assumed that the nerve was not caught, whereupon a double ligature was passed through by means of the first thread, and, being tied, the vein was cut across. But on inspection of the mass it became clear that the nerve was included in the ligature and had been cut through. The tumor was easily dissected up after this until a pedicle was formed containing the jugular vein from above. This being tied, the tumor was removed. Drainage, suture, and dressings were applied in the usual manner. The patient recovered without one untoward symptom. *Dec. 31st.*—The first dressing was removed, together with the drainage-tubes. *Jan. 3, 1887.*—She was discharged cured.

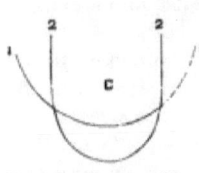

FIG. 45.—Outlines of flap-incisions.

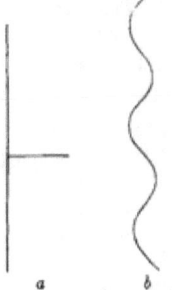

FIG. 46.—*a*, T-shaped incision. *b*, Undulating incision.

Having thus gone through the entire subject, we may sum up in the following points:

To accomplish a thorough and at the same time safe removal of a tumor located in the vicinity of large vessels, an adequate, that is, very ample, incision is absolutely necessary.

NOTE.—On the trunk and the extremities, straight incisions, with the addition of a transverse extension, will be found most convenient. Where a transverse cut is inopportune, considerable gain in space can be effected by *undulating* the line of incision.

In Scarpa's triangle, but especially about the neck, flap incisions are the most convenient.

Methodical dissection, guarded by as many preliminary double ligatures as necessary, will insure a steady and uninterrupted progress of the operation. Loss of blood will be minimal, and the flurry and haste incumbent upon profuse accidental hæmorrhage will not lead, as it always does, to the disregard of the rules of asepticism.

# SPECIAL APPLICATION OF THE ASEPTIC METHOD. 61

Aseptic canons are easily forgotten during frantic efforts to check dangerous hæmorrhage, although it is conceded that avoidance of suppuration is all the more important because of the injury to large vessels.

After thorough irrigation and cleansing, the *drainage* of the cavity is to be attended to. *It should be direct*—that is, should reach the surface on the shortest possible route, if necessary through a counter-incision—and care must be taken of not letting the square inner end of the tube impinge upon a large artery. Especially must this point be heeded where the tube consists of hard material, as perforation of the vessel by friction against the hard edge of the tube is possible.

NOTE.—There are cases on record where the innominate was ulcerated through by friction pressure of the margin of a tracheotomy cannula.

The inner end of the tube should be placed so as not to touch the vessels, the general direction of the mesial end of the tube being parallel with them. To secure this position the inner end of the tube should be fastened to a suitable part of muscle or fascia by a catgut stitch.

Change of dressings will be required, according to the size of the tumor, on from the third to the sixth day, when the tubes can be withdrawn.

### III. AMPUTATION OF LIMBS.

In performing a major amputation, the modern surgeon has to solve three problems:

*The first* is to avoid septic infection of the amputation wound, or, if sepsis of the limb be present, to eliminate it.

*The second* one is to limit hæmorrhage to an unavoidable minimum.

*The third* problem is to secure a good stump.

1. Aseptics and Antiseptics of Amputation.—To the adoption of aseptic and antiseptic measures must be ascribed the remarkable reduction of the rate of mortality after major amputations, now prevalent wherever such measures are practiced. Formerly one third of all cases were directly lost mainly through primary septicæmia, or pyæmia, or indirectly by secondary hæmorrhage due to ulcerative destruction. At present, deaths from acute and chronic blood-poisoning or secondary hæmorrhage are very rare, and limited to cases that come under the surgeon's knife in a neglected or septic state.

The total mortality, as computed from nearly 1,000 unselected hospital cases of various surgeons, treated on the new plan, is about fifteen per cent.

The author's personal experience embraces fifty-five cases of major amputation, mostly done in hospital practice. These were:

| | | |
|---|---|---:|
| Amputations of the thigh (1 Gritti's) | | 30 |
| " | " leg | 9 |
| " | " foot | 7 |
| " | " shoulder | 1 |
| " | " arm | 5 |
| " | " forearm | 3 |
| Total | | 55 |

The amputations were performed:

| | |
|---|---|
| For suppurating compound fracture in | 2 cases. |
| " phlegmon of soft parts in | 6 " |
| " acute and chronic osteomyelitis | 8 " |
| " spontaneous gangrene (non-diabetic) | 7 " |
| "      "            "      (diabetic) | 3 " |
| " articular suppuration (diabetic) | 1 case. |
| "       "         "       (pyæmic) | 1 " |
| "       "         "       (uratic) | 1 " |
| "       "     tuberculosis | 12 cases. |
| " congenital angio-lipoma of lower extremity | 1 case. |
| " malignant new growths | 8 cases. |
| " incurable ulcers | 5 " |
| Total | 55 |

Of this number were cured:

| | |
|---|---|
| By primary union | 19 cases. |
| " partial adhesion (open treatment and secondary suture) | 20 " |
| With suppuration | 8 " |
| Cured | 47 " |
| Died | 8 " |
| Total | 55 " |

The eight fatal cases were as follows:

CASE I.—Max Loffmann. Amputation of thigh at Mount Sinai Hospital for secondary hæmorrhage due to phlegmon of popliteal space after exsection of knee. Patient came on table collapsed, and died immediately after ablation (see page 259).

CASE II.—Gustav Leuber, aged forty-nine. *March 22, 1883.*—Syme's amputation of foot, at the German Hospital, for tuberculosis of tarsus. Died May 5, 1883, of general marasmus, due to pulmonary tuberculosis. Wound nearly healed.

CASE III.—Carl Frank, aged sixty. Senile gangrene of foot and leg; amputated at the German Hospital. On account of the collapsed and septic condition of the patient, twenty ounces of a six-pro-mille saline solution were transfused before commencing the amputation. The pulse rallied, and transcondylic amputation was done, but patient died immediately after the bone was sawed off.

CASE IV.—Louis Bourbonus, carpenter, aged twenty-nine. Acute progressive gangrenous phlegmon of hand and forearm. Septicæmia with petechial eruption. *February 24, 1880.*—Amputation of arm at the German Hospital. Patient died two hours after ablation.

CASE V.—Catharine Argast, aged fifty-four. Senile gangrene of fore part of foot. *September 18, 1882.*—Syme's amputation at the German Hospital. Marastic thrombosis of the femoral vein. Died October 23d of marasmus.

CASE VI.—Beckie Sternfeld, married, aged twenty-eight. Admitted to Mount Sinai Hospital, March 23, 1889, with puerperal pyæmia. A large number of abscesses were successively incised, among them one involving the right knee-joint. Though her case appeared to be hopeless, amputation of the right thigh was done October 28th, to rid her of the pain and inconvenience caused by the disorganized knee-joint. The wound was treated by the open method, and was doing well, though cicatrization was very slow. *November 18th.*—She died of general exhaustion **due to prolonged suppuration and chronic nephritis.**

CASE VII.—Lazar Schatel, commission **merchant, aged forty-six.** Spontaneous gangrene of leg due to endarteritis obliterans. **No diabetes. General atheromatosis.** General condition poor. *January 1, 1889.*—Amputation of thigh **in** lower **third.** Femoral artery occluded by old clot. Open treatment. *January 11th.*—Slight fever with delirium developed. Marginal necrosis of flaps. *January 31st.*—Apoplectic convulsion, followed by death. *Autopsy:* Chronic purulent lepto-meningitis. General atheroma.

CASE VIII.—Mrs. J. D. Brodek, married, aged sixty-five. Senile gangrene of **left** foot. No diabetes. Amputation of right thigh had been performed in June, 1887, **for** gangrene of right foot. *October 15, 1889.*—Amputation of left thigh in lower **third.** Occlusive treatment. Necrosis of flaps under septic symptoms. Wound **reopened.** *October 17th.*—Death from sudden heart failure, probably due to septicæmia.

The **author's total rate** of mortality would be 14·54 per cent.

Excluding the **hopeless and** moribund Cases I, III, IV, and **VI,** the death-rate will be reduced **to 7·84** per cent.

But one of the **patients (Case** VIII) **died of acute** septicæmia clearly chargeable to the operation. **Case** II died **of** tuberculosis; Case V (senile gangrene), of thrombosis due to general marasm.

Considering the large proportion of amputations of the thigh (thirty), and the fact that ablation was done twenty-nine times for acute septic processes under a vital indication, **during** a more or less pronounced state of **general sepsis,** the **final** results **may** be favorably **compared** with those achieved without antiseptics.

*In four cases complicated by grave diabetic symptoms the patients all recovered, the proportion of sugar in the urine either disappearing or being reduced to mere traces after ablation.*

NOTE.—It is well known to what a pernicious extent diabetic patients become disposed to tissue decay in the shape of ulcerative and gangrenous destruction, especially of the terminal parts of the extremities. On the other hand, suppurative or necrotic processes have a very marked tendency to aggravate the symptoms of melituria by causing a decided increase of the saccharine contents of the urine. Thus, whenever a diabetic patient becomes the victim of gangrene or a phlegmonous process, his system will not only have to contend with the inherent dangers of septic fever **and** exhaustion from more or less extensive suppuration, but his power of **resistance will be** dangerously sapped by an aggravation of the diabetes. It is not to be wondered **at, then, that surgeons will not** be very eager to resort in these cases to active operative **measures, but rather will content themselves** with palliatives, practically leaving the patient **to the resources of Nature.** Most patients of this order will then succumb to the combined **malignity of the septic** attack and the diabetes. **In** a few cases, however, the gangrene will **reach its limits,** the necrosed portions **will be cast** off, and cicatrization will commence. It **was** observed that whenever this took place, **a marked** reduction of the quantity of sugar contained in the urine occurred simultaneously with the limitation of **the** fever and suppuration. Upon this observation was based the plan first successfully executed **by** Koenig, to eliminate the source of sepsis by a high ablation of the affected member, carrying **the line** of section through healthy tissues.

The conditions of success are: First, **to** amputate where the nutrition **of the flaps is** adequate to insure against marginal necrosis. This means amputation above the knee In the lower extremity. Secondly, the insurance of **a** faultless asepsis together with the open wound treatment.

# RULES OF ASEPTIC AND ANTISEPTIC SURGERY.

To further a better understanding of the methods employed for the maintenance of the aseptic condition during amputation, it will be necessary to class all cases requiring ablation in three groups.

*a.* CLEAN CASES.—*The first group* consists, on the one hand, of cases where amputation is indicated for various reasons, such as deformities, tumors, etc., in which the skin of the member is unbroken, and no subcutaneous, acute, or chronic suppuration is present; on the other hand, of injuries requiring amputation, that come under treatment immediately after the accident.

These are called *clean cases*. They require the ordinary aseptic precautions, such as shaving, thorough scrubbing, and disinfection of the field of operation, and a careful protection of the hands and instruments of the surgeons from contact with non-disinfected parts of the patient's body. This is best accomplished by wrapping the whole limb, excepting the field of operation, into a swathing of disinfected towels, which should be fixed in position by safety-pins or a few turns of a roller-bandage. The patient's feet and hands, disinfection of which is difficult at best, should never remain unnecessarily exposed in amputations of the upper or lower extremity. If the operation is to be done near, or on the hand or foot, these must be, if time permit, subjected to a careful preliminary process of cleansing. It consists of a prolonged bath of warm soap-water, and subsequent packing in compresses moistened with a two-per-cent carbolic solution, and an external wrapping of rubber tissue to prevent evaporation. Large masses of epidermis will be soaked off in this manner, and can be removed by gentle friction with a brush or flannel rag in soap-water. This process must be repeated until the skin is perfectly clean, and does not shed epidermis. The part to be operated on is kept wrapped in a carbolized towel until anæsthesia is well under way, and the operation is about to begin.

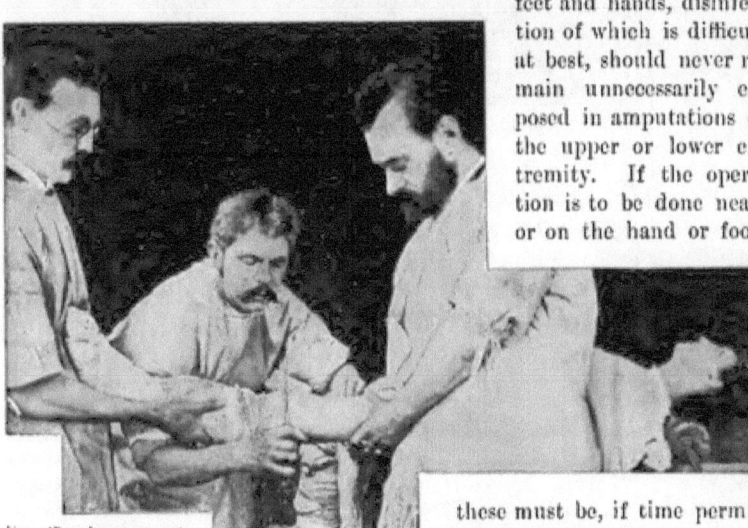

FIG. 47.—Arrangement of patient for amputation of thigh.

## SPECIAL APPLICATION OF THE ASEPTIC METHOD. 65

Esmarch's constrictor being applied, and the patient's body protected by rubber sheets, these and the parts of the limb not needing special disinfection are covered with disinfected moist towels. The parts of the assistants are distributed, and every one takes his place. Now the surgeon unwraps the field of operation, and, having once more rubbed it off with corrosive-sublimate lotion, begins to operate.

Frequent irrigation of the wound and especially rinsing of the hands of operator and assistants

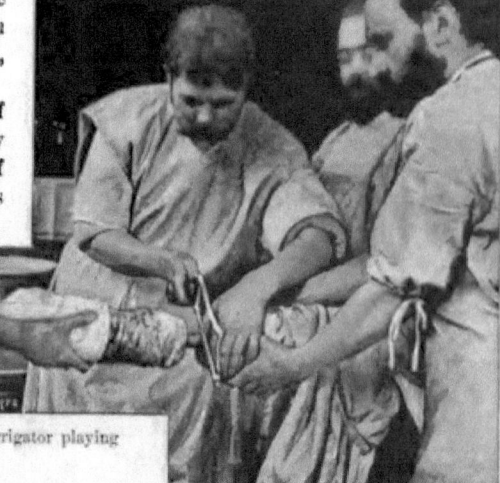

FIG. 48.—Section of femur. Irrigator playing from the left.

should not be neglected until the dressings are finished and the patient is ready for bed. The other precautionary detail mentioned in a previous chapter should also be carefully adhered to.

With the exception of the *saw*, most instruments required for amputation are easy to clean. *The saw is a frequent medium of pyogenic infection.*

CASE.—Arnold Bitter, mechanic, aged thirty-four, was amputated at the knee-joint eighteen years ago for a compound fracture of the leg. On account of insufficient covering, a large adherent cicatrix occupied the under and posterior side of the condyles, which were constantly ulcerated. Re-amputation of the thigh above the condyles, January 8, 1887, at the German Hospital. Drainage and suture. Fever developed on the second day, rising to 103° Fahr. on the third, wherefore the house-surgeon removed the dressings, but found nothing to explain the pain and fever. On the fifth day the author inspected the stump, and found firm union of the flaps between each other and to the sawn surface of the bone, the drainage-tubes still filled with fresh, sweet clots, but the extremity of the stump decidedly club-shaped and œdematous, the œdema being of the deep-going, firm variety, characteristic of acute osteomyelitis. The stump was nowhere painful on pressure, except at a point corresponding to the upper margin of the sawn surface of the bone. In a few days pus began to exude from the drainage-tube placed at the time of the operation through a counter-incision into the quadricipital bursa, and the patient's fever subsided. *Feb. 9th.*—The upper margin of the sawn surface was exposed and a narrow, sharp edge of necrosed bone was detected. This was chiseled away until healthy bone presented; fistula scraped, wound sutured. Primary union; patient cured, March 5th.

Apparently some filth had been detached from the teeth of the saw when it was drawn across the bone the first few times, and became lodged near the upper margin of the bone section, causing there a circumscribed acute osteomyelitis, ending in necrosis.

NOTE.—The proper way to cleanse a saw-blade is to scrub it thoroughly for five minutes in hot water with soap and a *stiff* brush, held across the blade, then to immerse it in carbolic lotion until used. It is best to do this as the last thing before the operation. *Wiping with a towel should be avoided,* as a number of linen fibers are detached thereby and remain adherent to the teeth of the saw. *Sterilization by boiling in water is still better.*

*b.* MILDLY SEPTIC CASES.—The second group contains *cases characterized by chronic suppuration,* due to tuberculosis of joints or bones, or to ulcerative processes of various kinds requiring amputation. Infection of the amputation wound through contact with hands or apparatus that have touched the ulcers or fistulæ, or through escaping secretions, occurs very easily in these cases, and special precautions have to be employed to avoid it.

A careful examination of the affected parts should be made several days or a week before the time appointed for the amputation. *Abscesses should be incised and drained,* retentions removed by counter-incision, and the amount of secretion reduced by all known means, as, for instance, frequent irrigation and change of dressings.

The field of operation should be prepared as indicated for the first group. Immediately preceding the operation the suppurating focus or ulcer should be irrigated and dressed in bed, and over the usual dressing a piece of rubber tissue should be tightly bandaged so as to overlap it on all sides, the margin of the gutta-percha adhering to the skin.

The patient being anæsthetized, Esmarch's constrictor is applied, and the rubbers are arranged in the proper manner to shield the patient's body from drenching with the irrigating fluid. After this the whole surface of the limb, with the exception of the field of operation, is wrapped in clean towels, the carbolized towel covering the site of the operation is removed, this and all hands are finally disinfected, the irrigator is started, and the amputation should commence.

It is not very difficult in these cases to exclude suppuration and to secure primary union by the exercise of a moderate amount of care and by intelligent attention to important details. Should infection occur on account of faulty management or the inherent difficulty of the case, the inevitable suppuration will be mostly of a benign character, and well-nourished and well-coapted portions of the wound may even heal by primary union.

*Where amputation has to be done through ulcerating or suppurating parts of a limb,* the surgeon has a still more difficult problem to solve. But even in some of these cases primary union can be achieved. Before commencing the operation, the skin surrounding the ulcer or sinus must be thoroughly scrubbed with brush, soap, and water, then the ulcer or sinus is repeatedly washed or injected with an eight-per-cent solution of chloride of zinc, and the granulations are thoroughly scraped off with the sharp spoon. Indurated or poorly nourished tissues are removed, and all *débris* is washed away with the irrigating stream of mercurial lotion. After this the amputation is done as usual, good care being taken to provide for ample drainage, either by means of rubber tubes, or *preferably by packing from the bottom with iodoform gauze.*

c. SEPTIC CASES OF GREATER INTENSITY.—To the third group belong all cases in which *an acute progredient septic process of spontaneous or traumatic* origin necessitates ablation of the affected limb under a vital

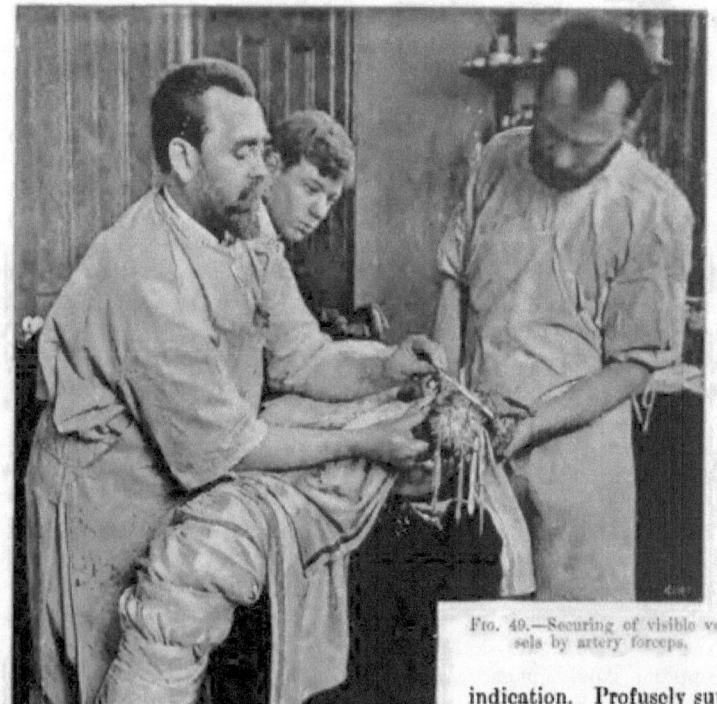

FIG. 49.—Securing of visible vessels by artery forceps.

indication. Profusely suppurating compound fractures, rapidly progressive phlegmons of the hand and arm, cases of embolic or other forms of spontaneous gangrene, compose this class, in which the surgeon has to contend not only with the local trouble, but also frequently with a deep and dangerous general intoxication of the system, due to the massive absorption of ptomaïnes and bacteria.

In many of these cases the processes determining phlegmonous destruction have progressed beyond the highest limit of amputation, and securing of an aseptic state of the wound is impossible. No amount of irrigation will here do any good, and the surgeon, having removed most of what is a source of further infection, has to trust to good luck and the power of resistance of his patient, aided by ample stimulation and other restorative measures. *In these cases the open after-treatment is in order.*

But, even in those instances where amputation can yet be done in healthy tissues, preservation of an aseptic state is an extremely difficult matter on account of several reasons. First of all, we have profuse secretion of pus or ichor, containing an extremely virulent culture of micro-organisms, a few individuals of which are sufficient to start up another phlegmon. Nobody who has not tried it can conceive the difficulty of keeping free from contamination in such cases. Another difficulty lies in the limits to our choice of the place of amputation. When we can go high up, far out of the reach of the infection, we should always do it without regard to so-called conservative considerations. *What is first to be conserved here is the life of the patient,* and before this view all objections ought to vanish.

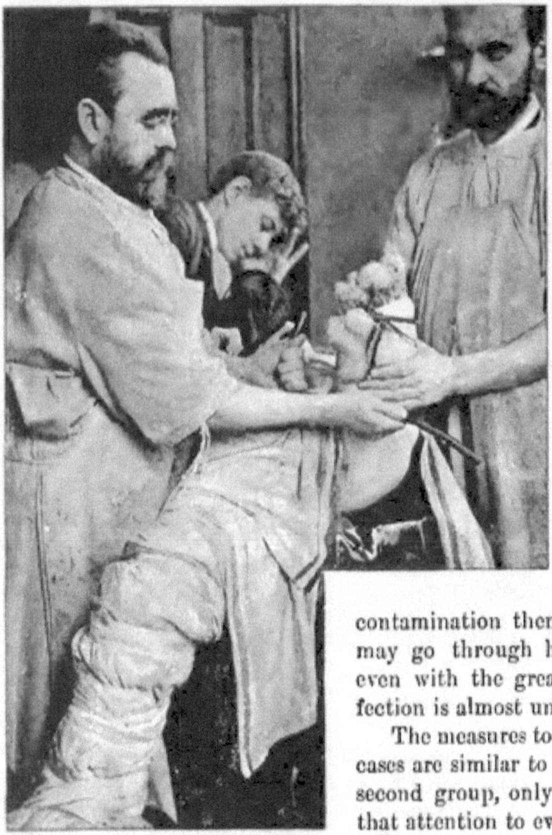

But, when the process has extended up beyond the knee or the elbow, how keep free from contamination then? True, the section may go through healthy tissues; but, even with the greatest care, contact-infection is almost unavoidable.

The measures to be employed in these cases are similar to those detailed for the second group, only with this difference: that attention to every step of the preparation should be more rigid; that, if possible, the filthy part of the preparation should be done by a separate person or persons; and, finally, that the judicious use of our strongest antiseptics for irrigation (1 : 500 to 1 : 1000 of corrosive sublimate) is justified. The lotion used for rinsing the hands must be repeatedly changed, and everything that

FIG. 50.—Compression of cut surface by sponges placed *over* the folded flaps. Removal of constricting band.

## SPECIAL APPLICATION OF THE ASEPTIC METHOD. 69

has come in mediate or immediate contact with the focus of infection must be rigidly rejected.

*Amputation wounds belonging to this group should not be sutured, but require loose packing and moist dressings (open treatment).*

Our first and second groups coincide with "*primary*" and "*secondary,*" the third with "*intermediate*" *amputations* of the old nomenclature.

2. **Hæmorrhage.**—Esmarch's apparatus and the animal ligature have undoubtedly had a great share in bettering the statistics of major amputation.

*a.* ARTIFICIAL ANÆMIA.—The most important and really blood-saving part of Esmarch's apparatus is performed by the constricting band, used instead of a tourniquet. The theoretical advantages of the use of the elastic roller-bandage, employed for evacuating the vessels of the limb, are offset by some serious drawbacks. It is an undeniable fact that the aërostatic pressure will effectually prevent the escape of considerable quantities of blood from a limb, the circulation of which has been suppressed by central constriction. Therefore, the expulsion of all the blood contained in a limb is not an absolute requirement of blood-saving in non-mutilating operations, as, for instance, joint exsections.

In amputations the blood contained in the removed limb is an absolute loss, but its quantity can be effectually limited to a very small amount

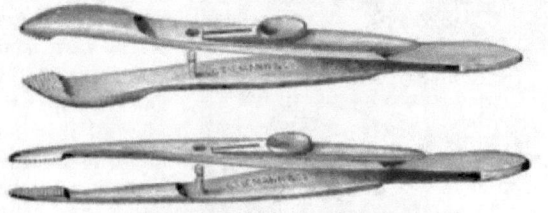

FIGS. 51, 52.—Esmarch's artery forceps.

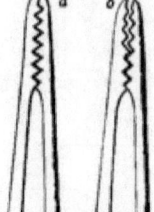

FIG. 54.—Showing the difference between *a*, a good, and *b*, a worthless, artery forceps. On compression, points of *a* remain in contact; those of *b* gap.

FIG. 53.—Hahn's artery forceps.

by previous vertical elevation of the limb. And this loss is abundantly repaid by the agreeable assurance, that no septic material or infectious cell-elements, detached from a malignant new-growth, are thrown into the general circulation with the blood and lymph which is expelled from the diseased limb by the elastic roller-bandage.

The retention of a certain quantity of blood in the vessels of the stump affords additional advantages of no mean value. By pressure upon the stump, the smaller and smallest arteries and veins each will pour out a minute quantity of blood, which will greatly aid the surgeon in finding and securing them before the removal of the constrictor. Thus all considerable ostia can be occluded, so that, on detaching the rubber band, no spurting vessels will be observed, and the capillary oozing will easily be controlled by compression of the wound, aided by digital pressure exerted upon the main artery of the limb. Compression should not be done by packing the wound full of sponges, and folding the skin-flaps over these. True that their elastic pressure will check hæmorrhage. But, on the other side, most of the small thrombi occluding the vessels, that are continuous with the clot occupying the outer meshes of the sponge, are torn away when the latter is removed, and renewed oozing results. The same objection must be raised against vigorous sponging of the wound-surface. Even after oozing has stopped completely, frequent sponging is apt to renew it, and thus to prolong the time required for stanching the hæmorrhage.

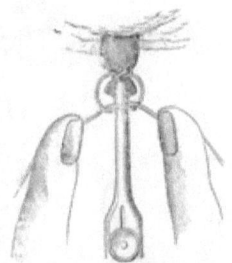

Fig. 55.—Manner of tying vessel. (Esmarch.)

A better way of employing compression is to fold the flaps over the wound, and then to arrange the sponges outside of them. This will insure the good effect of compression without the disadvantage mentioned above (Fig. 50).

As soon as all visible vessels have been secured, the wound is compressed, and the constrictor is removed while the limb is held vertically. The assistant who removed the constricting band applies digital compression to the main artery. Immediately after removing the rubber band, the skin of the parts that had been subjected to artificial anæmia is seen to flush up, and to remain vividly red for from five to ten minutes. This is the period of excessive hyperæmia, due to paresis of the vasomotor nerves. Hyperæmia is all the more lasting and intense, the longer and the tighter was the constriction. Attention should be devoted by the surgeon to learn the exact amount of tension of the rubber required to just stop arterial circulation.

The band should never be applied before the patient is relaxed, and it should not remain on longer than absolutely necessary.

NOTE.—The rubber constrictor exerts an enormous amount of constant and undiminishing pressure, hence it must be used with discretion. Applying it to the thigh held in flexion may lead to rupture of all flexors if the limb is straightened out afterward.

For a number of years, the author has discarded all specially made bands and apparatus recommended by authors and sold by dealers for the production of artificial anæmia.

*A piece of pure gum-elastic tubing, of the thickness of a man's index-finger or thumb, and of the length of one and a quarter yard, is all that is necessary.* Its application is illustrated in Fig. 56. The limb being held vertically for a few minutes, the elastic tube is put on the stretch, and thus coiled about the limb once or twice, its tension and the number of turns being determined by the relative thickness of the limb, the muscularity, and amount of adipose tissue underlying the skin. To estimate the tension

SPECIAL APPLICATION OF THE ASEPTIC METHOD. 71

required, the feel of the radial and dorsalis pedis arteries may serve respectively. As soon as their pulsation disappears, the constriction is sufficient.

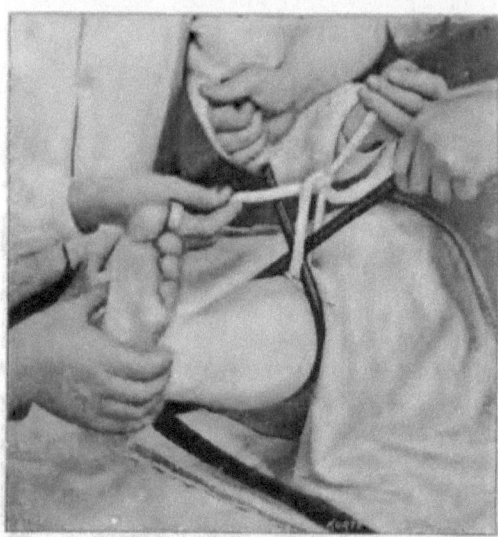

Fig. 56.—Manner of applying elastic constrictor (rubber tube) for the production of artificial anæmia.

When the required amount of constriction is secured, the ends of the tube are crossed, a short piece of cord or muslin bandage is passed under the crossing, and is firmly tied in a slip-knot. The ends of the tube being released, the rubber crowds up against the cord, and can not slip. (Fig. 57.)

This mode of constriction is very energetic, and deserves the preference for very large and muscular extremities.

*Another practical and more gentle way of applying elastic constriction is by means of an ordinary pure gum roller or Martin's elastic bandage.* It is especially suited for emaciated limbs and for operations on women of delicate frame, and children.

The manner of applying Martin's bandage is well illustrated in the accompanying cuts. As many turns of the bandage are superimposed tightly around the limb as necessary. The last turn is grasped in the left hand, and is pulled away forcibly from the limb, forming a bight, into which is thrust the remainder of the roller.

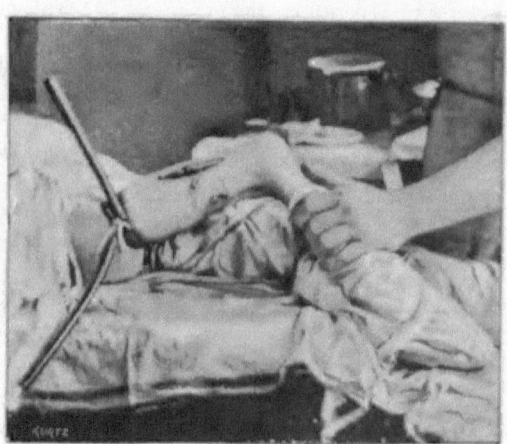

Fig. 57.—Elastic constrictor *in situ.*

As soon as the left hand releases the loop, it tightens about the roller, and holds it in place firmly and securely. (Fig. 58.)

*b*. LIGATURES AND FINAL HÆMOSTASIS.—The visible lumina of all cut vessels—veins and arteries—are tied with catgut, which is in every way preferable to silk. The objections raised against the new material have been entirely disproved by experience. The author never saw one case of secondary hæmorrhage from a vessel tied with catgut; and knows of two cases only, quoted on pages 5 and 57 respectively, where catgut ligatures slipped or gave way. In both, very brittle catgut was used, and the knot was not sufficiently tightened on account of the fear of breakage. Therefore it may be said that improper material was improperly applied in both of these instances.

In tying larger vessels it is very necessary to grasp and withdraw them from their sheaths for inspection.

*Arteries* will sometimes be laterally nicked just a little above the transverse section, and the ligature must be applied above the lateral opening.

*Large veins* must be also well inspected, as it may happen that the

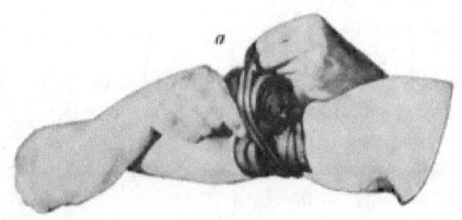
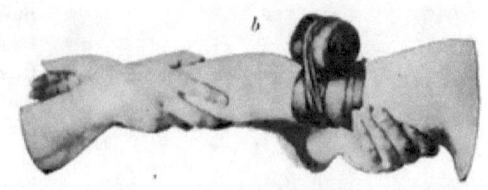

FIG. 58.—*a*. Applying of Martin's bandage as a constrictor. *b*. Martin's bandage *in situ*.

lumen of a hastily tied vein may be only partially occluded by the ligature. An ordinary artery forceps can not grasp at once the entire circumference of a principal vein, and the author has repeatedly seen only one half of the vein deligated in the shape of a dog's ear, the remainder of the vein con-

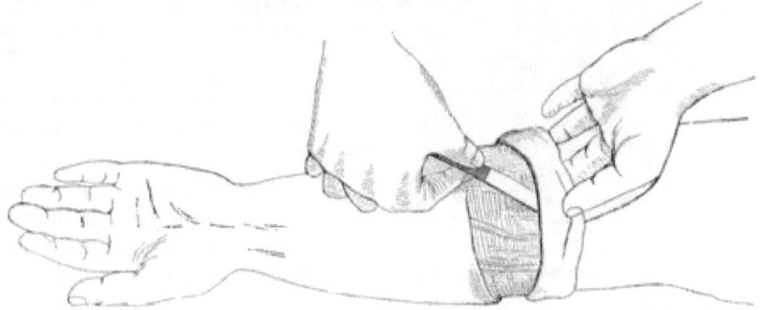

FIG. 59.—The wrong way of detaching the skin-flap. The knife should be held vertically. (Esmarch.)

tinuing to bleed in spite of the ligature. The best way to secure the entire lumen of a large vein is to grasp and withdraw it with one or two forceps

# SPECIAL APPLICATION OF THE ASEPTIC METHOD.

until its whole circumference is clearly visible, and then to twist it around its own axis, when it will be seen to form a neck which can be easily tied.

Atheromatosis of arteries is no valid objection to the application of the catgut ligature.

The grasping of vessels affected by it is difficult on account of their liability to slip before, and break after, being caught by the forceps.

FIG. 60.—Liston's bone forceps.

The ligature must not be tightened too much on an atheromatous vessel, or it may cut through it.

Vessels imbedded in sclerosed tissues must be secured by a circular stitch.

After the removal of the elastic constrictor, local compression of the wound is kept up until the marked hyperæmia of the limb begins to wane. Then, an assistant compressing the main artery, the wound is exposed. The glazing of clotted blood is removed by irrigation and gentle friction with the tips of the fingers, and the assistant is directed to release the compressed main artery. Then any additional vessels seen spurting should be secured. The hyperæmia of the limb will have ceased by this time, and with it the oozing.

NOTE.—Should a larger nutrient artery be divided at the time of the section of the bone, its bleeding can be readily stopped by the insertion of a short piece of stout catgut into the spurting orifice, where it can be left behind without any harm. The employment of wax for the same purpose is unsafe, unless the material is first sterilized by boiling.

FIG. 61.—Amputation wound of thigh, sutured and drained.

The statement that Esmarch's apparatus is not blood-saving, but, on the contrary, causes undue hæmorrhage, is misleading. It may be positively said that skillful management of the application of Esmarch's constrictor will enable the surgeon to perform major operations with an astonishingly small amount of hæmorrhage, and that loss of much blood after the removal of the rubber band is due to faulty manipulation.

3. **Securing of a Good Stump.**—In circular amputations, as well as in flap operations, an important object should be to gain abundant covering, and to bring about easy and natural apposition of the wound-surfaces without much external pressure.

In performing circular amputation, the assistant holding the mesial part of the limb can greatly influence the shape of the stump. As it is desirable to produce a wound of the shape of a hollow cone, multiple circular sections of not too great depth are commendable, while the assistant successively retracts each layer divided by the amputating knife until the periosteum is cut through and pushed well back. The soft parts are inclosed in a two- or three-tailed compress of sublimated gauze, and the bone or bones are sawed off, care being taken on the leg and forearm to complete the section of both bones simultaneously. After this the sharp edges of the bone are clipped off with bone-cutting forceps, and the vessels are attended to.

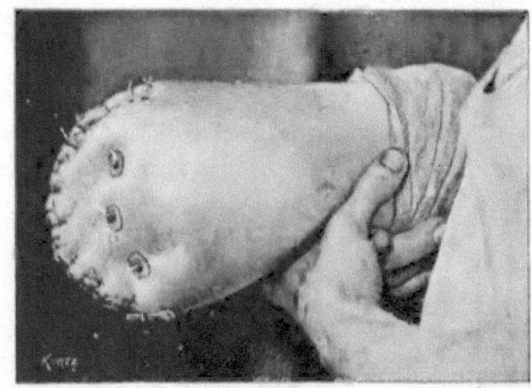

FIG. 62.—Amputation wound of leg, sutured and drained. Retentive button sutures.

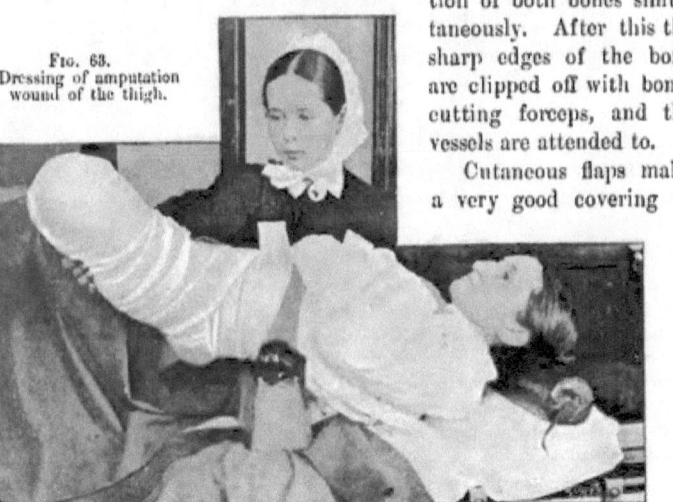

FIG. 63. Dressing of amputation wound of the thigh.

Cutaneous flaps make a very good covering to most stumps, and can be very easily adapted. As soon as the hæmorrhage is perfectly under control, suture of the wound can be commenced.

The author is using exclusively the interrupted suture, for reasons elsewhere mentioned.

If the case was unimpeachably aseptic, and no suppuration is expected, one medium-sized drainage-tube will suffice to carry away the first secretions. Otherwise abundant ways of egress must be provided in the shape of several properly distributed tubes. The protruding end of each tube is transfixed with a safety-pin, and cut off on a level with the skin. An ample dry dressing, consisting of a few layers of iodoformed and a generous mass of sublimated gauze is snugly bandaged to the stump, so as to reach at least twelve inches above the line of section.

FIG. 64.—Dressing of amputation wound of the leg.

If proper care was devoted to the stanching of the hæmorrhage, no great pressure will be required to check the oozing, which is, anyway, moderate after the use of corrosive sublimate for irrigation.

The idea of bringing about close apposition of the wound-surfaces by energetic pressure is not to be cultivated, as it will lead to frequent marginal necrosis of the flaps, frustrating complete primary union. Surface apposition should rather be accomplished by a proper fashioning of the wound and flaps, and the sutures should exert no traction whatever, but should merely secure contact of the cutaneous edges.

For securing contact of the deeper portions of an amputation wound, Lister's lead-plate, or button, sutures are very advantageous. (Fig. 62.)

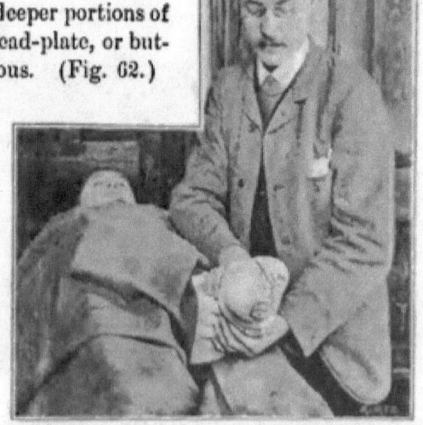

NOTE.—In former times, when carbolic lotions were employed for irrigation, oozing used to be quite free, and necessitated the use of a good deal of pressure, which was somewhat tempered by the interposition of thick layers of borated cotton between the dressing proper and the outer bandage. Flap necroses were then much more common than nowadays.

The sole office of the dressings is to lightly support the wound, and to absorb and render innocuous the secretions.

FIG. 65.—Amputation wound of the thigh fourteen days after the operation. Case of Mrs. Walther.

The author's custom is to make the first change of dressing on the fourth day after the operation, when the drainage-tubes can be withdrawn. Another lighter aseptic dressing is then applied, and remains undisturbed for a week. By the end of this time the drainage-tracks will have either healed completely, or their place will be marked by a small patch of granulations, requiring merely a borated-salve or simple adhesive-plaster covering.

This refers to correct cases only. Should septic fever develop or marginal gangrene be noted, frequent moist dressings are in order, and the rules appropriate for the treatment of suppurating wounds obtain precedence.

CASE: *Illustrating a Correct Course of Healing.*—Mrs. Pauline Walther, seamstress, aged fifty-one. Far-gone tuberculous destruction of knee-joint with fistula, the latter the result of a previous exploratory incision. *Feb. 14th.*—Amputation of thigh in middle third. Aseptic fever, with rise of temperature to 103° Fahr., on the two days following the operation. *Feb. 18th.*—Temperature, 99° Fahr. *March 1st.*—First change of dressings; drainage-tubes removed; wound redressed. *March 7th.*—Wound completely healed, except where one minute spot of granulations marks the former site of a tube. *March 12th.*—All firmly cicatrized; the stump can be lightly pounded without pain. *March 17th.*—Patient discharged cured. See Figs. 61 and 65.

### IV. OPERATIONS ABOUT NON-SUPPURATING JOINTS.

**1. Puncture and Irrigation.**—Chronic hydrops, or, as Volkmann calls it, catarrhal synovitis of the knee-joint, is often benefited or even cured by puncture and subsequent irrigation.

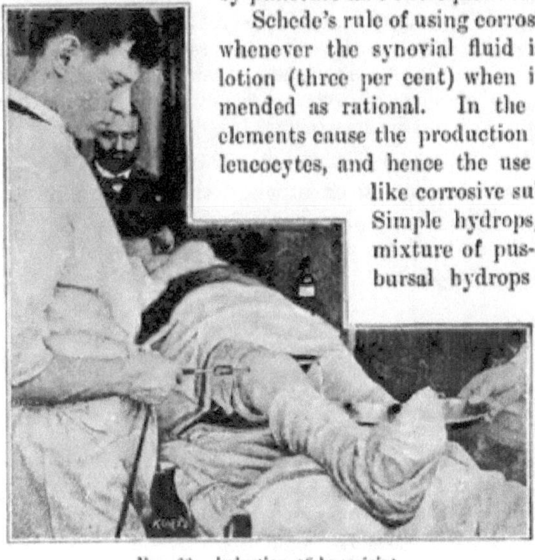

Schede's rule of using corrosive sublimate (1:1,000) whenever the synovial fluid is turbid, and carbolic lotion (three per cent) when it is clear, can be commended as rational. In the former case pyogenic elements cause the production of a certain amount of leucocytes, and hence the use of a strong germicide like corrosive sublimate is appropriate. Simple hydrops, where there is no admixture of pus-cells, is comparable to bursal hydrops or hydrocele, and is benefited by the application of an irritant substance like carbolic acid.

The manner of procedure employed by the author is as follows:

Two large-calibered trocars are rendered aseptic either by boiling the tubes for an hour in a five-per-cent solution of carbolic acid, or by heating them in a large alcohol flame to incandescence, after which

FIG. 66.—Irrigation of knee-joint.

## SPECIAL APPLICATION OF THE ASEPTIC METHOD.

they are dropped into carbolic lotion. Too much care can never be exercised in attending to the proper disinfection of the trocar-tubes, as their hollow shape renders their cleansing a difficult matter at best.

CASE.—Thomas Casey, hostler, aged twenty-three. Hydrops of right knee-joint of several years' standing. *March 14, 1887.*—Puncture and irrigation with Thiersch's solution and carbolic lotion. Dorsal splint. The trocars had received a rather superficial attention by boiling of too short duration. The following day high fever appeared with great distention of the joint. *March 15th.*—Aspiration yielded pus. *March 16th.* —Multiple incision and drainage. The fever not abating, although secretion was very scanty, the limb was suspended in a wire cradle, and weight extension was applied, so as to enable the house-surgeon to frequently irrigate the joint without disturbing the patient's rest. In spite of the most attentive treatment, new abscesses developed, and the patient's evident failing finally compelled amputation of the thigh, which was done, May 30th, by Dr. F. Lange. The patient recovered. Extensive tuberculosis of the head and shaft of the tibia was ascertained by examining the specimen.

After the usual preparation of the patient's limb, the trocars are thrust into the knee-joint from opposite sides, and the synovial fluid is let out.

To remove flocculæ of coagulated fibrin, Thiersch's solution is first used for washing out the joint cavity. The reason for this is the fact that carbolic acid hardens the fibrinous clots and makes them tough and unfit to pass the cannula. Corrosive sublimate, on the other hand, is poisonous, and dangerous quantities of it may be absorbed if irrigation be carried on sufficiently long to free the joint of all deposits of fibrin.

CASE.—John Schurz, mason, aged thirty, chronic hydrops of knee-joint. *April 8, 1886.*—At the German Hospital, double puncture and rather prolonged irrigation with corrosive-sublimate lotion (1 : 1,000) on account of the presence of large quantities of fibrinous deposit. *April 10th.*—*Mercurialism;* salivation and sharp colic, lasting for five days, with some fever, ending in recovery on appropriate treatment. Hydrops cured.

As soon as Thiersch's fluid is seen to escape clear from the efferent cannula, corrosive sublimate or carbolic lotion is substituted therefor, and the joint is thoroughly flushed with it. To prevent the retention of a dangerous amount of either of these solutions, the joint is flexed and emptied

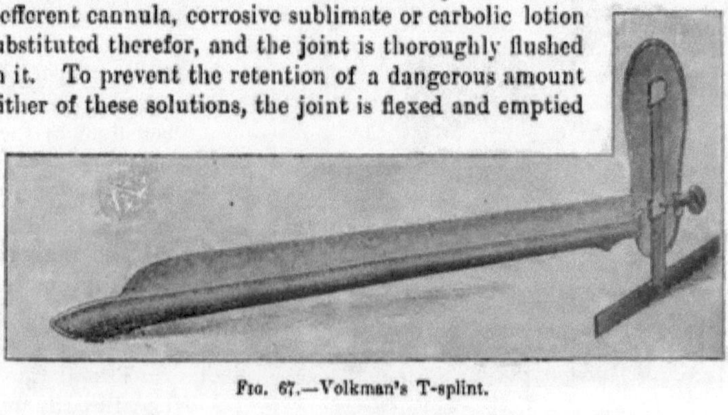

FIG. 67.—Volkman's T-splint.

by external pressure. The tubes are withdrawn, a small patch of iodoform gauze is attached with a strip of adhesive plaster over each puncture-hole, and the limb is placed on a dorsal splint. (Fig. 67.)

2. **Arthrotomy for Chronic Fibrinous Hydrops, for Vegetations, Tumors, and Floating Bodies of the Knee-joint.** *a.* HYDROPS GENU.—In cases where a thick coating of fibrinous deposit is lining the entire cavity of the knee-joint, simple puncture and irrigation will be found impracticable on account of the continuous clogging of the efferent cannula. To completely free the joint of these masses, immediate incision must be done. The internal aspect of the knee presents the most convenient place for this procedure. The skin and fascia are successively incised, and all bleeding vessels are carefully tied. On being exposed, the bluish capsule is cut into, and the incision is extended to about an inch in length. After this, irrigation by Thiersch's solution is practiced, and the joint is repeatedly flexed and extended to aid detachment and expulsion of the membrane, which can be hastened by sweeping the index-finger through all the recesses of the joint. The slight hæmorrhage following this manipulation will cease spontaneously, and the clots are washed out by a strong jet of irrigating fluid.

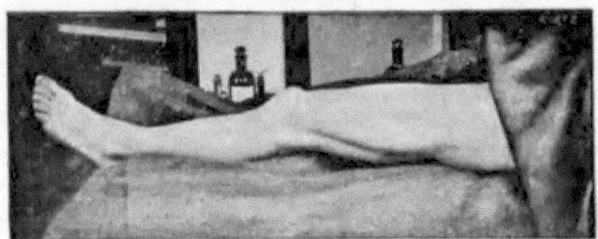

FIG. 68.—Arrangement of rubber sheets for operations about the lower extremity.

After the insertion of a short piece of medium-sized drainage-tube, which should reach just within the cavity of the joint, the capsular incision is closed by a few interrupted catgut sutures.

The fascia and skin are likewise united, the protruding end of the tube is transfixed with a safety-pin and trimmed off short, and the joint receives a final flushing with carbolic or mercurial lotion according to the indications mentioned in the preceding paragraph.

After this the wound is dressed and the limb is fixed upon a dorsal splint.

If the aseptic measures were sufficient, no reaction whatever will follow the operation. In cases where the hydropic fluid was limpid, no secretion of any account will be observed, and the tubes can be withdrawn at the first change of dressings, which is usually done on the fifth day after the operation. As soon as the wound is in progress of cicatrization, active movements and cautious use of the limb should commence, the joint being protected by a small aseptic dressing, held in place by Martin's elastic bandage.

CASE of John Schurz, page 77, who was discharged cured June 29, 1886, with partially restored and constantly improving mobility.

*Passive movements are unnecessary and very painful.* Restoration of the mobility should be hastened by cold or warm douching and subsequent massage, and its final establishment left to the active efforts of the patient himself.

Cases in which large quantities of firmly adherent membrane were removed and some hæmorrhage followed, especially if the hydropic fluid was very turbid, will develop a moderate secretion of serous bland pus, that may continue for some time. Some fever will also occur, to subside as soon as the dressings are changed and the joint is washed out again.

It will commend itself to apply in these cases a fenestrated plaster-of-Paris splint, and to repeat irrigation once or twice daily in the beginning, diminishing the number of washings *pari passu* with the disappearance of the secretion. As soon as the discharge shall have become serous and scanty, the tube can be withdrawn and the case treated as above explained.

CASE.—Fred. Scheeker, laborer, aged twenty-six, had been suffering for several years from a painless, massive, hydropic distention of the right knee-joint, that could not be traced to a traumatism. Considerable lateral mobility was the main cause of his seeking relief at Mount Sinai Hospital. *Dec. 7, 1885.*—Double puncture and irrigation were done, but had to be abandoned on account of large masses of dense fibrin. Immediate incision and clearing of the joint were practiced. Fever and some secretion being noted, the dressings were changed December 10th, and, the limb being put up in a fenestrated plaster splint, irrigation with corrosive sublimate was employed twice—later on, once—daily. *Dec. 20th.*—Normal temperature was noted. *Feb. 1st.*—Irrigation discontinued and splint removed. *Feb. 20th.*—Patient discharged cured, with increasing flexion (twenty degrees).

*b.* VEGETATIONS.—The favorite seat of vegetations in the knee-joint is that lax part of the capsule situated below the inferior margin of the patella, which is overlaid by a thick cushion of loose fat and the ligamentum patellæ proprium. They are rarely pedunculated, their common appearance being that of a yellowish or purple coxcomb, and their direction transverse. The functional disturbance produced by them is sometimes very slight, but occasionally extremely severe, especially when it happens that their margin is caught and jammed in between the articular surfaces. Hæmorrhage with acute synovitis and an effusion may follow this accident.

The diagnosis of vegetations, sufficiently massive to cause functional trouble, is not difficult to the careful examiner. Frequently the patients themselves will point out the kernel-like slipping bodies of soft consistency. They are easily distinguished from free floating bodies by the fact that on manipulation they never disappear entirely from their seat of predilection, to reappear in a distant part of the joint.

Topical treatment is generally powerless against this complaint, although the constant use of a Martin's bandage may mitigate the trouble by confining somewhat the motion of the joint, and thereby diminishing the chances of contusion of the growths by jamming.

In aggravated forms, arthrotomy and excision of the vegetations is proper. With strict attention to the cautelæ before mentioned, the joint is incised, and, the patella being tilted upward by a sharp retractor, the mass is grasped with a pair of mouse-tooth forceps, and is bodily excised. Should it extend across the entire width of the patella, another lateral incision will have to be made on the opposite aspect of the knee, to enable the surgeon to complete the excision.

If much hyperæmia of the growth be present, as shown by its purplish color, hæmorrhage may be rather free. In such a contingency the raw surface should be seared with the thermo-cautery.

Toilet of the joint cavity is followed by suture, and a small drainage-tube is inserted to serve as a safety-valve. The subsequent treatment coincides with that given for simple hydrops after puncture and irrigation.

Case I.—Miss Lena C., aged fourteen, vegetations occupying the internal inferior margin of the patella. The patient had frequent attacks of sudden, very sharp pain in the knee, followed by effusion. Various plans of local treatment had been employed unsuccessfully for about a year. *Dec. 5, 1881.*—With the assistance of Dr. B. Scharlau, the family attendant, incision of knee-joint on its inner aspect was done. A series of yellow, smooth bodies presenting, they were excised with forceps and curved scissors. Drainage, suture, and plaster-of-Paris splint. Some fever, due to constipation, but no inflammation followed. *Dec. 9th.*—A laxative being administered, a copious stool was had, whereupon the temperature at once fell to, and remained at the normal standard. *Dec. 12th.*—The tube was removed. About New Year's the patient commenced to walk about, and shortly after was discharged cured. In the spring of 1886 circumscribed swelling of the synovial membrane in the vicinity of the cicatrix was noted. It subsided upon the use of an elastic bandage, which was ultimately abandoned. In January of 1887 the patient was still perfectly well.

Case II.—Frank Mann, clerk, aged twenty-five, well-defined painful vegetations to be felt near the lower margin of the knee-pan, on both sides. Duration of trouble, six months. Functional disturbance very marked. *April 8, 1886.*—Double incision of knee-joint at the German Hospital. Excision of a deep-red, transversely situated, coxcomb-like growth from the lower rim of the patella. A good deal of oozing necessitated searing of the denuded surface of the capsule with the thermo-cautery. Drainage; plaster-of-Paris splint. Eventless course of healing. The tube was removed on the tenth day. Patient discharged cured, with good motion, May 20, 1886.

c. Floating Bodies of the Knee-Joint:

Case.—E. Behrmann, painter, aged thirty-eight. Large floating body of the knee-joint, with chronic hydrops. *May 15, 1886.*—Arthrotomy at the German Hospital. Previous to the incision the floating body was fixed by finger-pressure near the line of section, but disappeared in the joint cavity when the last stroke of the knife opened the capsule. The author swept through the joint with a well-rinsed finger, and found the body in the bursa of the quadriceps muscle. By means of bimanual manipulation, the body was brought down to the aperture, and was readily extracted. Irrigation with corrosive-sublimate lotion, drainage, suture, and fixation upon a dorsal splint followed the extraction. Normal course of healing. *June 15, 1886.*—The patient was discharged cured with good function of the knee.

d. Suturing of the Fractured Patella.—Although not perfect, yet the functional results achieved by the ordinary forms of treatment employed in cases of transverse fracture of the patella are generally so good, that arthrotomy, for the sake of wiring or otherwise suturing the patellary fragments, is rarely if ever justified at a time immediately following the injury. Hamilton has shown that even a considerable degree of diastasis of the fragments is not incompatible with a very fair functional ability of the limb, provided that the intervening ligamentous band be strong, the action of the quadriceps vigorous, and the lateral extensions of the quadriceps tendon uninjured.

It seems, then, rational, in cases of patellary fractures, first to employ the usual methods of treatment by rest and appropriate bandaging, and thus to await the result. It never can be predicted with accuracy, and may turn out to be very satisfactory after all.

Should the result be unsatisfactory, either through failure of union or subsequent rupture of the new-formed ligament, arthrotomy and secondary suture may properly be taken into consideration.

On account of the presence of large quantities of blood and serum, found shortly after the accident effused into the joint and its vicinity, primary arthrotomy for patellary fracture is a more risky undertaking than the secondary operation. The slightest error in the use of the aseptic apparatus may cause irreparable damage, and may cost the patient's limb or life. Especially dangerous are those cases in which open ulcers or abrasions, or other secreting wound-surfaces due to the primary injury, are located near the field of operation, be they however small or superficial. Pyogenic infection and suppuration of the knee-joint are here nigh to inevitable. Anchylosis is the most favorable issue that can be expected in case of suppuration; very often, however, the limb will have to be sacrificed.

The conditions for the successful performance of the secondary operation are, as far as the chance of avoiding suppuration is concerned, infinitely better. The effusions due to recent traumatism are mostly absorbed, the parts have recovered their physiological equilibrium, and faults of aseptic technique are easier to avoid and not as hard to remedy as in recent cases.

The circumstance can not be urged as a serious drawback, that a few weeks after the accident, the fracture-planes are found covered with new-formed connective tissue or a cicatrix, and that this must be first removed before suture can be applied.

More difficulty may be encountered in overcoming the retraction of the quadriceps. But even such high degrees of retraction as are occasionally observed in complete failure of union, or met with in old secondary rupture, representing a diastasis of several inches, can be managed so as to permit suture and bony union of the fragments.

The mode of procedure is well illustrated by the following history:

CASE.—Mrs. Lizzie P., housewife, aged twenty-eight, an extremely obese woman, contracted in 1884 a transverse fracture of the *left* patella, which was attended to by her family physician, and was treated by rest and bandaging. It healed with a seemingly satisfactory ligamentous union, which, however, gave way a few weeks after the completion of the treatment, resulting in a wide gap between the fragments. Measurement gave a hiatus of two and a half inches in extension, five inches in flexion at a right angle. Her gait was rather uncertain, causing many falls, one of which produced, May 2, 1887, a transverse fracture of the *right* patella. This recent fracture was treated by approximation with two broad strips of adhesive plaster, bandaged on and laced, the limb resting on a T-splint. *May 25th.*—The old patellary fracture was united by operation at the German Hospital. The limb having been rendered anæmic by constriction, the joint was laid open by a transverse incision, and the cicatricial tissue investing the fracture-planes of the knee-pan was cut away, and the bone scraped free from all adhering connective tissue, until the corresponding surfaces of the patella were clean and smooth. After this four equidistant holes were drilled through each fragment, while the bone under treatment was held immovably fixed by an assistant in the grasp of a lion-jaw forceps. The drilling of the apertures in the upper fragment was much easier than of those in the lower one. By the aid of a flexible silver probe, a double thread of thick catgut (No. 4) was drawn through the corresponding drill-holes, the ends of each suture being temporarily secured in the grip of an artery forceps. The most difficult part of the operation consisted in the approximation of the fragments. The quadriceps tendon was exposed by a longitudinal incision of six inches in length, and, the upper fragment being forcibly drawn downward with bone-forceps, a number of alternating lateral notches were cut into the muscle and tendon, until the fragment yielded to moderate traction. The first suture nearest the edge of the patella was tightened—not tied—by an assistant until the fragments were brought in contact, whereupon the second suture was firmly knotted. After this the fourth suture was tightened and the third one tied; finally, the two outermost sutures were attended to. The ends of the catgut were trimmed, and three short drainage-tubes were inserted in the three angles of the wound. During the whole operation a stream of a 1 : 2,500 solution of corrosive-sublimate lotion was played on the exposed tissues. Before the closure of the wound, it was finally flushed with a 1 : 1,000 mercuric solution, and the application of a number of external catgut stitches completed the process. The knee was enveloped in an ample dry dressing and a plaster-of-Paris splint, enforced by a few lateral strips of white-wood veneering. Finally, the constricting elastic band was removed, and the extremity suspended in the vertical position, which was abandoned twenty-four hours after the completion of the operation. *June 3d.*—Splint removed; dressings changed; drainage-tubes withdrawn. *June 17th.*—Wound healed throughout. Silicate splint applied. *June 20th.*—Patient commenced to walk on crutches. *July 2d.*—She was discharged cured. *July 13th.*—The union of sutured patella was found firm, the operated limb much more useful than its mate. Flexion could be carried to a right angle. The course of healing of the case was feverless throughout.

### 3. Arthrotomy for Irreducible or Habitual Dislocation, and for Deformity due to Fracture.

—Dislocations that are irreducible from the outset, or have become so through neglect, can be corrected by means of aseptic arthrotomy.

CASE I.—Henry Köhler, aged nine. Dislocation of basal phalanx of thumb upon dorsum of metacarpal bone, of six weeks' standing. *December 29, 1879.*—Repeated unsuccessful attempts at reduction under chloroform. Immediate arthrotomy. Dissection of abnormal adhesions, and excision of a shred of interposed capsular tissue, followed by ready reduction. Suture and catgut drainage. Primary union. *Jan. 10th.*—Patient discharged cured with improving function.

CASE II.—John Becker, aged twelve. Fresh compound dislocation of terminal phalanx of the ring-finger on the dorsum of the middle phalanx. *March 29, 1884.*—Ether was administered at the German Hospital, and, after careful disinfection

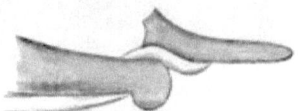

FIG. 69.—Explaining relation of parts in John Becker's case of phalangeal dislocation.

## SPECIAL APPLICATION OF THE ASEPTIC METHOD.

of the patient's hand, reduction was repeatedly attempted without success. The small transverse laceration of the integument of the volar aspect of the finger did not give the least advantage as to examining the interior relations of the displacement, hence a lateral incision was made on the radial side. It was then ascertained that the tendon of the flexor digiti profundus was displaced upon the dorsum of the middle phalanx, and was interposed between the articulating surfaces. Reduction could only be accomplished after a free division of all resisting bands of torn capsular ligament, caught between the flexor tendon and the articulating surfaces. Suture and catgut drainage; fixation of the finger on a small volar splint. *April 5th.—* Primary union. In May the function of the injured joint became nearly normal. (Fig. 69.)

CASE III.—John U. Kena, publican, aged thirty-nine. Admitted to German Hospital February 9, 1889, with old irreducible subcoracoid dislocation of right shoulder-joint. A number of unsuccessful attempts at reduction had been made by several medical men. *February 11th.—*Ether being administered, reduction by manipulation was again unsuccessfully tried. *Arthrotomy by a vertical anterior incision* revealed fracture by divulsion of the minor tubercle. The head of the humerus was dissected out of its newly-formed adhesions, and was replaced in the glenoid cavity. Drainage by a posterior button-hole incision. Suture. Uninterrupted primary union of wound. Patient was discharged cured, with improving mobility, on March 16th.

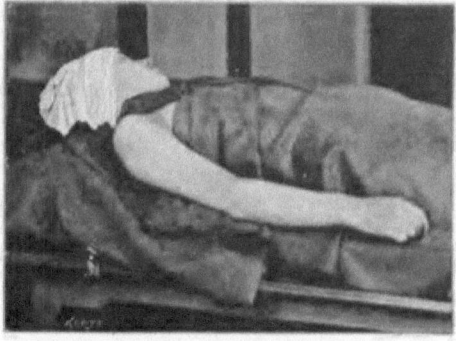

FIG. 70.—Arrangement of rubber sheets for operations about the upper extremity.

*Condyle fractures of the elbow with posterior or lateral displacement of the forearm are a common injury with children.* What with the great difficulty of an exact diagnosis in the presence of a large effusion, and the great differences of opinion of the authors as regards the proper manner of treatment, no wonder that, after elbow-fractures, cases of gunstock deformity and partial dislocation with inability to flex the elbow are not at all rare.

The author's conviction is that in many instances exact reposition and retention are utterly impossible unless the fragment is cut down upon and sutured or nailed to its original seat. The insertions of the muscles of the

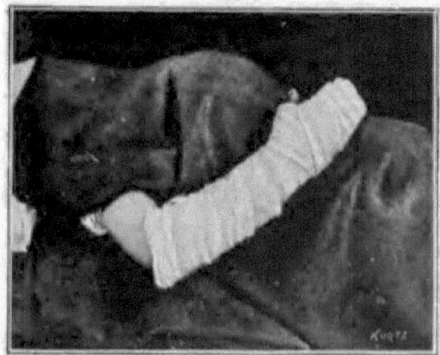

FIG. 71.—Dressing for wounds of hand and forearm.

forearm about the epicondyles must exert a great influence upon the displacement of the fragments, hence it seems that flexion would be the better position to counteract the tendency to displacement. But all assertions made to that effect, that, in spite of the presence of a large swelling, reduc-

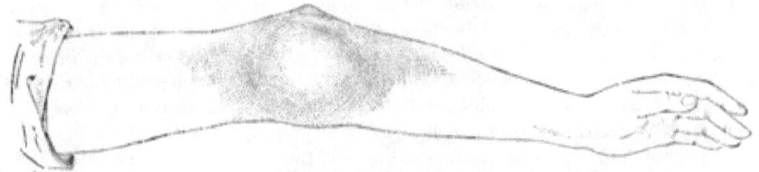

Fig. 72.—Anterior view of gun-stock deformity due to elbow fracture.

tion can always be accomplished and retention maintained, have appeared to the author as a hollow pretense or self-deception.

A very guarded prognosis in elbow-fractures is, on the part of the physician, a sign of wisdom and discretion.

Where very limited motion and an unfavorable position result in spite of careful treatment, the only means of correction is arthrotomy with subsequent partial or total exsection.

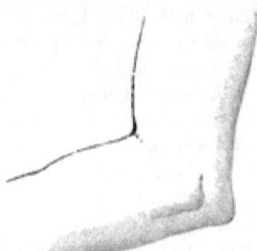
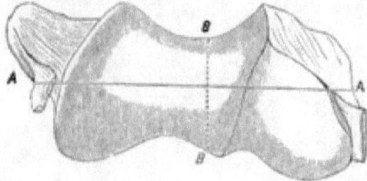

Fig. 73.—Lateral view of Bernhard Loebel's elbow.

Fig. 74.—Normal aspect of lower end of humerus. A A. Transverse diameter. B B. Line of fracture. In Bernhard Loebel's case.

CASE I.—Bernhard Loebel, aged two. October 27, 1886, injured his elbow by falling off a chair. The arm was put up by a physician in the flexed position in plaster

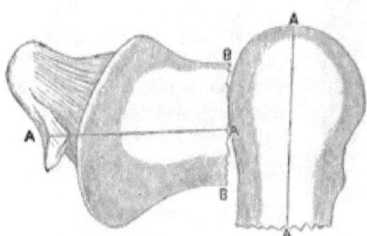

Fig. 75.—Showing relative positions of fragments in Bernhard Loebel's case.

Fig. 76.—Anterior view of lower end of humerus in Bernhard Loebel's case.

of Paris, and remained in this dressing for a fortnight. *Dec. 7, 1886.*—The elbow joint showed very marked gun-stock deformity. It was held at an angle of about one hundred and forty degrees. Flexion could be carried to about one hundred and ten degrees; extension not beyond the angle first mentioned. The forearm was displaced inward and backward, and the tendon of the triceps described a well-pronounced concave line. An abnormal mass of bone could be felt in the bend of the elbow externally, behind and below which the head of the radius could be made out with some difficulty. A posterior incision midway between the abnormal mass of bone and the olecranon opened the joint, and the periosteum was raised by means of the knife and elevator on both sides of the incision until the lower end of the humerus could be turned out for inspection. It was found that the deformed callus consisted of the external epicondyle, capitellum, and a small portion of the trochlea that had been broken off obliquely, and was tilted and pulled forward by the action of the flexors so as to present its articular aspect forward, part of the fractured surface looking backward. In this position bony union had taken place. The elongation of the outer half of the articular end of the humerus accounted for the gun-stock deformity; the presence of the large mass of bone displaced forward by tilting of the fragment explained the inability to flex. The lower end of the humerus was pared off horizontally with the knife, care being taken to remove a little more from the external than from the inner half of the lower end of the humerus, in order to preserve the "carrying point." The capsule and skin were united by suture. One drainage-tube was inserted. The arm was put up in extension in a couple of lateral pasteboard splints. No fever followed. *Dec. 14th.*—First change of dressings. In anæsthesia the tube was removed, and the arm was flexed to an acute angle and put up in this position in two lateral pasteboard splints. *Dec. 19th.*—Passive motion was practiced in anæsthesia, and the arm was fixed in the straight position. *Dec. 23d.*—Passive motion without ether. Fixation at an acute angle. *Dec. 29th.*—Free passive motion to normal limits. Splints abandoned and active movements commenced. *March 3d.*—Outline of elbow almost normal. Flexion and extension normal.

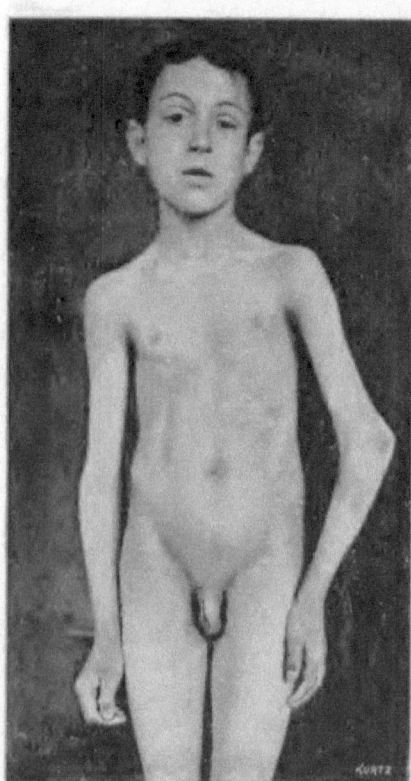

FIG. 77.—Gun-stock deformity due to T-fracture of the lower end of the humerus. Willie H.'s case.

CASE II.—Willie H., aged eleven. Very pronounced gun-stock deformity due to fracture of the elbow-joint sustained two and a half years ago. The treatment had been conducted by a surgeon of good repute. Flexion could be carried to a right angle, extension to about one hundred and thirty degrees. Fig. 77 shows the boy's arm in full extension. *June 17, 1887.*—Arthrotomy done at Mount

Sinai Hospital revealed a very curious condition of things. The broken-off external condyle and capitellum occupied a position similar to that observed in the preceding case. The ulna was dislocated backward and inward from the fragment representing the trochlea, which was attached by callus to the anterior aspect of the lower end of the humerus. Apparently a T-shaped fracture of the lower end of the humerus had taken place. The articular surface had a most grotesque shape. The cartilaginous surfaces of the trochlea and sigmoid incisure were coated with a dense mass of connective tissue. The broken-off coracoid process was attached to the fragment of the trochlea. The articular surface was pared off to approximate the shape of a normal humerus, and the wound was drained, sutured, and the arm put up in a pasteboard splint. Normal union by primary adhesion of the wound took place, but an annoying complication, consisting of *paralysis of the forearm and hand*, was noted. This untoward event was probably caused by the fact that the pad of Martin's bandage, used for producing artificial anæmia, had been placed *over the inner aspect of the arm*, exerting undue pressure over the nerves. *June 19th.*—The compressive dressings were removed, the drainage-tube was withdrawn, and the wound redressed. *July 2d.*—The patient was discharged from the hospital with healed wound. Local treatment of paralysis by galvanism and massage was commenced. *July 22d.*—Flexion and extension of forearm and fingers re-established. *Aug. 1st.*—Function of elbow becoming normal. *Aug. 19th.*—Muscular power fully restored. (See Fig. 78.)

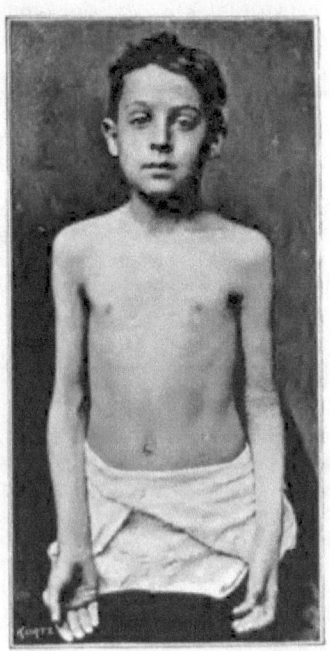

Fig. 78.—Result after exsection of elbow-joint for gun-stock deformity. Willie H.'s case.

*Habitual luxation of the shoulder-joint*, a very annoying and rebellious complaint, may also be cured by arthrotomy and partial exsection of the redundant capsular ligament. (See case on page 8, Note 2.)

## V. OPERATIONS FOR DEFORMITIES.

1. **Knock-Knee and Bow-Leg.**—Operative exposure of the medullary tissue of the long bones is a dangerous procedure unless suppuration can be excluded from the wound. By the successful employment of the aseptic method the danger of osteomyelitis can be virtually excluded.

McEwen's osteotomy is one of the safest and most useful procedures of the newer surgery. It has almost entirely displaced purely orthopedic methods.

For knock-knee, after division of the soft parts by a short longitudinal incision, the cancellous tissue of the lower end of the femur is divided by a properly shaped chisel, called osteotome. For bow-leg, the osteal section is carried through the upper end of the shaft of the tibia and fibula, or through the lower end of the femur, or both. The operation is done under artificial anæmia; and the dressings are applied, and the limb is put up in a contentive dressing—preferably plaster of Paris—before the removal of the constricting elastic band. New-formed bone is thrown out into the gap caused by the correction of the position of the bones, and by the end of three or four weeks firm union in a normal position is the result.

CASE.—Leopold Heymann, clerk, aged nineteen. Very marked bow-legs, the distance between the internal condyles of the femora being three and a half inches. *November 15, 1883.*—Double osteotomy of the thighs at Mount Sinai Hospital. Plaster-of-Paris splints. *Dec. 14th.*—Change of dressings. Wounds healed by primary union; bones firmly consolidated. The knees were in contact, but the curvature of the tibiæ, which represented a great part of the deformity, was still very marked. Undoubtedly osteotomy of the shin-bones would have given a better result. The patient declined further operative interference.

### 2. Bony Anchylosis in a vicious position.

CASE I.—Lina Frieberger, aged fifteen. Bony anchylosis of right and pseud-anchylosis of left maxillary joint, probably due to acute osteomyelitis of right ascending ramus. The teeth were in absolute apposition, and no solid food could be taken. Marked facial hemiatrophy. In childhood a suppurating affection of the right cheek was noted. *April 3, 1886.*—Exsection by chisel and mallet of the left maxillary joint (hemiatrophy of the same side). The operation did not relieve the functional trouble; the joint was found pseud-anchylosed, the cartilages gone, and the capitellum nearly absorbed. The wound healed by primary intention. *April 29th.* — Exsection of right maxillary joint, which was found firmly anchylosed. The

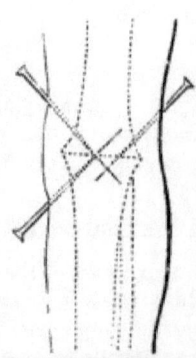

FIG. 79.—Arrangement of nails in Maggie Schweizer's case.

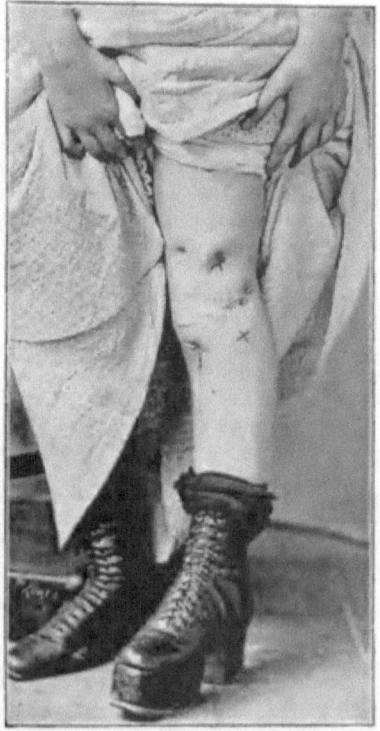

FIG. 80.—Final result in Maggie Schweizer's case. Cross-marks indicate places where nails were driven in. (Page 88.)

semilunar incision was obliterated, the capitellum, coronoid process, and temporal bone forming one solid mass. Immediately after its removal the teeth could be separated to the distance of an inch and a quarter. Primary union. Perfect restoration of function noted in January, 1887.

CASE II.—Maggie Schweizer, aged fifteen. Bony anchylosis of knee-joint at a right angle, in consequence of infantile acute osteomyelitis of tibia, with suppuration of knee-joint. *January 22, 1886.*—At the German Hospital, excision of the patella and of a wedge-shaped piece of bone, with preservation of the epiphyseal lines of femur and tibia. Transverse cutaneous incision, as for knee-joint exsection. Division of the bones by the saw, after peeling off of the periosteum. The sawed surfaces were brought together, and their fixation was secured by three steel nails, which were driven diagonally through the tibia and femur in the horizontal plane—that is, from the lateral aspect of the extremity. The locking of the femur and tibia was so firm that the limb could be raised and handled like a solid staff. The application of the dressings was thereby made a very easy procedure. Full plaster-of-Paris splint. No reaction and no fever were observed. *Feb. 23d.*—First change of dressings. The nails and two drainage-tubes inserted at the operation were removed. The bones were found firmly united. Over a small aseptic dressing a light silicate-of-soda splint was applied, and the patient was directed to walk on crutches. *March 15th.*—Discharged cured with light silicate splint. *May 10th.*—Presented herself to author, walking excellently with the aid of a raised sole. Shortening, two and a half inches.

### 3. Deformed Callus.

CASE I.—William Paradies, laborer, aged thirty-eight. Deformed callus of the lower end of the tibia following a supra-malleolar fracture of the leg. Radiating pain issuing from the site of the deformity, due to pressure on the integument, which was tightly stretched over the protruding edge of the upper fragment. *March 7, 1887.*—The deformed bone was exposed and chiseled away on a level with the surface of the distal fragment. Suture; no drainage. Primary union. *March 21st.* —Patient discharged cured from the German Hospital.

CASE II.—Ernst Langer, carpenter, aged forty-five. Deformed callus of fibula. *August 29, 1885.*—At the German Hospital, incision and exsection of the callus by chisel and mallet. Apposition and fixation of the fragments by a strong catgut bone-suture. Primary union. Discharged cured, September 26, 1885, with firm consolidation.

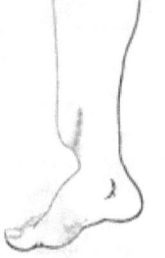

FIG. 81.—Deformed callus of lower end of tibia. (William Paradies.)

### 4. Club-Foot and Pes Valgus.

On account of its simplicity and the excellent results reported both from abroad and at home after its practice, Phelps's operation seems to deserve extended trial. It consists in the combination of tenotomy of the tendo Achillis with a free division of all the soft tissues situated on the mesial side of the planta pedis, the incision penetrating down to the bone and, if necessary, into joints. The idea of dividing all resisting tissues underlies the plan of procedure. The incision includes the tendons of the tibialis posticus, flexor digitorum communis longus, the belly of the flexor digitorum brevis, of the abductor hallucis, the plantar fascia, the long plantar ligament, the deltoid ligament, the nerves, and, if unavoidable, the vessels. The incision need not be a very long one. It commences just in front of the tip of the inner malleolus, and extends downward, according to the age of the patient, for about an inch.

All the parts named above can be easily reached from the wound with a tenotomy knife, unless they are in the direct line of section, when they are divided with the scalpel. Preservation of the integrity of the plantar artery is very desirable, on account of the avoidance of saturation of the

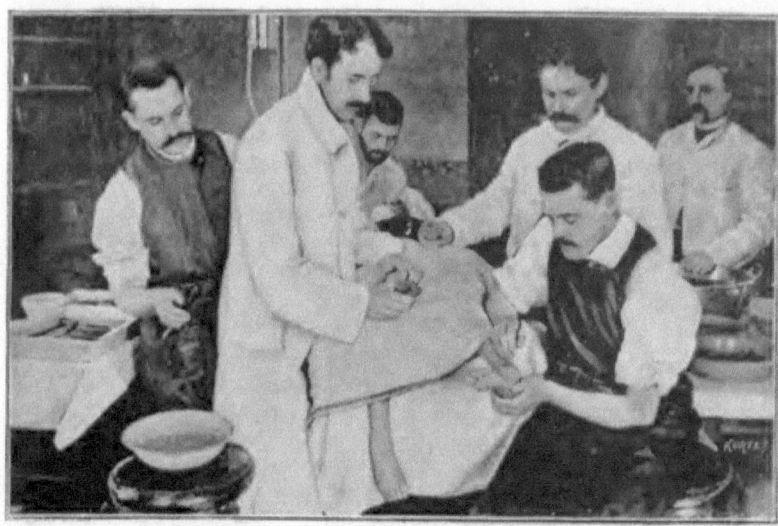

FIG. 82.—Group illustrating an operation about the foot or ankle.

dressings with blood. The operation being done with the aid of Esmarch's band, all the tissues can be readily identified as they are gradually exposed step by step. The internal plantar artery can thus be seen and doubly tied. The main trunk of the artery sweeps in a long curve outward to the external side of the sole, and is out of the line of section. Should it be divided accidentally, and the blood soil the dressings at once, it is proper to remove them, to re-apply Esmarch's band, to enlarge the incision, and to find and deligate the cut ends of the vessel. In extreme cases of adults, where the bones have acquired a definitely vicious shape, linear osteotomy of the neck of the astragalus must be added to the tenomyotomy performed in the planta.

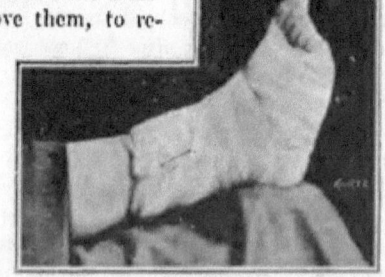

FIG. 83.—Dressing for wounds of ankle and foot.

The author was surprised to see the ease with which even great deformities could be corrected after the division of all tissues mentioned above. Of course, the wound is a wide gap, which is widened still more by the corrected position. Its healing is accomplished by the "organization of the moist blood-clot" (Schede's method). As soon as the wound has been well cleansed by irrigation, a piece of rubber tissue, previously kept immersed in a five-per-cent solution of carbolic acid for twenty-four hours, is placed over the gap. This is covered with a few strips of iodoform gauze and an ample dressing of sublimated gauze. While the foot is held in the correct position by an assistant, the surgeon applies over the aseptic dressing a plaster-of-Paris splint. While the plaster is setting, the foot is held with force in a somewhat overcorrected position, which will allow for the slight giving way of the aseptic dressing. Then Esmarch's band is removed, and the feet are held in the vertical posture for an hour or two after the operation. After disappearance of passive hyperæmia they are placed on a pillow in the horizontal posture.

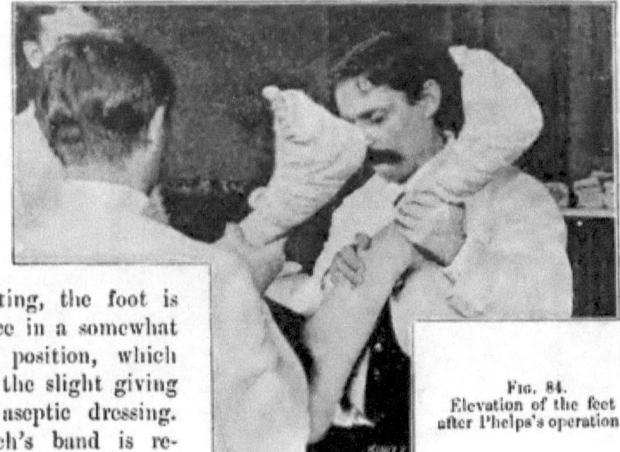

Fig. 84. Elevation of the feet after Phelps's operation.

In a fortnight or so the plaster of Paris is cut away, and a silicate splint is applied. As soon as it is dry the patient is allowed to walk with the aid of crutches. In about four weeks after the operation the silicate shoe is split on top, and the dressings are removed. In many cases the wound will be found cicatrized over by this time. Should this not be the case, however, the aseptic dressing and silicate shoe must be reapplied. When the wound is perfectly healed, the silicate splint can be replaced by a well-fitting laced shoe.

NOTE.—The silicate shoe must not include more than about one third of the leg, in order not to prevent treatment of its debilitated muscles by massage and electricity.

The fear that the severed tissues will not grow together properly is unfounded. Schede had the opportunity of ascertaining by autopsy the exact re-establishment of the physiological relations of the cut tissues. The best proof of the fact is, however, the restoration of the function of the cut parts.

The results exhibited by Phelps at a meeting of the New York State Medical Society at Albany were so excellent as to deserve the utmost attention. They surpassed everything the author has seen accomplished for the cure of this deformity.

CASE.—Harry Epstein, school-boy, aged twelve, suffering from chronic interstitial nephritis as a consequence of scarlatina. General condition poor, on account of lack of exercise, due to disability from club-feet. The patient was walking on the outer edge of the plantæ. The urine contained granular and hyaline casts, and twenty per cent of albumen. *March 14, 1887.*—At Mount Sinai Hospital, double Phelps's operation was done under chloroform, which was borne excellently, the operation lasting forty-five minutes. No fever, no reaction followed. *March 28th.*—The plaster shell was cut away, and the patient commenced to hobble about in the ward on crutches. *April 10th.*—The old water-glass splints were removed, and were replaced by a new set, which were worn until June. After this the patient was fitted with a pair of lacing shoes.

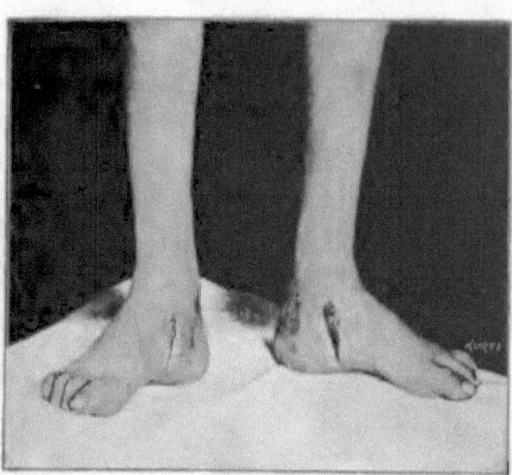

FIG. 85.—Appearance of wounds four weeks after Phelps's operation. Harry Epstein's case.

CASE II.—Aaron Meyer, oysterman, aged twenty-nine, far gone and very painful pes valgus of both feet. *Oct. 12, 1885.*—At Mount Sinai Hospital, exsection of a bony wedge by chisel and mallet from the internal aspect of the head of the astragalus, the scaphoid, and calcaneum of the right foot. Area of the base of the wedge about one square inch. The remnants of the neck of the astragalus and calcaneum were divided entirely by the osteotome, and the foot was broken into shape by manual force and put up in an aseptic dressing and plaster-of-Paris splint. *Nov. 1st.*—Dressings removed, wound presenting a strip of shallow granulations. *Dec. 1st.*—Discharged cured. *Feb. 1st.*—Foulis's operation on the left foot, which showed a lesser degree of deformity than the right foot before operation. The talo-navicular joint was incised, and its entire cartilaginous covering was removed by scraping with a scoop. *Feb. 21st.*—First change of dressings; primary union. *Feb. 27th.*—Patient discharged cured. In March, 1887, patient presented himself for examination. Firm anchylosis of the talo-navicular joints of both sides, and very good function had been secured, the patient attending to his accustomed business.

## VI. PLASTIC OPERATIONS.

Aseptics have greatly improved the results of plastic operations, and especially erysipelas has been almost entirely banished from facial wounds made for plastic purposes. In performing any operation about the face it is necessary for the surgeon to protect himself and the patient from two sources of infection. One is the oral and nasal secretions, the other the patient's head, notably his hair. The latter should always be enveloped in a cap extemporized from a good-sized towel or compress wrung out of corrosive-sublimate lotion. The accompanying illustrations show the manner of folding the towel about the head. It should be firmly fastened by a narrow roller-bandage encircling the forehead and occiput. Whenever vomiting occurs, a careful cleansing of the soiled skin and a change of towels are indicated.

Where there is no great tension to be overcome, fine catgut (No. 0) makes excellent suturing material for facial wounds after plastic operations.

Where the tension is great (which, however, should be reduced to a minimum by the proper shaping of flaps and free dissection), silver wire, or silkworm-gut well soaked in carbolic lotion, will be well employed for retentive purposes. Sutures of coaptation are best made with fine catgut.

Hare-lip pins were never used by the author, as they are unnecessary, and offer no advantages over the suturing material more generally employed by surgeons.

Where the wounded surfaces can be completely closed by suture, no dressings whatever are needed. A thick layer of iodoform dusted over the line of union will soon unite with the oozings into a paste, which on becoming dry will form an excellent and unirritating protection to the wounds and suture-points. Daubs of collodion, or the application, after hare-lip operations, of strips of adhesive plaster to the face, are especially unpleasant and irritating to infants. They create uneasiness, and excite the little patients into crying fits, and the distortion of the face resulting from frequent crying is certainly not conducive to the uninterrupted rest and union of the wounds.

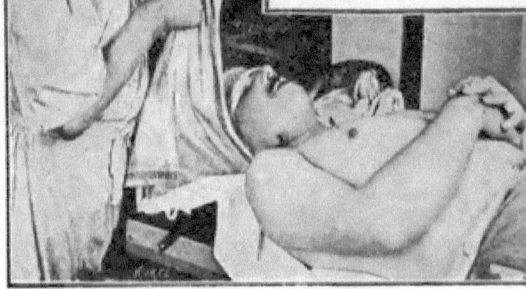

FIG. 86.—Applying aseptic cap. First step.

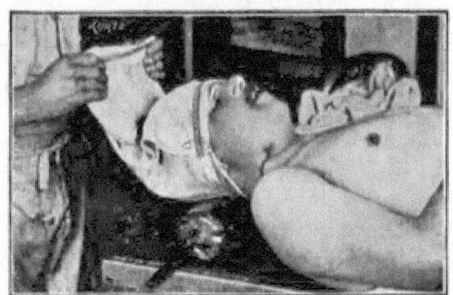

FIG. 87.—Applying aseptic cap. Second step.

SPECIAL APPLICATION OF THE ASEPTIC METHOD. 93

Retentive sutures should never be removed too soon—that is, before the seventh day. The smaller catgut sutures will be absorbed by that time.

Fig. 88.—Aseptic cap *in situ*. Cancer of lip.

Where an uncovered defect is unavoidably left behind, on account of lack of integument or some other reason, Schede's procedure is the best means of preventing suppuration. A strip of rubber tissue is laid over the defect, and is suitably inclosed in an aseptic dressing. The blood-clot, which will form under the rubber tissue, will, if it be well protected from desiccation and decomposition, rapidly become organized.

*In plastic operations performed about the soft and hard palate* the condition of the teeth should be well attended to previous to the undertaking. Decaying teeth should be removed, and an unwholesome state of the gums and mucous membrane should be corrected by the diligent use of the tooth-brush and a 1:1,000 solution of permanganate of potash as a mouth-wash.

**Urethroplasty** will fail almost invariably if ammoniacal urine is permitted to pass over the line of union. Acid urine is not deleterious to the wounds. Where chemical examination has established the presence of ammoniacal decomposition of the urine, frequent washings of the bladder and the urethra with weak solutions of permanganate of potash (1 : 4,000 or 5,000) and the internal administration of boracic acid will suitably prepare those organs for the operation. To prevent the soiling of the wound by ammoniacal urine, a soft Nélaton catheter should be passed into the bladder and fixed by

Fig. 89.—Dressing for excision of the upper jaw.

a proper bandage to prevent its escape. Daily antiseptic irrigation of the bladder should be continued all the time while permanent catheterism is used. As soon as the wound is firmly united, catheterism may be stopped.

*Perineal plastic operations* on the female require a previous thorough disinfection of the vulva and vagina by mercurial irrigation, which should be kept up during the entire time of the operation. Here, too, dressings are annoying and unnecessary. Catheterism, temporary confinement of the bowels, and frequent irrigation, with subsequent dusting with iodoform powder, will afford all the security needed against infection.

Aside from the care for the production and maintenance of the aseptic condition during and after the operation, another important requirement must be fulfilled. *This is a thorough and complete apposition of the entirety of the wounded surfaces by several tiers of catgut sutures, and a correct union of the mucous membranes of the vagina, and of the rectum if necessary.* A slovenly manner of suturing will lead to the formation of hollow spaces, which will become filled by blood-clot; and, if the sutures of the mucous membranes be also inexact, contact of the vaginal or rectal discharges with the unprotected clot will lead to its inevitable putrescence, and to partial or general suppuration. *An exact, deep and superficial suture is the best protection of perineal operative wounds against infection.*

NOTE.—The stitches holding the mucous membrane together should never pass through the epithelium. They should be entered and brought out just below the epithelial lining. This will prevent inversion of the edges, and the stitch-holes will be also protected from infection by the ridge of protruding mucous membrane.

On account of the great vascularity of the face, facial wounds will often heal without suppuration, even if very indifferent asepticism was observed.

*Not so in other parts of the body, notably about the extremities*, where suppuration is much more easily produced, and is generally followed by sloughing of the flaps. *Strict asepticism, avoidance of tension by sutures and of pressure by dressings, are imperative conditions of success in plastic operations done on the extremities.*

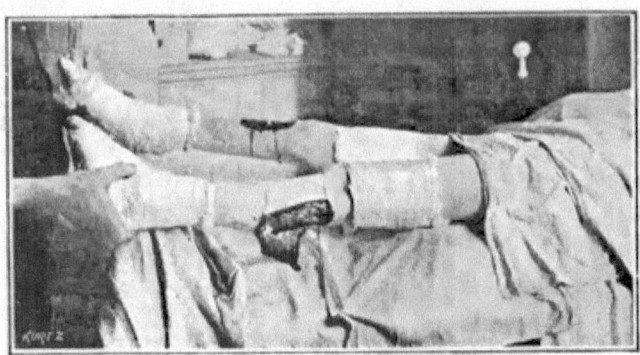

FIG. 90.—Maas's operation. Primary plaster-of-Paris dressings. On the right leg, the defect to be covered; on the left leg, flap detached from calf.

CASE I.—Abraham Strecker, aged seven. Circular, extensive skin defect of the right leg, due to old compound fracture: extensive ulceration of frontal part of the cicatrix; œdema of the foot, caused by contraction of the circular cicatrix. *Dec.* 7,

## SPECIAL APPLICATION OF THE ASEPTIC METHOD.

*1885.*—At Mount Sinai Hospital, plastic repair of the frontal part of the defect by Maas's procedure. Each thigh and foot was first incased in a plaster-of-Paris splint, then the cicatrix was disinfected with an eight-per-cent solution of chloride of zinc and pared off

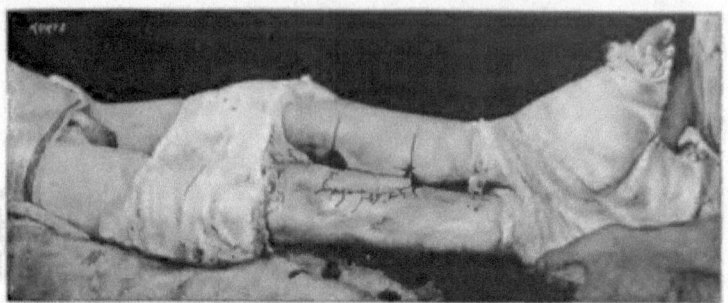

FIG. 91.—Maas's operation. Secondary plaster-of-Paris dressings fixing relative position of extremities. Flap attached to its new habitat.

with the scalpel. After this a properly shaped, generous skin-flap was raised from the posterior aspect of the left leg. Now the extremities were superimposed in such a manner as to bring the flap over the vivified surface of the right leg, wherewith it was brought in contact on its raw surface. A secondary plaster-of-Paris dressing applied over the primary plaster splints secured the limbs and the flap in their new relative position. The exposed raw surface of the pedicle of the flap was wrapped in an envelope of rubber tissue to prevent its desiccation; the flap was lightly attached to its new habitat by a few catgut sutures. The edges of the flap were dusted with iodoform, and the defect of the calf was inclosed in an aseptic dressing. With the exception of a small portion of the end of the flap which necrosed, primary union throughout was achieved. *Dec. 21st.*—The pedicle of the flap was cut, and the limbs were released from their confinement. Rapid cicatrization of the remnant of the original and of the defect of the calf followed, and, January 30, 1886, the boy was discharged cured. The œdema of the foot had disappeared.

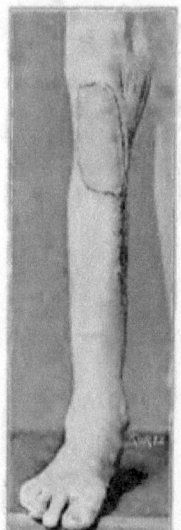

FIG. 92.—Maas's operation, final result. Cicatrix is marked with ink.

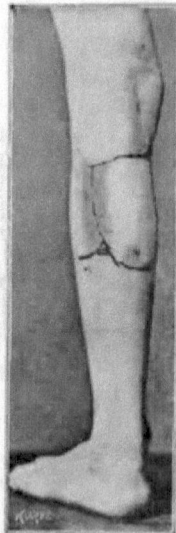

FIG. 93.—View of cicatrix of the place whence the skin-flap was taken.

CASE II.—Adolph Carstens, schoolboy, aged eleven. *Feb. 17, 1887.*—At the German Hospital, Maas's operation for a large skin defect of the anterior aspect of the tibia, due to severe traumatism. The case was managed exactly like the foregoing one, with this additional circumstance, however, that it became necessary to pare off an area of the anterior aspect of the tibia by chiseling, corresponding to the size of the flap, in order to remove the condensed cicatricial tissue underlying the extensive elevated ulcer. Thus, a well-vascularized base was secured for the skin-flap. *March 3d.*—The pedicle was divided, and, April 10th, the patient was discharged cured.

## VII. ASEPTICS OF THE ORAL CAVITY.

Long after the principles of the aseptic treatment of external wounds had become recognized, the proper management of the wounds of the normal openings of the respiratory, digestory, and uro-genital tracts was still a mooted question. It was a comparatively easy thing to produce in these regions an aseptic condition for the time of the operation. But how to protect the wounds from the inevitable soiling by the continuous discharges pertaining to these several apertures, was first shown by Billroth, who successfully employed iodoform as an effective preventive of putrefaction in the oral cavity.

If a fresh wound of the oral cavity is rubbed off with iodoform powder and packed with gauze saturated with iodoform, this dressing will become matted together with the tissues of the raw surface, and will form an effective protection against infection by septic influences. The secretions will innocuously pass over the surface of the gauze, and the penetration of active germs to the wound will be prevented by the air-tight and closely adherent packing.

The course of oral wounds treated in this manner differs widely from that observed under other forms of treatment. Diphtheritic and phlegmonous processes, formerly so common in wounds freely communicating with the mouth, have become things of great rarity. The terrible odor which could not be kept down by however frequent irrigations with any kind of deodorizing lotion until the necrosed layer of tissues was cast off, is now generally absent. By the time that the packing of iodoformed gauze becomes loose, healthy and vigorous granulations will have sprung up, and the wound will progress toward uninterrupted healing without pain and without fever.

As long as the packing is firmly adherent, it should not be disturbed. Its forcible extraction would certainly cause a good deal of pain, and would be followed by hæmorrhage and inflammation. The superficial layers of iodoformed gauze, becoming soiled by secretions or food, can be daily renewed.

Another important point to be observed in operations about the oral cavity is the control of hæmorrhage. The abundant blood-supply of this region is apt to be the source of copious hæmorrhage, dangerous in itself, but especially perilous on account of the possibility of the entrance of blood into the air-passages.

This accident may, on the one hand, cause instant death from suffocation; on the other, it may produce catarrhal or septic pneumonia by decomposition within the bronchi.

Hæmorrhage from oral wounds can be controlled in two ways. They may be employed separately or combined.

The first one is by preliminary ligature of one or both lingual arteries; the second, by the exclusive use of the actual cautery and galvano-caustic wire loop.

Where the operation must needs extend to the floor of the mouth, deligation of the lingual arteries will be insufficient, and the use of the actual cautery point or loop often impracticable. In such a case, *preliminary tracheotomy* and the employment of a *tampon cannula* will be the only safe means of preventing the entrance of blood into the bronchi.

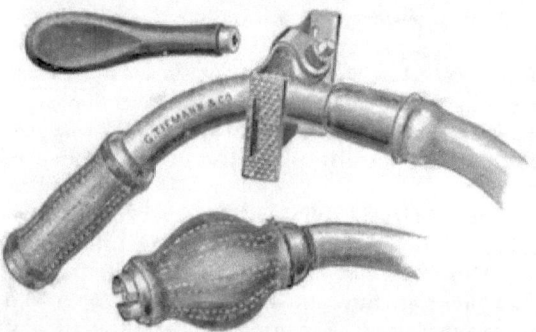

Fig. 94.—The author's tracheal tampon cannula.

Although Whitehead's speculum is an excellent instrument to render the oral cavity accessible, yet it will be unsatisfactory in operations to be done on the floor of the mouth. Here section or even partial excision of the lower jaw may be unavoidably necessary to afford ample space for complete excision of a malignant tumor, and to make accurate hæmostasis practicable.

Where most or all attachments of the tongue to the inferior maxilla must be severed, a strong loop of silk should be drawn through the stump of the tongue near the epiglottis, to be brought out by the mouth and attached by a strip of adhesive plaster to the cheek. This precaution will enable the nurse or attendant to instantly clear the epiglottis should the stump of the tongue ever slip back upon and occlude the entrance to the larynx.

In the more extensive cases of oral surgery, especially after removal of the tongue, nutrition will have to be carried on for some time by the stomach-tube, which can be left in for several days, or can be daily introduced by the mouth or nostril.

Early operations for cancer of the tongue will give better results in every way than late ones. But even of the latter it can be said that, as a rule, the patient's life will be prolonged by them, and will be made more tolerable.

Every oral operation should be preceded by a careful preparation of the mouth by extraction of carious teeth and frequent washings with a germicide lotion, preferably a 1 : 1,000 solution of permanganate of potash. Present stomatitis should be first got rid of by all means.

CASE I.—Mr. David S., wholesale butcher, aged fifty-four. Strong smoker. On the inner aspect of the right cheek, opposite a carious and sharp-edged molar, where an opaline mucous patch had existed for some time, an elevated ulcer of the size of a silver dollar had established itself, and was steadily extending. The submaxillary lymphatic glands were intumescent. *April 30, 1884.*—Extirpation of the growth from a transverse incision extending backward from the angle of the mouth. The outer skin was saved and brought together by a line of stitches. The intumescent submaxillary glands were also removed. Uninterrupted recovery followed, but a small fistula remained behind, corresponding to the middle of the incision ot the cheek, which, however, closed after a few applications of the thermo-cautery. The contraction of the cheek was successfully overcome by the insertion and wearing of wooden wedges, which were abandoned in the fall of 1884. During the summer a relapse of cancer had developed in the deep-seated submaxillary glands of the right side and in the submental gland. *September 25, 1884.*—The glandular swellings were extirpated from both mentioned regions. The complete removal of the submaxillary glands necessitated excision of two inches of the deep jugular vein. The wound healed by the first intention; the patient took his first walk twelve days after the operation. He remained free from the disease until September, 1885, when a rather rapid swelling of the submaxillary glands of the *left* side was observed. Apparently the infection had extended to the opposite side of the neck by way of the diseased submental gland. The original site of the epithelioma in the cheek remained intact by relapse. *October 22, 1885.*—An attempt was made to remove the glandular swelling of the left side of the cheek, but it had to be abandoned on account of the wide extension and infiltrating character of the new growth. *January 31, 1886.*—Patient died of extension of the disease to the cerebrum.

Had the first operation been undertaken at an earlier date, the respite secured to the patient would have been much longer.

Case II.—Katie Johs, aged thirteen. Mucous cyst of the left under side of the tongue, deeply imbedded in the lingual tissues, and extending back to the hyoid bone. *March 24, 1883.*—Deligation of the left lingual artery from an external incision above the hyoid bone. Whitehead's speculum being inserted, the tongue was transfixed and secured by a strong fillet of silk. By this it was withdrawn, and the cyst was easily extirpated from its bed by means of scissors and forceps. Care was taken not to grasp the cyst with the mouse-tooth forceps, which served only to hold aside the muscular tissue of the tongue. Minimal hæmorrhage was observed. The wound was stitched with fine silk throughout its entire length, a few threads of catgut being inserted into its upper corner for drainage. Both wounds healed by primary union, and, April 7th, the patient was discharged cured from the German Hospital.

Case III.—Adolph Böttger, cooper, aged forty-two, a strenuous smoker and hard drinker, had contracted an epithelioma of the right anterior margin of the tongue, extending well forward to the gums of the canine tooth, and involving the intervening part of the floor of the mouth. No intumescence of the lymphatic glands could be made out. *August 28, 1883.*—At the German Hospital the right lingual artery was deligated, and the right half of the tongue was excised by the aid of forceps and scissors. A morphine injection had been administered before the operation, and anæsthesia by chloroform was not carried to insensibility. Hæmorrhage was very moderate. In excising the floor of the mouth the bleeding was somewhat profuse, and a large number of spurting vessels had to be tied. The resulting wound was packed with iodoformized gauze. No fever or inflammation followed, and the power of deglutition was re-established on the third day. The patient left the bed on September 9th, and October 9th was discharged cured. In February, 1884, the disease returned on the inner aspect of the gums. *March 10th.*—Three inches of the alveolar process of the horizontal part of the lower maxilla were excised, together with the entire cicatrix. Cure was delayed by necrosis of the remaining portion of the body of the jaw. *April 30th.*—The sequestrum was extracted. *May 20, 1884.*—Patient was discharged cured.

## SPECIAL APPLICATION OF THE ASEPTIC METHOD.

*May 17, 1886.*—The patient returned with a far-gone relapse, starting from the left submaxillary stump. *May 19th.*—Exsection was performed. Violent delirium tremens set in immediately after the operation, followed by death in collapse.

CASE IV.—Fritz Osterwald, shoemaker, aged sixty-three; strong **smoker; cancer** of the right margin of the tongue well back near the anterior pillar of the fauces, with considerable involvement of the floor of the mouth. *February 2, 1886.*—Deligation of the left lingual artery, followed by excision of the corresponding half of the tongue and floor of the mouth in morphine-chloroform anæsthesia **at the German** Hospital. Access was gained to the oral cavity by a semicircular **incision following the under** side of the lower jaw, from which the attachments of the muscles were raised together with the periosteum. The mucous membrane was cut through, whereupon the tongue and floor of the mouth could be drawn out from under the maxilla and turned out upon the front of the neck. Hæmorrhage was rather free in spite of the preliminary ligature of the lingual artery; and, though the patient was not fully anæsthetized, alarming asphyxia suddenly took place, apparently due to the occlusion of the glottis **by a** blood-clot. Efforts to dislodge **this were** unsuccessful, therefore hasty tracheotomy had to be performed, resulting in re-establishment of respiration. After this the excision was completed without further mishap. More than half of the tongue was removed up to the epiglottis, together with the left side of the floor of the mouth and the anterior faucial pillar. The wound was packed with iodoformized gauze. Nutrition was carried on by stomach-tube. No fever followed, but, February 15th, symptoms of iodoform mania necessitated the removal of the original packing, which was replaced **by corrosive-sublimate** gauze. *Feb. 18th.*—The restless patient was taken to his home, **whence he was transferred to** Bellevue **Hospital,** where he **died** a maniac on February 28th.

The foregoing case illustrates **the** dangers **from the entrance of blood** into the larynx, and the greatest drawback of iodoform **when used on elderly** individuals—namely, its tendency to produce acute **mania.** From this instance the author learned the lesson of never risking a rather bloody operation in the oral cavity without preliminary tracheotomy and **the use of a** tampon cannula. The anxious moments spent in opening the **suffocating** patient's trachea will never be forgotten.

CASE V.—Victor Jeggi, silk-weaver, aged fifty-three, **a very** moderate smoker, admitted August 20, 1885, **to the German** Hospital with lingual cancer, involving nearly one half and principally **the right side of** the tongue. No glandular swelling. *Aug. 22, 1885.*—Both lingual arteries **were** deligated, and two thirds of the entire length and width of the organ were **excised with very** little hæmorrhage in mixed (morphine-chloroform) anæsthesia. The wound **was packed** with iodoformed gauze. Deglutition returned on August 28th. The **wound healed very rapidly, so that,** September 5th, patient could **be discharged** nearly **cured.** He presented **himself,** February 21, **1886,** with a relapse in **the** floor of the mouth, but delayed operation until March 30th, **when the** disease had assumed formidable proportions. Preliminary tracheotomy being done, **the** author's tampon canula was inserted. **The** middle portion of the lower jaw was **excised,** and the remnant of the tongue was **removed** together **with** the entire floor of the mouth by means of the thermo-caustic knife. **The stumps of** the severed arteries did not retract (atheromatosis), and **were** successively **tied. The** wound was packed with iodoformized gauze, and nutrition was carried on by the stomach-tube. *April 2d.*—The patient vomited, and undoubtedly some of the ejecta **found** their way into the bronchi. *April 3d.*—Catarrhal **pneumonia** set in with a chill and a temperature of 104° Fahr. *April 6th.*—The critical condition changed for the better, and by April 15th the patient left the bed. To avoid vomiting produced by the frequent introduction of the stomach-tube, this was carried in through the nostril and left *in situ* with evident comfort to the patient. The wound contracted rapidly, but in the middle of May relapse appeared in the pharynx, which ended the patient's existence in June, 1886.

The presence of the tampon cannula in the trachea, effectually shutting off the possibility of the entrance of blood into the air-passages, made this otherwise very bloody and formidable operation comparatively easy and safe.

CASE VI.—Mr. Joseph T., wholesale liquor-dealer, aged sixty, a smoker, had been suffering for twelve years from opaline patches of the tongue, two of which, situated on the left side of the organ, developed, toward the end of 1886, into epitheliomata. The otherwise well-nourished patient suffered also from chronic interstitial nephritis, as evidenced by the presence of albumen and hyaline and fine granular casts in the urine. *Feb. 10, 1887.*—The left lingual artery was deligated under chloroform anæsthesia. The tongue was secured by a strong fillet of silk, and was withdrawn from the mouth. A straight Peaslee's needle was then carried into the bottom of the deligation wound, and was thrust through the middle of the base of the tongue just in front of the epiglottis into the oral cavity. One end of a platinum wire was passed through the eye of the needle, withdrawn through the wound and disengaged. The same needle was reintroduced by the wound into the oral cavity, emerging this time just alongside of the left anterior pillar of the fauces. The other end of the wire was brought out by the needle through the external wound. Thus, one half of the base of the tongue was included in a loop, and, the wire being connected with a galvanic battery, was singed through without loss of blood. After this the tongue was divided longitudinally by the thermo-cautery in two unequal halves, and finally was severed from its connections with the floor of the mouth by the same instrument. A few spurting arteries had to be tied off during this last step of the operation, which was completed within the time of forty minutes. The hæmorrhage was really insignificant, to which circumstance is to be mainly attributed the rapid recovery of the patient. The oral wound was packed with iodoformized gauze, and the external incision was dressed in the normal manner. The temperature remained normal throughout, and feeding by tube was discontinued on the third day. The mouth was irrigated every hour with a 1 : 1,000 permanganate of potash solution, until February 18th, when the packing came away. The wound appeared clean, and rapid contraction was manifest. *Feb. 25th.*—The external wound was firmly healed. *March 8th.*—The oral wound was closed.

NOTE.—In preparing iodoformized gauze for use in wounds of the oral cavity of elderly subjects, care must be taken not to sprinkle too much of the chemical upon the gauze. The surplus of iodoform should be rinsed out of the meshes of the fabric, which should be tinged just a very faint yellow color.

## VIII. LARYNGEAL OPERATIONS.

1. **Tracheotomy.**—The belief that tracheotomy is an easy operation is by no means justified by the author's experience. Occasionally, on a slender neck, and when there is competent assistance to be had, it is a simple enough procedure. But in most cases, especially on children, it calls for the best qualities of an experienced and cool surgeon.

The necessity of tracheotomy having become manifest, three requirements are to be fulfilled. *First,* infection of the wound has to be avoided ;

*secondly*, unnecessary hæmorrhage has to be guarded against; and, *thirdly*, the trachea has to be properly incised, and the cannula properly introduced and secured.

The risks of the operation are not inconsiderable, hence intubation of the larynx, a much simpler, easier, and more physiological procedure, must be declared to be far preferable to tracheotomy where its application is proper, as in croupous laryngitis.

For the removal of foreign bodies and in cases of tumor of the larynx, tracheotomy will remain the proper measure.

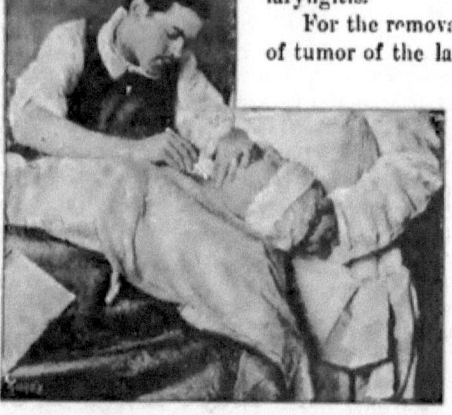

Fig. 95.—Arrangement of the patient for tracheotomy.

Avoidance of infection of the wound from within or without is an ever important matter in all laryngeal operations. But it is especially important, and also more difficult, in cases where the operation is done in the presence of an infectious process, as, for instance, diphtheritic croup, where the extension of the septic condition to the external wound signalizes a very grave complication of the otherwise precarious state of the patient.

The aseptic rules laid down in preceding parts of this work obtain to their full extent in laryngeal operations. Infection from within must be guarded against by careful cleansing of the external wound and rubbing iodoform powder into all its recesses before incising the trachea. As soon as the cannula is inserted, the external wound must be well mopped out with a sponge soaked in corrosive-sublimate lotion. Then it is dusted with iodoform, and lightly packed with iodoformized gauze. In all cases of croup the external wound should not be sutured, as sutures favor retention. A small slit compress of iodoformized gauze is slipped in under the flange of the cannula before its fastening by the two lateral pieces of tape. By slipping in over the gauze compress a slit piece of rubber tissue or oiled silk, the dressings and the patient's shirt will be protected from soiling by the sputa. A narrow roller bandage passed several times over and under the outer opening of the cannula will give additional security against accidents.

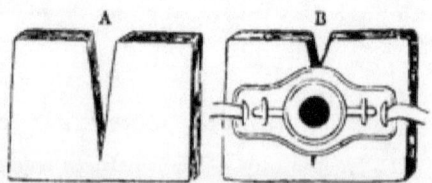

Fig. 96.—A, Slit compress. B, Same *in situ*.

NOTE.—Unruly children will sometimes attempt the forcible removal of the cannula. In 1880 the author performed tracheotomy on a boy twelve years old, who, on regaining consciousness, at once tore out the cannula from the wound, breaking its fastenings to the flange, which remained attached to his neck. The family attendant, an elderly gentleman, attempted the re-introduction of the instrument. Finally, during the violent struggles of the patient the cannula slipped into place, whereupon respiration, which had been labored before, suddenly ceased altogether. The author reached the bedside by this time, and at once removed the cannula from the asphyxiated child's neck, restoring respiration. It was found that the cannula had been introduced upward into the oral cavity, instead of downward into the trachea. Another tracheal tube was properly introduced, and peace was once more restored, but the boy died subsequently of septicæmia, due to the wide extent of the diphtheritic affection of the pharynx.

*Hæmorrhage*, always characteristic of an overhasty and bungling operation, can be guarded against by observing the rules laid down in the chapter on the technique of surgical dissection. Nothing will retard the performance of tracheotomy as effectively as the disregard for hæmorrhage. And every drop of blood spilt unnecessarily will proportionately diminish the chances of recovery, not to mention the danger of suffocation from the entrance of blood into the lungs.

NOTE.—The author once assisted a colleague who in his anxiety to open the trachea cut the isthmus of the thyroid gland. The formidable hæmorrhage following this step only increased the doctor's haste. He plunged the knife into the pool of blood and fortunately opened the trachea. The patient aspirated a large quantity of blood, and would have surely been suffocated but by the timely turning of his body face downward. The patient, a boy of seven years, recovered.

As soon as the skin, platysma, and superficial fascia have been amply divided, the two groups of longitudinal muscles situated in front of the larynx are exposed. Sharp retractors are inserted and the bleeding vessels are attended to. A faint white mark indicating the median line where the muscles meet, is incised, and the muscles are taken up and raised by the retractors as the wound deepens.

Thus far everything is easy. The most difficult part of the operation consists in the proper treatment of the isthmus of the thyroid gland.

The surgeon must decide whether to approach the trachea from above or below the isthmus, and this decision depends upon the length of the neck and the size of the isthmus. In long, slender necks, the trachea is easily exposed below the isthmus; in short, fat necks, with a massive isthmus, the upper operation is more appropriate.

*a*. SUPERIOR TRACHEOTOMY.—Having chosen the upper operation, the surgeon must find his way to the upper part of the trachea, situated just behind the isthmus, without injuring the thyroid capsule and its complicated plexus of large and turgid veins. To accomplish this, Bose's method affords an easy way.

The deep cervical fascia divides into two layers just above the superior margin of the thyroid gland, these two layers forming the main body of the thyroid capsule. The point of division corresponds exactly with the upper margin of the cricoid cartilage, which can be easily identified by touch. The nail of the left index-finger is placed against the margin of the cricoid, the pulp of the finger looking downward, whereby the thyroid gland is protected, and the fascia is opened by a short transverse incision directed against the upper edge of the cartilage. As soon as this is done, a blunt hook can be introduced through the transverse slit behind the thyroid gland, which then can be drawn down with some force, exposing the two or three upper rings of the trachea. The author never saw this method fail, and, in employing it, never was compelled to cut the cricoid cartilage for want of space to limit the incision to the trachea. (See Fig. 97.)

*b*. INFERIOR TRACHEOTOMY.—When the lower operation is decided on, the two layers of the deep cervical fascia are successively incised *between two forceps*, and thus the trachea will be readily exposed.

*Incision of the trachea* should be done by the scalpel used for the first part of the operation, and rather by cutting than by puncture, as the latter may injure the posterior wall of the cylinder. Before cutting it, the trachea should be allowed first to adjust itself in its normal position, so that the incision should be placed exactly in the median line.

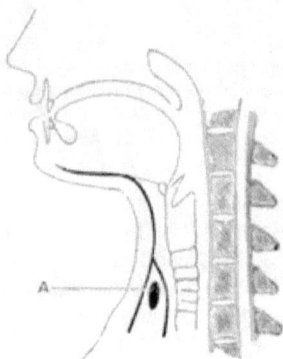

FIG. 97.—Diagram showing relations of deep cervical fascia. A, Thyroid body. Just above it, corresponding to cricoid cartilage, bifurcation of deep cervical fascia.

Grasping of the trachea while the incision is being made, but especially haste in opening the organ, may lead to very serious mistakes. It may happen that the trachea is not incised at all, or, what is still worse, the incision is placed laterally or even posteriorly on the tilted wind-pipe.

CASE I.—Mary R., aged five. *May 4, 1882.*—Tracheotomy performed by a colleague for laryngeal croup. The cannula could not be kept back in the wound, and the patient was found by the author suffocating, the instrument lying on the outside of the neck. Examination showed that the tracheal incision was placed to the left side and posteriorly, the trachea being twisted and bent while the cannula was *in situ*. An anterior tracheal incision was made, and in this the tube was retained without trouble. The child died of pneumonia.

CASE II.—Hermann Mollenhauer, aged two and a half. Croupous laryngitis. *March 27, 1881.*—With the assistance of the family attendant, Dr. Hase, superior tracheotomy, on account of imminent suffocation. The trachea was exposed without trouble, but in cutting it open too hastily it tilted around its axis, and the point of the knife shaved off a segment of the first tracheal ring. The tilting of the trachea was not noticed at first on account of the necessary haste; but, as soon as it was discovered, the trachea was properly incised, and the child ultimately recovered.

As soon as the proper number of rings are divided, the lips of the incision should be taken up by two small, sharp retractors. (See Fig. 18, page 40.) Hasty crowding in of the cannula is reprehensible, and may cause serious or fatal mischief by detaching and pushing membrane down into the deeper parts of the tracheal tube. Drawing asunder the tracheal wound will afford ample opportunity for free breathing, for ejection of blood and membrane or mucus, and will give the surgeon a welcome chance to inspect the trachea and to extract semi-detached membrane or a foreign body. It will also solve the question whether tracheotomy has accomplished its end or not by the relief from dyspnœa.

The *apnœa*, or seeming cessation of breathing, often observed immediately after the incision of the trachea, is apt to alarm beginners. It is due to the habituation of the patient to exist on a very small allowance of oxygen. The first deep and free breath taken through a newly-made tracheal incision gives the patient more oxygen than ten or fifteen labored inspirations could give before the operation.

As soon as the cannula and dressings are in place, the patient is brought to bed, and a sponge, hollowed out in cup shape by the curved scissors, is attached with a safety-pin or two to a suitable piece of bandage, is wrung out of hot carbolic lotion (one per cent), and is tied down loosely just over the orifice of the cannula. It should be cleansed at frequent intervals in the same lotion. Close attention to the cleanliness of the interior of the cannula is a constant duty devolving upon the nurse. It should be done by chicken or pigeon wing-feathers dipped in carbolic lotion. The little patients should be encouraged to drink as much as possible, preferably milk.

The first dressings can remain undisturbed for three days; on the fourth day they and the cannula are changed. The patient is laid out flat on a table as for tracheotomy, and everything possibly needed should be at hand and readily arranged in a pan. Two sharp retractors, thumb-forceps, scissors, a clean cannula, and a change of dressings will be needed. The bandages are cut, and they and the cannula are simultaneously removed with the outer compress of gauze. The deeper packing should remain unchanged till it becomes detached. The fresh cannula is slipped in at once, and usually without much difficulty if the procedure be not unduly delayed.

The packing of iodoformed gauze will become loose on about the fourth day, and should then be removed. If the wound is found clean and granulating, no repacking will be required.

As soon as the patient can breathe freely through the fenestra of the outer tube, the external opening of the cannula being occluded, the instrument should be removed, as it is apt to cause pressure-sores and troublesome granulations within the trachea.

The author's experience embraces forty-four tracheotomies performed for various reasons. Twenty-two were done for croupous laryngitis on children. Of these, five recovered; seventeen died. The superior operation was employed seventeen times; the inferior, five times.

One of the children died of suffocation caused by the ill-advised action of the father, who inflated the patient's bronchi through the cannula with a large quantity of burnt alum. The others died of extension of the process to the lungs, or of septicæmia.

Of the remaining twenty-two tracheotomies done on non-croupous cases, three concerned children, nineteen referred to adults.

The following table will elucidate the causes for which the operation was performed:

## SPECIAL APPLICATION OF THE ASEPTIC METHOD.

|  | Recovered. | Died. |
|---|---|---|
| Asphyxia from entrance of blood into trachea | 1 | 1 |
| "     "   malignant goitre |  | 2 |
| "     "   foreign body in right bronchus | 1 |  |
| "     "   arterial hæmorrhage into a cervical abscess | 1 |  |
| "     "   chloroform |  | 1 |
| Dyspnœa from cicatricial stenosis of bronchus |  | 1 |
| "     "     "     "     " larynx | 1 |  |
| "     "     "     "     " pharynx | 1 |  |
| "     "   lymphosarcomata of neck |  | 1 |
| "     "   sarcoma of tonsil and neck | 1 |  |
| "     "   laryngeal tumor | 3 | 2 |
| "     "   foreign body in trachea |  | 1 |
| "     "     "     "     " larynx | 2 |  |
| Preliminary tracheotomy | 2 |  |
| Total | 13 | 9 |

Of the two cases operated on for the entrance of blood into the larynx, one recovered (see Case IV on page 99); the other, where hæmorrhage came from a suicidal gunshot wound of the base of the skull, died of the cerebral injury.

In two cases the operation was done for threatening asphyxia by growing malignant goitre. Both died: one from collapse; the other from coma, produced by acute alcoholism or traumatic delirium (see Cases I and II on page 113).

In one case asphyxia, caused by hæmorrhage into a cervical abscess, necessitated the operation. The patient recovered (see Case III on page 231).

In two cases tracheotomy was done without success for deep-seated stenosis of the air-ducts.

One concerned a man of forty, in whose left bronchus post-mortem examination revealed a syphilitic cicatricial stenosis. The other bronchus was found compressed by acute swelling of a bronchial lymphatic gland.

The other case was that of Fred. Peckary, aged one, who exhibited symptoms of a growing tracheal stenosis, *principally obstructing expiration*. The case came, March 6, 1886, under the author's care by the kindness of Dr. Boldt. Tracheotomy was done at the German Hospital without relief. The child died of pneumonia March 10th. On autopsy a brass trousers-button was found imbedded in old cicatricial tissue between trachea and œsophagus, midway between the cricoid cartilage and the bifurcation. An open communication existed between the two tubes. The button was held in place by a rim of cicatricial tissue in the œsophagus, and projected downward with its free lower margin like a valve into the lumen of the trachea. Thus inspiration found no impediment, but on expiration the valve was raised, and expiration-stenosis was the result.

In one case syphilitic stricture of the fauces indicated the operation. Patient survived.

In five cases the trachea was opened on account of the presence of laryngeal tumors. Three survived, and two died of septic pneumonia, due to aspiration of the intensely fetid secretion of the ulcerated tumors.

Preliminary tracheotomy was done twice successfully before extirpation of cancerous tumors.

In one case the trachea was opened on account of acute asphyxia occurring during chloroform anæsthesia.

CASE.—Undersized boy, aged nineteen. *November 12, 1889.*—At Mount Sinai Hospital removal of an enormous congenital teratoma of the occipital region under chloroform. The growth had become sarcomatous, and extensive involvement of the cervical glands of both sides was present. The patient had to be placed in the prone position, and this and his generally weak state, together with the encroachment on the trachea by the glandular swellings, produced asphyxia toward the end of the operation. As artificial respiration did not seem to produce any effect, tracheotomy was performed at once, and respiration was restored. While the pedicle of the tumor was being detached, it was noted that respiration had again ceased. The cannula was found outside of the tracheal wound, from which it was allowed to slip by the assistant intrusted with the narcosis. It is fair to state that death was very likely due to exhaustion or collapse induced by the shock of the formidable operation upon the much emaciated patient. He was a lad of nineteen, but looked like a very sickly child of ten.

In one case increasing stenosis, caused by the presence of a disproportionately small tumor, indicated the operation.

CASE.—Julius Meyer, peddler, aged thirty-nine. Previous history pointed at the lodgment of a foreign body in the œsophagus with dysphagia, which spontaneously disappeared. Gradually, however, increasing dyspnœa supervened. The laryngoscope demonstrated the presence of a small irregular tumor in the larynx, the size of which did not seem to explain the intense dyspnœa. Tracheotomy was done December 18, 1886, at Mount Sinai Hospital. On incising the trachea above the thyroid body, a granuloma occupying the posterior and lateral aspect of the larynx just below the vocal chords was exposed. Surrounded by this mass was found the point of a *wooden skewer*, one inch in length, its ends being imbedded in the mucous membrane. The cricoid cartilage was divided, the body was extracted, and the granuloma was excised. *Dec. 27th.*—Tracheal tube was removal. (For continuation, see Case III on page 108).

The following histories of the removal of foreign bodies from the air-passages conclude the series of the author's non-croupous cases of tracheotomy :

CASE I.—Clara V., aged five and a half. *May 22, 1887.*—A foreign body entered the larynx of the patient, causing intense fits of coughing and transient attacks of choking. A number of unsuccessful attempts at endolaryngeal removal of the body were made the same day. Finally, the body became lodged in the right bronchus, where its presence was made out by the sibilant noise heard near the bifurcation and the absence of normal respiration sounds over the entire right lung. A short, hacking cough, moderate dyspnœa, and noisy respiration served as constant reminders of the impending danger. *June 14th.*—During a coughing spell, suddenly an alarming asphyctic attack set in, followed by dysphagia, aphony, hoarse, croupy cough, and distressing dyspnœa. Marked laryngeal stridor and diminished respiration sounds over both lungs pointed to the lodgment of the foreign body in the glottis. Inferior tracheotomy being performed, the dyspnœa at once disappeared. The foreign body, a headless and armless miniature doll of porcelain, five eighths of an inch long and three eighths of an inch wide, was found firmly wedged in the glottis, whence it was extracted through the wound without difficulty. The wound was treated openly, and the child recovered. (See Fig. 98.)

CASE II.—Josephine O., aged seven. *July 7, 1889.*—A large white bean was aspirated and lodged in the left bronchus, where its presence was diagnosticated by the entire absence of respiration sounds over the left lung. July 10th fever set in. Author saw the patient July 12th at Clayton, N. Y., with well-developed pneumonia of the left lung. Removal to New York was advised, and was fortunately accomplished without the necessity of operating in the railroad car. On July 14th, without previous warning, the bean was dislodged by an access of cough, and the child was immediately asphyxiated. Dr. E. D. Walker, being in constant charge of the patient, at once incised the trachea, and thus the immediate danger was averted. In the mean time, the size of the bean being steadily increased by the accumulation of blood-clot around it, asphyxia became very profound, respirations had ceased, radial pulse could hardly be felt, and death was imminent. The author arriving, the child was put on the table, the wound was enlarged, and it was found that the bean was resting on the spur of the bifurcation. A soft rubber drainage-tube, a little smaller in diameter than the trachea, was passed down to the foreign body, was firmly pressed down upon it, and aspiration by mouth being practiced, was sucked up to the external wound. There it became detached and fell back into the windpipe, to be brought in sight again by the next movement at expiration, when it was luckily seized with a forceps and brought out whole. The deep collapse was overcome by artificial respiration and general stimulation, and the child recovered in spite of the pneumonia.

FIG. 98.—Miniature doll, removed from larynx by tracheotomy. Exact size. (Clara V.)

2. Laryngofissure.—Fission of the larynx for the removal of tumors or a foreign body was performed five times by the author. In one case of recurrent diffuse papilloma a very good final result was secured. In another one, done for epithelioma, speedy relapse followed. In the third case the presence of a foreign body and inflammatory granuloma required the step. The body and new-growth were removed, but the perichondritic inflammation maintained for a very long time such an intense swelling of the laryngeal mucous membrane that the tracheal cannula had to be worn until June, 1887.

CASE I.—Mrs. C. Lehmann, twenty-four, epithelioma of both vocal cords. *April 11, 1884.*—At the German Hospital, laryngofissure and extirpation of both vocal cords and the adjacent mucous membrane were done. *April 15th.*—Cannula removed. *April 30th.*—Wound healed. Relapse manifesting itself soon afterward, excision of the larynx was done in the summer of the same year by Dr. F. Lange, who took charge of the service at the General Hospital after the expiration of the author's term.

CASE II.—David Popplewell, machinist, aged forty-two; recurrent papilloma of the larynx, that had been treated endolaryngeally by Dr. Gleitsmann, who kindly directed the patient to the author. *July 9, 1885.*—Laryngofission at the German Hospital. Removal of the posterior half of right vocal cord; excision of several disseminated papillomata and searing of their base by the thermo-cautery. *August 5th.*—External wound healed; voice much improved.

CASE III.—Julius Meyer, peddler, aged thirty-nine. Recurrent stenosis after tracheotomy (see case on page 106) done, December 18, 1886, for the removal of a foreign body and granuloma from the larynx. *January 27, 1887.*—Laryngofissure. Moderate return of the new-growth about the defect of the mucous membrane in which the end of the wooden splinter had been found imbedded. The probe was introduced into this aperture, and penetrated downward and backward to a distance of three fourths of an inch, thin pus exuding from the sinus. Intense swelling and hyperæmia of the entire mucous membrane and submucous tissue were noted. Perichondritis was diagnosticated, and a tracheal tube was left inserted in the wound. The patient readily recovered from the operation, but subsequently could not get along without a cannula till June, 1887.

CASE IV.—P. Lewin. Laryngofissure and extirpation of tubercular tumor of larynx. Subsequent laryngoplasty. See "New York Medical Record," April 6, 1889.

CASE V.—Fanny Kupfer, aged sixty. Chronic suppurative perichondritis with partial necrosis of thyroid cartilage. Laryngofissure October 25, 1889, at Mount Sinai Hospital. Laryngeal fistula closed by laryngoplasty February 14, 1890. Discharged cured, March, 1890.

3. **Extirpation of the Larynx.**—There is no doubt in the author's mind that partial or total extirpation of the larynx for malignant new-growths, *if done early*, is the correct treatment, and will be successful in direct proportion to the readiness and thoroughness with which it is done. This view is in full accord with the accepted principles of the treatment of malignant neoplasms of all other regions of the body. The large rate of mortality recorded so far after extirpation of this organ is due in a great measure to the fact that the step was resorted to mostly in hopeless and desperate cases, in which endolaryngeal therapy had utterly failed to give relief. How the precious opportunity is lost of rendering substantial aid, or even securing durable relief, was illustrated by a famous case which not long ago engaged the interest of the whole civilized world. Generally the course of events is as follows: A laryngeal tumor of doubtful character being noticed, its cure is attempted by necessarily imperfect and superficial endolaryngeal methods at removal, which, however adequate for the treatment of benign new-growths, are unquestionably unreliable so far as the eradication of a malignant neoplasm is concerned. The repeated employment of caustics or the application of forceps will not only fail to eradicate all the elements of evil, but, on the contrary, will serve as a stimulus to the rapid extension of the malady to the unaffected parts of the organ. False hopes thus raised by professional ignorance or cupidity are not only doomed to disappointment by unmistakable recurrence, but the chances of relief are marred by the further dissemination of the infectious elements in the diseased organ itself and to the pertinent lymphatic glands.

The earlier the operation is done after due establishment of the diagnosis, the less mutilating it need be. Unilateral extirpation of the larynx is far less dangerous than the total removal of that organ, and, as a number of successful cases testify, even a fair degree of phonation, together with unimpaired deglutition, may be preserved by it.

CASE I.*—Paul Hahn, barber, aged fifty. *November, 1879.*—Increasing dysphagia. Dr. E. Gruening diagnosticated an elevated ulcer of the size of a half-dollar coin, occupying the depression bounded by the right side of the base of the epiglottis, the right side of the base of the tongue, and the right wall of the pharynx, a site corresponding to that of the glosso-epiglottic and aryteno-epiglottic folds, and more particularly to that of the sinus pyriformis. The mucous covering of the epiglottis was seen to be thickened and congested. The cervical glands did not appear to be affected. No evidence of syphilis could be elicited, either from the history or from the physical examination of the patient, excepting a moderate degree of onychia, characterized by roughening of the finger-nails. In the course of the treatment it became evident, however, that this latter trouble was due only to the fact that, in pursuing his trade, his fingers were much exposed to the action of soap-lather.

Anti-syphilitic treatment was instituted and continued for some time with apparent benefit, the patient regaining to a certain extent the ability to swallow. The improvement was, however, merely temporary; the dysphagia returned, and the patient soon began to suffer from the inanition thus engendered.

Preliminary tracheotomy was performed January 18, 1880, at the German Hospital. *March 5, 1880.*—Unilateral exsection of the larynx was done with the able assistance of Drs. Gruening, Bopp, Lefferts, and Dr. Degner, the house-surgeon, to whom great credit is due for the skill and patience exhibited in the difficult and tedious after-management of the case.

An incision was carried from the median line of the hyoid bone along its upper margin outward to the extent of three inches, exposing the right lingual artery, which was ligated. A second incision was carried downward from the starting-point of the first, in the median line, to the opening for the cannula, exposing the anterior surface of the hyoid bone and larynx, and the flap thus formed was dissected up with all the underlying soft parts and turned outward. Trendelenburg's tampon-cannula had been fitted into the trachea. The right half of the hyoid bone was then exsected, a double ligature placed around the superior laryngeal artery, and the same divided. The crico-thyroid ligament was cut across, a pair of bone scissors inserted into the larynx, and the thyroid cartilage divided in the median line. Trendelenburg's tampon cannula did not fulfill the requirements owing to a leak in the inflated bladder, so that blood managed to find its way into the trachea. An attempt to make it serviceable by winding layers of moistened gauze around the cannula was unsuccessful, and during the rest of the operation it became necessary to fill out the lower part of the larynx with small sponges. The interior of the larynx was now exposed and showed an oval tumor, of about the size of a pigeon's egg, situated in the substance of the right false vocal cord, involving the posterior half of the true vocal cord and the small cartilages belonging to it. The right half of the thyroid and the whole of the arytenoid cartilage were now dissected up and removed, together with the whole epiglottis. The pharynx being thus exposed to view, its entire right side was seen to be diseased, and was removed, together with the right tonsil and the lower half of the right pillars of the palate. The base of the tongue, likewise involved, was dissected up on the right side with the scalpel, on the left with the thermo-cautery. The hæmorrhage was insignificant, and the patient rallied promptly after the operation.

One of Tiemann's excellent soft-rubber tubes was introduced into the œsophagus, the wound thoroughly cleansed with a ten-per-cent solution of zinc chloride, and the whole cavity packed with moistened balls of carbolized cloth. The edges of the horizontal incision were then united by catgut sutures.

* "Archives of Laryngology," vol. i, No. 2, June, 1880.

The œsophageal tube was remarkably well tolerated, and the patient's nourishment was satisfactorily effected through it during the whole course of the treatment.

The dressing was changed once every twenty-four hours.

On the fifth day after the operation the patient was well enough to sit up in a chair for an hour. Three days later he could ascend a flight of stairs in being removed to another room, and a week later he spent most of his time out of bed. By the 1st of April, twenty-six days after the operation, he took a walk in the garden, and his weight had increased by 6½ pounds.

The large cavity contracted rapidly, and finally became a canal, bounded on one side by the remaining half of the larynx, on the other by a smooth cicatrix uniting the skin with the mucous membrane of the posterior wall of the pharynx.

On the 29th of April the patient made a first attempt to speak. When the tracheal tube was closed, he could converse with a hoarse, dull voice, quite audible, and easily understood at a distance of from two to three yards. His ability to swallow has in a measure been recovered, but he preferred to use the œsophageal tube, to which he had become accustomed. By the 5th of May he had gained 14½ pounds in weight.

The patient continued well until February, 1881, when he contracted an acute pleurisy, to which he succumbed rather suddenly on account of fatty heart. The specimen of the larynx gained at the post-mortem examination showed absence of any sign of a relapse.

The tumor was found to be an adeno-sarcoma.

CASE II.*—Henry O., porter, aged fifty-seven. Rebellious hoarseness of five months' standing, with increasing difficulty of deglutition. Marked loss of flesh and power. *March 16, 1885.*—When the patient was directed to the author by Dr. S. W. Gleitsmann, a deep-seated, nearly immovable, hard, glandular swelling of the size of a hen's egg was noted in the left submaxillary triangle. Endolaryngeal inspection revealed the presence of a smooth, pale tumor, the size of an almond, commencing in the left glosso-epiglottidian fold and extending through the substance of the left vocal cord into the ary-epiglottidian fold, to terminate in the arytenoid cartilage with a knob-like protuberance. *March 18th.*—Chloroform being administered, the diseased glands were removed. The sterno-mastoid was found partly involved, and this, together with a piece of the internal jugular vein of about one and a half inch in length, was removed in one mass. Then inferior tracheotomy was performed. The wound healed kindly, except where the tracheal tube was located, and April 27th, under chloroform, the left half of the larynx was removed. A tampon cannula, made by George Tiemann & Co. after the author's directions, was inserted and suitably distended so as to prevent the entrance of blood into the trachea. After this an incision, commencing at the upper notch of the thyroid cartilage and extending to the lower margin of the cricoid cartilage, laid bare the larynx in the median line. To this was added another incision, commencing in the upper angle of the first cut and extending horizontally to the anterior margin of the left sterno-mastoid muscle. The crico-thyroid ligament was split to admit a strong pair of bone-pliers for the division of the thyroid cartilage; but it was found impossible to perform this act, as the strongly inclined position of the cartilage did not permit an effective handling of the instrument. Therefore, access was gained through an incision in the thyro-hyoid ligament from above, and in this manner an exact division of the calcified cartilage was successfully effected. After this the epiglottis was cut through lengthwise, the left half of the crico-thyroid ligament was divided, and the superior thyroid artery was included in a double ligature and cut through. The most difficult part of the operation consisted of the dissection of the lateral portions of the larynx and pharynx, closely adherent to the carotid artery

* "Annals of Surgery," January, 1886, p. 20.

by cicatricial tissue, caused by the extirpation of the submaxillary glands. Shallow incisions, running parallel with the course of the carotid artery, were cautiously made one after another, and the difficult task seemed almost completed when suddenly a powerful jet of arterial blood welled up from the bottom of the wound. The bleeding point was easily secured in a pair of artery forceps, and then it was ascertained that the trunk of the superior thyroid artery (doubly ligated further below prior to this) had been cut away on a level with its inosculation into the carotid. A catgut ligature was applied around the main trunk above, another below the artery forceps, and when the instrument was removed a round hole in the side of the carotid became visible. The remaining adhesions, corresponding to the lateral portion of the pharynx on the left side, could now be easily dissected out. The tampon cannula was removed, and it was found that no blood whatever had entered the trachea. A soft tube was inserted into the œsophagus, the wound was loosely packed with iodoformed gauze, and an ordinary tracheal cannula was left in the lower angle of the tracheal wound. Finally, the horizontal incision was closed by a number of catgut sutures. The duration of the operation was one hour and three quarters—the anæsthesia throughout undisturbed.

Microscopical examination of the new-growth by Dr. L. Waldstein gave the diagnosis of alveolar sarcoma.

The subsequent course of the wound was very satisfactory and free from fever or suppuration, the patient's only complaint being a rather profuse secretion of saliva. Nutrition was carried on by the œsophageal tube, the patient consuming considerable quantities of milk, eggs, and an emulsion composed of beef-tea and crushed boiled beef; finally, a generous supply of good whisky.

From May 10th on, the œsophageal sound was introduced twice daily for purposes of nutrition. On May 13th the tracheal cannula was abandoned. On the same day the innermost layers of the iodoformed gauze packing became detached, and were replaced. The entire wound was found to be in a vigorous process of granulation, and was considerably contracted.

*May 15th.*—The patient swallowed a small quantity of coffee.

*May 27th.*—Sutures were removed; wound firmly united. Increase of body weight four and a half pounds. *May 31st.*—Patient was discharged cured from the hospital, good deglutition being noted. *June 12th.*—Removal of a small, suspicious gland from the left supraclavicular space. *March 13, 1886.*—Removal of an enlarged lymphatic gland from left suprahyoid region. Since then the patient remained well, attending to his laborious occupation. He could speak with a very audible hoarse intonation. The right vocal cord performed its function normally. In March, 1887, relapse appeared in the cicatrix about the insertion of the stump of the epiglottis, for which subhyoid pharyngotomy was performed, April 22, 1887, at the German Hospital. A portion of the cicatrix, together with a section of the base of the tongue, was removed. The external wound was united by three rows of superimposed catgut sutures. Deglutition was hardly disturbed by the operation; the external wound healed by adhesion. and, May 2d, patient was discharged cured.

In both of the preceding cases decided alleviation of the patients' wretched condition and an undoubted prolongation of life were achieved.

## IX. GOITRE.

The aseptic method and an improved technique of dissection have materially reduced the formidable perils of the surgical treatment of goitre, justly dreaded by old-time practitioners.

In goitre encroaching upon the trachea, the question must be first decided whether the growth is cystic or parenchymatous. If cystic, various forms of treatment offer a fair chance of cure. The cyst can be tapped and injected with tincture of iodine, like a hydrocele; or it can be exposed by dissection, incised, and its walls sutured to the skin, like the sac in hydrocele operated on by Volkmann's method (Schinzinger).

CASE.—Lena Kaiser, aged thirty-five. Cystic goitre of the thyroid body. It was as large as a child's fist, and the source of much discomfort to the patient on account of the severe dyspnœa it produced. *November 23, 1882.*—At the German Hospital, exposure of the capsule of the goitre. A plexus of much-distended veins was included in two sets of double mass ligatures, between which the capsule was cut into. The parenchyma of the gland was divided, and the sac of the cyst being exposed was incised and attached to the skin by two continuous sutures. The cavity was packed with carbolized gauze. *December 22d.*—Patient was discharged cured.

Where the presence of a number of contiguous cysts is made out, their enucleation will be appropriate. The procedure is not difficult, and offers the additional advantage of the possibility of primary union and a speedy cure.

CASE.—Hannah S., servant, aged thirty-one. *January 16, 1886.*—At Mount Sinai Hospital, extirpation of four contiguous cysts of the thyroid body. Flap incision; the thyroid capsule was cut into between two rows of mass ligatures; after this the cysts were shelled out without difficulty. The wound was drained and sutured. Primary union. Patient was discharged cured February 21st.

*Parenchymatous goitre* may be treated with some hope of success by the methodical injection of tincture of iodine in cases in which the tumor is soft and vascular. Should this plan fail, or when the tumor is very dense and hard, excision must be performed.

Total removal of the thyroid gland is apt to produce a deep alteration of the general condition denoted "*myxœdema,*" or "*cachexia strumipriva*" (Kocher), characterized by idiotism, loss of sexual power, and general dense œdematous infiltration of the subcutaneous connective tissue ending in death. Hence, a portion of the glandular tissue ought to be always left behind to perform its function, so necessary to the healthy state of the nervous system.

The principles laid down for the safe removal of tumors (page 52) should guide the surgeon in exsecting thyroid swellings. Hæmorrhage from the large veins of the capsule is to be avoided by the timely use of Thiersch's spindles and of double ligatures. Dissection should be systematic and deliberate, and especial care should be devoted to the preservation of the recurrent laryngeal nerve, which will be found behind the lateral lobe of the thyroid gland in the groove separating the trachea from the œsophagus.

CASE.—Rosa Rosenfeld, cook, aged twenty-four. Parenchymatous hyperplastic goitre of the body and right thyroid lobe, causing severe dyspnœa. *October 9, 1884.*— At Mount Sinai Hospital, extirpation of the right lobe and body of the gland from a spacious flap incision. A pedicle was formed toward the left lobe, and, being first ligatured, was cut off. In dissecting up the right lobe, which was found to be insinuated between the trachea and œsophagus, the recurrent laryngeal nerve was separated and drawn aside. Drainage, suture, and aseptic dressings. The wound healed, with the exception of the drainage-tracks under the first dressing, which was changed on October 19th. Some hoarseness due to paresis of the right vocal cord persisted for five months, but ultimately disappeared.

*Tracheotomy for goitre* is one of the most formidable tasks the surgeon may be called upon to perform. It was twice the author's duty to undertake this procedure for extreme dyspnœa caused by malignant tumor of the thyroid gland. One case was complicated by mitral insufficiency and acute broncho-pneumonia, and ended fatally. In the other one the supra-sternal portion of a very large fibro-sarcoma of the thyroid gland had to be first extirpated before access could be had to the trachea. This case also ended lethally.

CASE I.—Rosa Guttmann, widow, aged thirty-six. Large and growing originally parenchymatous, later sarcomatous, substernal goitre of five years' standing. Mitral insufficiency and severe acute broncho-pneumonia. Dr. S. Kohn, who referred the patient to the author, diagnosticated paralysis of the right vocal cord. *November 11, 1879.*— Patient was admitted to German Hospital in a very exhausted condition. After copious stimulation tracheotomy was performed. Only a very small amount of ether was administered for the cutaneous incision. Division of the goitre by the thermo-cautery was tried, but had to be given up on account of the slowness of the process and the great hæmorrhage from the enormously distended veins. The expedient of at once taking up and firmly retracting the divided tissues by large, four-pronged, sharp hooks, proved more efficacious in checking hæmorrhage. With a few rapid strokes the trachea was exposed and opened, and, a large-sized soft catheter being introduced, respiration became well established. But a few minutes afterward patient expired.

CASE II.—Elizabeth K., aged sixty-two. A very fat woman, with a small pulse, suffering from extreme dyspnœa due to the presence of a very large and hard supra- and infra-sternal fibro-sarcomatous goitre. *August 23, 1882.*—Extirpation of the supra-sternal part of the swelling with subsequent tracheotomy, for which a specially constructed cannula with a long tube was used. Relief of dyspnœa. Copious stimulation was employed by the family attendant to such an extent that in the night of August 24th the patient became boisterously drunk, and died in a soporous condition under the symptoms of acute alcoholism.

## X. AMPUTATION OF THE BREAST.

In preantiseptic practice the rate of mortality observed after amputation of the breast, mainly due to accidental wound complications, was nearly as high as that of major amputation of the limbs.

The notable depression of the death-rate that has taken place since is directly due to cleanlier methods.

The absence of a proportionate decrease of the death-rate, caused by relapse of the malignant growths for which the operation is performed, is to be attributed to the tardiness of the general practitioner in advising and urging early removal, and the unwillingness of the patients to heed timely advice.

In view of the fact that over ninety per cent of all mammary tumors are carcinomatous, the benefit of the doubt belongs to the view which urges to removal. *A probatory incision at least should be insisted on in every case of solid chronic intumescence of the breast that remains uninfluenced by proper local and general treatment directed against syphilis or chronic inflammatory mastitis.*

*Partial operations* are admissible only where the youth of the patients, the smoothness and mobility and slow progress of the tumor justify the assumption of a benign growth, such as adenoma or adeno-fibroma, or where probatory puncture leaves no doubt of the presence of a simple retention cyst.

In these cases the operation proposed by T. G. Thomas is very appropriate, and gives satisfactory results both as to the completeness of the removal and the cosmetic effect. The incision is laid in the pectoro-mammal fold, and the breast-gland is raised from the pectoral fascia sufficiently to enable the surgeon to incise it on its posterior aspect. After the enucleation of the tumor the breast is replaced, and, the wound being drained, the skin is united by an exact suture. The cicatrix remains hidden under the overlapping breast.

CASE I.—Miss C. L., governess, aged twenty. Adenoma of left breast of the size of a hen's egg. *December 12, 1884.*—At Mount Sinai Hospital, Thomas's operation. *December 22d.*—First change of dressings. *December 24th.*—Discharged cured. *December 12, 1886.*—No relapse; very fine linear cicatrix.

CASE II.—Miss Tillie G., aged sixteen. Adeno-fibroma of left breast of the size of a small apple.

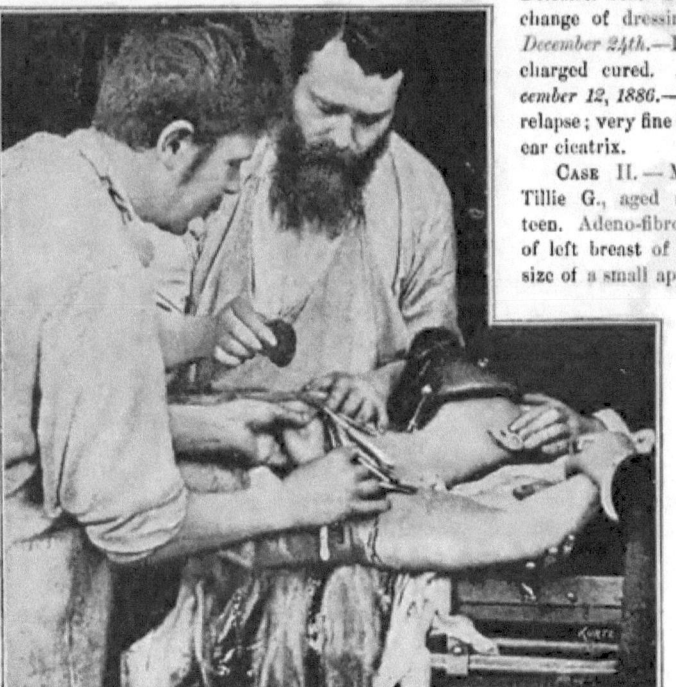

FIG. 99.—The mammary gland being detached from below, the surgeon inserts his left hand under the breast to complete the upper section.

*December 20, 1886.*—Thomas's operation at Mount Sinai Hospital. *December 30th* — Dressings changed. *January 4, 1887.*—Wound firmly united.

Whenever amputation of the breast is performed for malignant tumor, *the operation must be radical,* or at least as radical as possible. *No regard whatever should be paid to cosmetic considerations, the object of the measure being the extirpation of a deadly disease, which, if not eliminated, is sure to kill.* A wide berth should be given to the visible limits of the disease, and the knife should take away at least an inch and a half of apparently healthy skin. *The pectoral fascia, axillary fat and glands must be invariably removed in mass,* whether intumescence is to be felt or not. If the axillary vein be attached to degenerated lymphatic glands, the attached segment must be included in two ligatures, and the intervening piece cut out together with the adherent mass. (See case of Betty Lowy, page 59.)

The technique of breast amputation is simple. After marking by a shallow cut the extent of the two semi-elliptic incisions that should include the part to be removed, the inferior margin of the breast-gland is exposed. The pectoral fascia being incised, the mamma is gradually dissected up from the thorax till its upper limit is reached. The surgeon's hand is slipped in under the breast, and the upper incision completes its detachment, except where the lymphatic vessels, passing along the pectoral fold from the breast to the armpit, form a sort of a pedicle. The bleeding vessels are secured as they are cut, and the pectoral wound is covered with a towel wrung out of corrosive-sublimate lotion, to remain under its protection during the removal of the axillary contents. The incision is extended well up the arm into the axilla, and the skin is dissected up for about an inch to each side of the cut. The fascia is divided where the incision can be made boldly upon the edge of the pectoral muscle anteriorly, and the latissimus dorsi posteriorly. Proceeding from this latter incision, the loose connective tissue is divided by blunt dissection with a thumb-forceps and the handle of the scalpel, until the axillary vein is exposed to view. With this the most important step of the operation is accomplished. Seeing the vein will prevent its accidental injury, and from

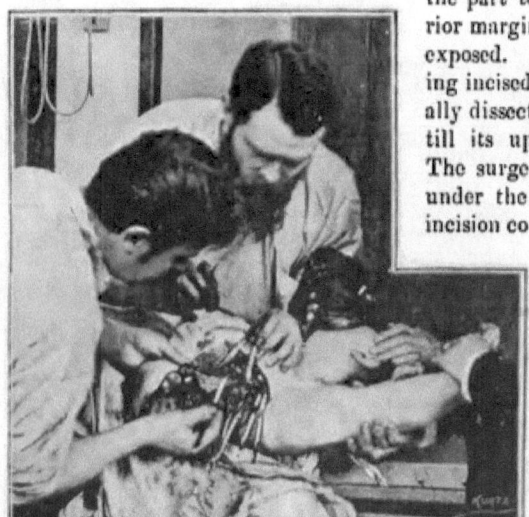

FIG. 100.—Removal of axillary contents. The surgeon holding the detached breast serving as a handle.

this on, in most cases, dissection will be directed *away from* instead of *toward the vein*. The loose fat can be easily detached from all its lateral adhesions.

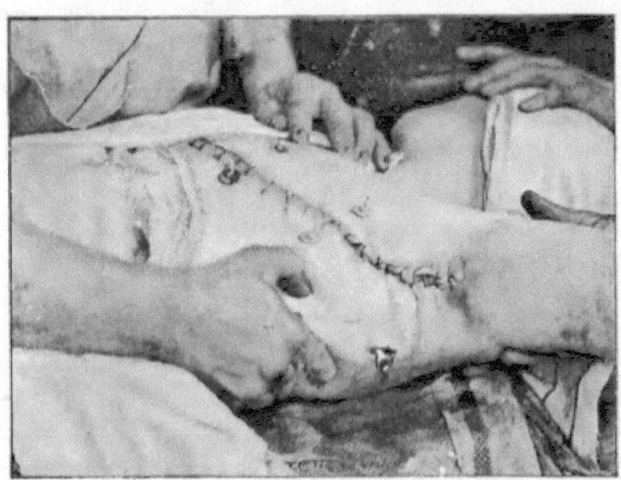

FIG. 101.—Sutured wound after amputation of breast. Counter-incision through latissimus for purposes of drainage.

The vessels and nerves which traverse the adipose tissues can be distinctly felt and seen as they are successively approached. If necessary the long thoracic artery and vein, and sometimes the subscapular vessels, should be

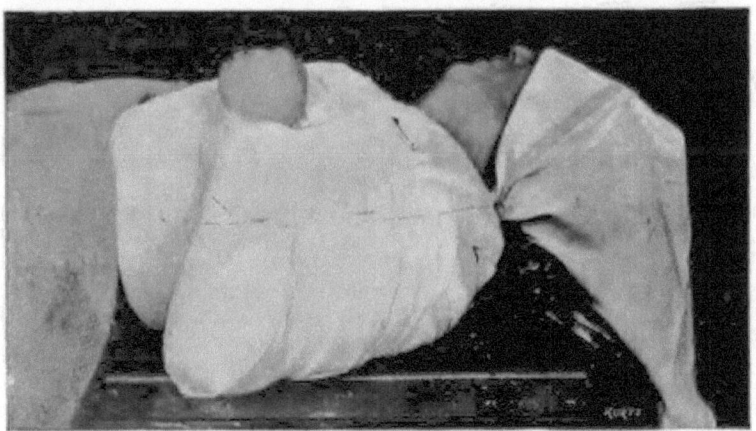

FIG. 102.—Completed dressing after breast amputation.

taken up and cut between two forceps. The nerves ought to be preserved. During the dissection of the axillary contents, the breast serves as a suitable handle. Breast and axillary contents are removed in one mass. Thus the intervening lymphatic ducts are certainly taken away together with the mammary gland and the axillary lymphatic glands. After due irrigation, a counter-incision is made on the external aspect of the latissimus-dorsi muscle. The knife should divide the skin and fascia only; then a dressing-forceps is thrust through the muscle into the most dependent part of the axillary wound, when it is made to grasp the end of a stout drainage-tube, which is drawn out through the counter-incision, to be transfixed with a safety-pin and clipped off even with the skin.

After this the pectoral wound is united. **Lister's** button suture, **or a** quilled suture, or any **other of the** known forms of retentive suture, is applied to relieve tension. After another irrigation, the fine catgut sutures of coaptation are put in until the wound **is closed**. The wound is once more flushed **out** with mercuric lotion, and is covered with the dressings, care being taken **to** make them the thickest about where the drainage-tube issues forth. **The** dressings are secured by roller-bandages, and the arm is either included in the turns of the bandage, the ulna first being well padded, or, being **left** out, is supported by an extra sling.

Ordinarily, the dressings are changed **and the** tube **is removed on the** fourth day after the operation, **when** the retention sutures are also extracted should they not have been absorbed by this time. A smaller dressing secures **the parts against injury. Five days later** another change of dressings may take place, when the drainage opening will be **found** closed by a plug of granulations. After this a covering of cerate or lead plaster, with a little pad of cotton secured by a strip of adhesive plaster, **will** be all that is necessary until cicatrization is complete.

It is remarkable how soon the arm regains its power of abduction in cases that remain free from suppuration.

Of seventy operations for tumors of the mammary gland, sixty-eight were done on women mostly past middle life; two were performed on men. The male cases were as follows:

CASE I.—A. B., aged seventeen. Growing adenoma of right mammary gland. *August 4, 1883.*—Extirpation **of the tumor**; axilla was not interfered with. Uninterrupted primary union.

CASE II.—George Eckert, blacksmith, aged sixty. Large, very hard epithelioma of the right mammary gland, starting from the nipple, which was unrecognizable in the ulcerated mass. Axillary glands involved. *April 27, 1886.*—Amputation of breast and evacuation of axilla at the German Hospital. Large portions of skin and of the pectoralis major and minor muscles had to be removed. Primary union followed, except where the skin could not be brought together. *June 7th.*—Discharged cured.

In five cases of adenoma of young women, the tumor alone was removed.

In five instances (Mary Hauser, adeno-cystoma; Emma Bockhold, cysto-sarcoma; Albert Baron, adenoma; Sarah S., cysto-adeno-fibroma; Frida Meissner, adeno-fibroma), **the mammary gland alone was** amputated, the axillary space remaining intact.

The remaining fifty-eight cases consisted of **fifty-two cancers, five** sarcomata, and one instance of tuberculosis. In each **of these the entire breast and all the axillary contents were removed.**

| | |
|---|---|
| Cancer | 53 cases. |
| Sarcoma | 6 " |
| Adenoma | 7 " |
| Adeno-fibroma | 2 " |
| Adeno-cystoma | 1 case. |
| Tuberculosis | 1 " |
| Total | 70 cases. |

Of this number, *sixty-one times healing by primary union was observed.* *Five cases suppurated* in consequence of infection of one or another kind at the time of the operation; three cases healed by granulation, as it was impossible to cover the defect caused by the operation. A fourth granulating case died of erysipelas, contracted outside of the author's care (Julie Schmalz, scirrhus) while the wound was not yet healed.

Of the cases healed by primary adhesion, one died of continuous thrombosis of the axillary and innominate vein, with subsequent embolism of the pulmonary artery. The catastrophe took place shortly after the first change of dressings, made eight days after the operation.

Case.—Clara Hahn, spinster, aged thirty-two. *November 30, 1883.*—Amputation of left breast, with evacuation of axilla for small-celled adeno-carcinoma; suture; no drainage. *December 14th.*—First change of dressings; entire wound absolutely healed. On Christmas eve the patient was selling crockery over the counter. *April 4, 1885.*—Typical amputation of *right* breast at the German Hospital for the same affection, together with excision of relapsing cancer in the shape of a small node in the cicatrix of the left side. Patient was doing excellently till April 12th, when the first dressings were changed, and the wound was found faultlessly healed. Immediately after the dressings were completed, the patient became faint and cyanosed; breathing labored, pulse scarcely to be felt; the left deep jugular vein was permanently distended. Hydropericardium and hydrothorax developed with œdema of both arms, and the patient died April 20th, sixteen days after the operation, having had normal and later subnormal temperatures throughout. Autopsy revealed continuous *thrombosis of left axillary and anonyma vein,* the thrombus extending into the right auricle and the pulmonary artery; bilateral hydrothorax, hydropericardium, and a hæmorrhagic infarction of the connective tissue in the posterior mediastinum.

The only unusual circumstance that attracted the author's attention immediately before the second and fatal operation was the fact that, a hypodermic injection of morphia being administered, extensive ecchymosis appeared shortly afterward at the site of the injection, suggesting a morbid alteration of the patient's vascular system.

Thrombosis and embolism were observed in another case, which, however, ended in cure.

Case.—Mary Lier, school-teacher, aged fifty-seven. Suffering from old pulmonary emphysema and chronic bronchitis. Face slightly cyanosed. Scirrhus of right breast; nipple retracted, discharging dark, tar-like serum. *November 14, 1878.*—With the kind assistance of Dr. F. Lange, amputation of right breast and evacuation of the axilla were performed. Anæsthesia by ether was very bad. Feverless course of healing. *November 19th.*—Drainage-tube was removed. *November 23d.*—Apoplectiform seizure, followed by aphasia and agraphy, which, however, gradually disappeared. *December 25th.*—The wound was entirely healed, and patient could again speak Bohemian, her mother-tongue. Gradually she regained her German and English, and in 1882 the author heard from her as being able to write again.

One of the suppurating cases died of acute catarrhal pneumonia and carcinosis of the lungs, twenty-two days after the operation, the wound doing well at the time under process of granulation.

CASE.—Mary Volkmer, housewife, aged forty-seven. Soft adeno-cancer of both breasts, the large tumor of the left mamma causing much distress. *March 17, 1881.*— At the German Hospital, amputation of left breast and evacuation of the axilla were done. Wound was united in part only on account of extensive loss of integument. Suppuration of axillary space followed, but the fever resulting therefrom subsided directly after drainage was re-established. Nevertheless, patient appeared to be very ill. *April 8th.*—Catarrhal pneumonia set in, to which she succumbed. *April 9th.*— On post-mortem examination general carcinosis of lungs and liver and catarrhal pneumonia were found.

In computing the three fatal cases, that of Julie Schmalz, who died of erysipelas contracted under the care of another physician before perfect cicatrization had taken place, can justly be excluded. Accordingly, of the remaining sixty-seven cases, two died directly in consequence of the operation ; none, however, on account of septic processes established in the wound. Thus, the author's rate of mortality from accidental wound infection in amputation of the breast would be 0 ; from other causes beyond the influence of the surgeon, a trifle less than three per cent (2·98).

## XI. ABDOMINAL OPERATIONS.

### 1. *General Remarks.*

The relation of aseptics to the surgical treatment of the peritoneal cavity is in some quarters a subject of hot controversy to this day. On one side we see the advocates of a more or less complicated antiseptic apparatus, including the spray, achieving very good results, and basing success upon the strict enforcement of their cantelæ. But, on the other hand, we notice a most successful laparotomist maintaining that antiseptics are unnecessary, or even harmful, and that he is accustomed to flush the peritoneal cavity with "water from the tap," teeming with millions of bacteria, and yet his results vie with those of the most scrupulous Listerian. Both sides to the controversy have abundant and incontrovertible facts to support their positions, and the contradiction seems to be hopelessly insurmountable. It certainly is extremely bewildering to the student and beginner. Yet this contradiction is unreal, and let us say, on one side, also disingenuous.

The physiological peculiarities of the peritonæum, most notably its enormous absorbent power, endow it with the quality of neutralizing the deleterious effects of limited quantities of pyogenic or septic micro-organisms—a quality not possessed to such an extent by any other part of the human organism.

Grawitz[*] has brought experimental proof of the fact that the normal peritonæum will at once absorb into the circulation moderate quantities of

---

[*] "Charité Annalen," xi. Jahrg., page 770.

active pyogenic cocci, where they will be widely scattered through the blood and perish.

NOTE.—This fact goes very far to explain Lawson Tait's position, who, however, although disclaiming antiseptics, devotes most scrupulous care to *asepticism*—that is, to the cleansing of hands and instruments. His instruments are few, and selected with a view to simplicity. *His sponges are put into carbolic lotion for disinfection.* The water used for the immersion of his instruments is sterilized by boiling. Most of the bacteria contained in his " water from the tap " are innocuous—that is, non-pyogenic; and those that have the power to cause suppuration are too few to produce serious trouble. They are simply absorbed and killed off by the great germicide, the blood.

The limit of the quantity of pyogenic cocci required to produce acute purulent peritonitis varies with the size and state of health of the animal used in the experiment. A large dog's peritonæum would resist a much greater quantity of infectious pus than that of a small dog or rabbit. And a healthy animal would neutralize more septic material than a debilitated one of the same kind and weight.

The presence in the peritoneal cavity of a larger quantity of stagnant bloody serum than can be readily absorbed within an hour, will suffice to produce purulent peritonitis on the addition of a very small number of cocci.

If the fluid is absorbed or artificially removed by drainage *before* the cocci have a chance to vastly multiply, no peritonitis or only adhesive forms of the inflammation will develop.

Therefore, it is rational to employ drainage in cases where large surfaces, denuded of peritonæum, have to be left behind in the abdomen.

Denudation of the surface layer of the peritoneal endothelium by caloric, or mechanical or chemical influences, is the main factor in causing the development of purulent peritonitis. It favors exudation of serum and diminishes or destroys the power of absorption inherent to the normal peritonæum. Should even a minute quantity of pyogenic cocci be introduced into the peritoneal cavity under these circumstances, purulent peritonitis may readily develop.

The practical conclusions to be drawn from the preceding facts are:

1. Although the normal peritonæum will tolerate a greater quantity of infectious material than most surgical wounds, yet all precautions regarding the cleansing of hands, instruments, sponges, and other apparatus used for laparotomy should be employed, as septic infection of the peritonæum is much easier to prevent than to cure.

2. Unnecessary denudation of the uppermost layer of the peritonæum should be avoided as much as possible.

3. Corrosive solutions, as, for instance, of carbolic acid or mercuric bichloride, are not to be used on the peritonæum. As soon as the peritoneal cavity is opened, Thiersch's solution should be employed for rinsing the surgeon's hands, immersing the instruments, sponges, towels, and, if necessary, for irrigation.

4. A careful toilet, that is, removal of all exuded serum or blood, should precede closure of the abdominal wound.

5. Where large denuded surfaces have to be left behind, and a good deal

of oozing is to be expected, drainage, or plugging with strips of iodoform gauze, or both, must be employed.

NOTE.—If the drain-tube is brought out from a dependent part of the peritoneal cavity, as for instance through Douglas's cul-de-sac, the secretions will escape spontaneously by the operation of the law of gravity. Whenever the drainage-tube is brought out above the symphysis, the serum collecting at the bottom of the cavity must be removed either by hourly mopping out with a stick, armed with a pad of absorbent borated cotton, or by exhausting with a long-nozzled syringe, introduced to the bottom through the hollow of the drain-tube.

6. Should it become evident that the mode of drainage employed is insufficient to remove a copious gathering of secretions, febrile symptoms, tenderness, and tympanites developing on the first few days after the operation, a saline purge may be employed in preference to the accustomed opium treatment (Tait). Its object would be to favor rapid absorption of the effused serum in an analogous manner seen with the administration of cathartics for the rapid removal of hydropic accumulations from the abdominal cavity.

7. If purulent peritonitis be undoubtedly established, reopening and irrigation of the peritoneal cavity with hot Thiersch's solution may be taken into consideration, provided that the patient's general condition should warrant such a procedure.

## 2. *Herniotomy.*

In the main, the success of herniotomy depends upon the condition of the strangulated gut at the time of the operation. With aseptic precautions, as long as the gut is not necrosed, herniotomy is fraught with very little danger. From the moment that intestinal gangrene has set in, the preservation of asepticism becomes extremely difficult. Contact alone with the decayed gut is infectious. Laceration of the friable intestinal wall is very likely to occur on employment of the least amount of force, and usually leads to further contamination by escaping intestinal contents. In addition to this, the general condition of patients with intestinal necrosis is mostly wretched. Systemic intoxication, and the tendency to heart-failure induced by constant vomiting, vastly increase the perils of anæsthesia and hæmorrhage, and the prognosis is thereby rendered all the more doubtful.

The free exhibition of anodynes, especially in the shape of hypodermic injections, in the presence of strangulated hernia, is very often followed by fatal consequences. The most acute symptoms are blurred or blotted out entirely, and *a false sense of security is apt to lull the apprehensions*, and to betray patient and physician into undue procrastination.

Out of the fifty-one cases of herniotomy performed by the author both for strangulation and for the radical cure of the complaint, eleven died. Eight out of this number exhibited necrosis of the gut, and all of these died. Of the remaining three, one, whose gut was sound, died of acute nephritis, presumably due to the use of ether as an anæsthetic ; the other one of general tuberculosis of the peritonæum ; the third of acute sepsis due to paralytic distention of the gut induced by peritonitis.

CASE I.—A. Schlesinger, aged seventy-three, strangulated left inguinal hernia of twenty-four hours' standing. *April 12, 1885.*—At Mount Sinai Hospital, the hernial sac was exposed under other anæsthesia. A knuckle of gut could be felt within the sac, containing a cubic, friable body that was easily crushed, whereupon the gut was replaced in the abdominal cavity without any difficulty. The wound was sutured and dressed. Duration of the operation, twenty minutes. The wound healed by primary adhesion, but uræmic symptoms, with suppression of the renal secretion and vomiting, developed on the second day. The scanty urine was found containing blood and a large amount of albumen. *April 22d.*—The patient died in uræmic coma.

Inquiry elicited the fact that, preceding the day of the patient's illness, he had largely consumed of a dish of potato soup. The toothless old man had bolted some of the potato, a piece of which having made its way into the hernia caused strangulation.

The other fatal case, not due to necrosis of the gut, was as follows:

CASE II.—Mrs. Henrietta Bolz, housewife, aged sixty, an ill-nourished, emaciated person, who said that she had been suffering from belly-ache and constipation for two months, and that she has had severe and continuous fever that caused her present emaciation. She also noted that she had lost most of her hair. Forty-eight hours previous to her admission, irreducible femoral hernia of the right side was diagnosticated by a medical man. Vomiting, no fever, and great tenderness over the abdomen were found, and it was deemed proper to explore the hernia. Accordingly the operation was done, May 7, 1887, at the German Hospital. After incision of the sac, this was found to contain a portion of adherent omentum, together with a very much congested knuckle of small gut. The strangulating band was incised, the gut withdrawn, and, being in a viable condition, was replaced. The protruding portion of omentum was liberated, tied, and cut off. In replacing it, extensive adhesions of the stump to the parietal peritonæum could be felt inside of the abdominal cavity. The sac was excised and the wound closed and dressed in the usual manner. *May 12th.*—Change of dressings. The wound was found united, but the general condition of the patient had remained the same as before the operation. Gradually considerable ascites developed, the patient continuing to complain of much colicky pain; the vomiting and lack of appetite, together with rebellious constipation, seemed to justify the assumption of a general morbid condition of the peritonæum, namely, either tuberculosis or a neoplasm. *May 26th.*—The peritoneal cavity was reopened at the site of the cicatrix left by herniotomy, and extensive tubercular degeneration of the entire peritonæum, with dense infiltration of the omentum and almost universal agglutination of the intestines, were found. The parietal peritonæum and the gut were literally covered with a mass of miliary white nodules. With a view to relieving the obstruction caused by the multiple adherence of the bowels, a protruding part of the thick gut was attached to the wound by a number of catgut stitches, and the external incision was packed with iodoformized gauze. *May 28th.*—The bowel was found well united with the parietal peritonæum, and an artificial anus was established by incising the gut and sewing the mucous membrane to the skin. Sufficient stools followed, but the patient died, March 31st, of exhaustion.

CASE III.—I. F., jeweler, aged fifty-four. Strangulated ventral hernia of large proportions in a very obese subject. *July 1, 1887.*—Laparotomy in the patient's home. Several feet of very much congested and extremely dilated small intestine were replaced with much difficulty. The edges of the hiatus, located just above the umbilicus, were pared, and closed with eight button-sutures. Duration of the operation, two hours. Patient did not rally well, continuing to moan and to vomit bilious matter. The temperature rose to 104° Fahr. during the night, and the man died under symptoms of intense sepsis and peritonitis twenty hours after the operation. Some peritonitis was present

## SPECIAL APPLICATION OF THE ASEPTIC METHOD.

at the time of the herniotomy, and might have been stimulated by the operation, which, however, was clearly obligatory. No autopsy.

Forty (including those subjected to the radical operation) of the author's total fifty-one herniotomized patients recovered.

*a.* **Herniotomy for Strangulation.**—If gentle and not too prolonged efforts at reduction, first without then with anæsthesia, do not succeed, herniotomy should be done forthwith. The mode of procedure is as follows:

If fecal vomiting be observed, it is advisable to wash out the stomach with an œsophageal tube, to prevent the entrance of fecal matter into the air-passages during anæsthesia.

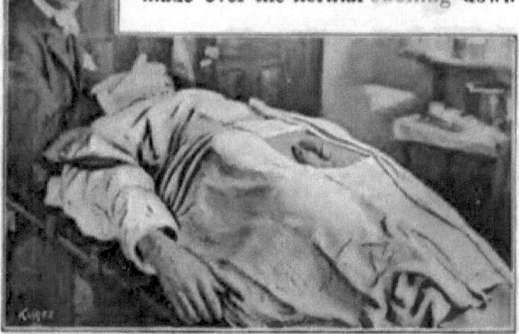

Fig. 103.—Patient ready for herniotomy (or for any other operation about the genital region).

The patient's inguinal region is shaved and scrubbed off with soap and hot water, and is disinfected with mercuric lotion. Towels wrung out of corrosive-sublimate solution are arranged about the field of operation, and a free incision is made over the hernial swelling down upon the sac. *The incision should extend well above the inguinal or femoral ring, and should freely expose the place where the hernia emerges from the abdominal wall.* By doing this the surgeon will be enabled to divide the constricting band under the guidance of the eye, and without the necessity of inserting the probe-pointed knife into the inguinal or femoral canal, a circumstance that may, even in the hands of a cautious and expert surgeon, lead to cutting or laceration of the intestine, especially if it be very brittle, or necrosed, or adherent.

CASE IV.—Philip Trumann, aged two years and three months, was presented to the author December 11, 1881, with a soft, fluctuating, scrotal swelling of the left side, which, however, could not be by pressure reduced in size. Congenital hydrocele was diagnosticated nevertheless, as the tumor showed transparency. Puncture with a hypodermic needle brought out intestinal contents. There were no signs of strangulation, therefore cold applications were ordered, and the child's mother was told to return the next day. By December 12th all symptoms of strangulation, with rather high fever and inflammation of the swelling, had developed. Herniotomy was done at the German Dispensary. In opening the sac, the gut was inadvertently incised. It was found that local peritonitis of the sac, with extensive fresh adhesions, presumably due to escape of fecal matter through the puncture-hole, had taken place. The gut was detached everywhere by the finger-tips, the parts were well disinfected by free irrigation with a two-per-cent solution of carbolic acid, and the slit in the intestine was closed with a Lembert suture of catgut. The strangulating band was then cut, and, the intestine being replaced, the wound was sewed up, drained, and dressed. Uninterrupted recovery followed. *January 12, 1882.*—The patient was discharged cured.

The sac is carefully opened between two forceps, and, if possible, at a place where there is no adhesion to the gut. After free division between two thumb-forceps, a careful inspection of its contents, gut or omentum, or both, should be made. This will be very much facilitated by taking up the edges of the incision made into the sac with a number of artery forceps, which will serve as handles to unfold it to a funnel, which can be easily looked over. (Fig. 105.)

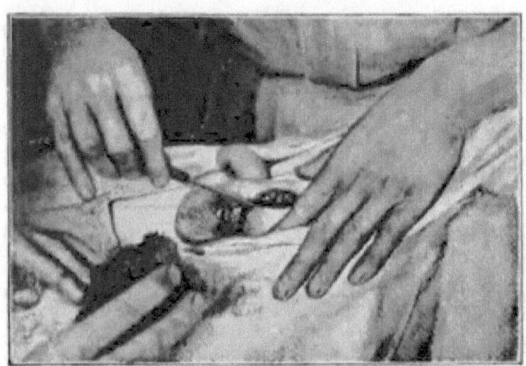

FIG. 104.—Herniotomy. Cutaneous incision.

Generally the gut will appear deeply congested, purplish, or brownish red. As long as it is turgid, and is seen to contract on pinching, it may be assumed to be viable.

But it still remains to be ascertained whether the points of strangulation be alive or not. *To do this the strangulating band or bands must be first cut to a sufficient extent.*

Attempts to withdraw the gut before the strangulation is *completely* removed may lead to very serious consequences, especially where necrosis of the strangulated portion of the intestine is present.

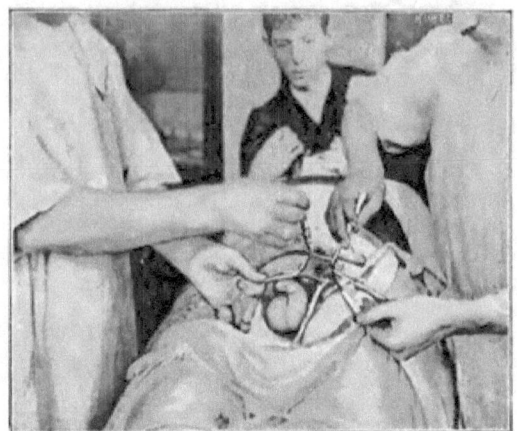

FIG. 105.—Herniotomy. The opened hernial sac is held apart for inspection by a number of artery forceps.

CASE V.—J. Schrank, saloon-keeper, aged fifty-nine. Left inguinal strangulated hernia of five days' standing. Herniotomy, March 8, 1886, at the German Hospital. The sac contained a large mass of adhering omentum, and a knuckle of deeply congested small intestine. It was thought that the strangulating band, corresponding to the internal abdominal ring, had been sufficiently incised, and a very gentle and unsuccessful attempt was made to withdraw the gut. The tip of the index was reinserted as a guide, and, the constriction being **completely divided, the gut was easily** withdrawn. At the same moment a considerable quantity **of fecal matter was seen to** escape. It was found that necrosis of the neck of the strangulated knuckle of gut **had** taken place, and that it had been torn or cut during the preceding efforts at liberation. The intestine was still further extracted, and was attached to the **skin** by a few silk sutures. After careful disinfection, the neck of the sac was loosely packed with strips of iodoformized gauze, and the wound was inclosed in a moist dressing. The collapsed patient died two hours after the operation.

In cases like the preceding one, the classical practice of invaginating **the** tip of the index into the inguinal canal **or femoral ring, for** the purpose of cutting the strangulating band, is dangerous, **as it may lead to** injury of the brittle gut.

The author has **found the** gradual division of all **tissues** from without inward much **safer,** although it must be admitted that the division of the fibrous tissues located above the place of strangulation is extensive, and often practically converts herniotomy into laparotomy.

With a few exceptions, the **author** has always employed open division of the strangulating bands of **tissue,** and never had reason to regret it. In some of the complicated cases he was thereby enabled to **at** once gain a very clear insight into the relations of **the** hernia, and **in a** great measure the ultimate success of the operation was attributed to that advantage.

CASE VI.—Fred. Bormann, laborer, **aged thirty-three, had been treated at the German** Hospital without success during several **days for internal intestinal** obstruction marked by the usual symptoms. On closer inspection, slight œdema of and somewhat **indistinct** resistance at the right inguinal region was noted. *January 17, 1884.*—An incision was made exposing the external inguinal ring, which **was seen** to be normal. The incision **was** further extended, and, when most of the fibrous layers surrounding the inguinal canal had been divided, a small but well-defined tumor could be seen and felt occupying **the inner aspect of** the abdominal wall near **the** internal orifice of the inguinal canal. The abdominal **wall was** completely divided, and then **a** small **hernia,** located between the parietal **peritonæum** and **the** abdominal **wall,** was exposed. **The** sac being incised, a knuckle **of small gut** was found contained **within** it. The place **of** strangulation **was at** the neck **of the sac. This** was completely **slit** open, the gut **was** reduced, and, the neck of the sac **being closed** by a purse-string ligature, it was cut away entirely. The incision in **the** abdominal **wall** was closed by three tiers of catgut **sutures.** Primary union followed. *February 16th.*—Patient was discharged cured.

CASE VII.—Mr. M. S., aged thirty-six. Left inguinal hernia, that had been repeatedly incarcerated, **but** was reduced **each** time. April 8, 1885, it came down again, and, after prolonged and very energetic efforts, the physician in charge succeeded in replacing it, but the symptoms of strangulation, notably vomiting and absence of alvine evacuations, persisted. *April 12th.*—Herniotomy at Mount **Sinai** Hospital. No external tumor **could be seen, but** on palpation **a dense resistant** swelling could **be felt in the** inguinal region within the abdominal wall. **The region of the** external abdominal ring was freely exposed by an ample incision, and **the abdominal wall was divided** above Poupart's ligament. The hernia **which** had **been reduced in mass was then** reached, and was pushed out through the inguinal **canal.** The remaining portion **of** the intervening abdominal wall was **divided,** together with the place of strangulation, and, the sac being tied and cut away, the abdominal wound was closed with three tiers of strong catgut sutures. The wound healed kindly. *May 15th.*—Patient was discharged cured.

It may be said, then, that open division offers great advantages, especially with regard to the avoidance of injury to necrosed or very brittle gut, and that its only drawback—the increased size of the incision—is vastly overbalanced by the security gained therefrom. If the gut be found necrosed, it can be safely withdrawn from the ample aperture, and establishment of an artificial anus can take place after securely packing the neck of the protruding knuckle of intestine with a sort of embankment of iodoformized gauze. This packing of gauze serves as a diaphragm against infection of the peritoneal cavity.

Out of twenty-four cases of herniotomy done for strangulation, undoubted gangrene of the gut was present at the time of operation in six. In two of these the necrosed part of the gut was injured within the inguinal canal by the unavoidable manipulations in liberating the intestine. In those cases where external or open section was used, the integrity of the much-decayed gut was preserved. In these latter cases the gangrene extended to the free part of the gut, and was taken notice of before dissolving the strangulation. In the former cases, however, in which the gut was inadvertently injured, gangrene was limited to the exact locality of the constriction, and was diagnosticated only after the mishap.

The practical lesson to be drawn from this experience is that open incision of the inguinal canal should be done whenever very acute strangulation has existed for more than four or six hours.

All the patients upon whom necrosed gut was found died either of collapse, shortly after the completion of the operation, or of peritonitis due to infection extending from the place of strangulation.

On one of them resection of the necrosed part of the gut was practiced, with subsequent suture. The patient died of peritonitis.

Case VIII.—Catharine Ihle, housewife, aged sixty-one, a very fat woman, having a large incarcerated *umbilical hernia*, was operated September 24, 1881, at her rooms in the presence of the family attendant, Dr. Arcularius. Open section of constricting bands, circumscribed necrosis of the neck of the protruding mass of transverse colon. Exsection of six inches of thick gut and of a triangular piece of meso-colon, and subsequent enterorrhaphy with fine catgut; closure of abdominal cavity. Peritonitis developed during the following night, and, September 25th, patient died with enormous tympanites.

Immediate exsection of the necrosed gut has little to commend it. The dangers of infection of the peritonæum are almost insurmountable, the comprehensive preparations required for enterorrhaphy are usually not made, and, the work being extemporized, generally lacks exactitude. In addition to this, the general condition of the patients is commonly so bad, that undue prolongation of anæsthesia itself would be very dangerous. *Therefore, in these cases, the establishment of an artificial anus is the proper thing to do.* (See Enterorrhaphy, page 158.)

## SPECIAL APPLICATION OF THE ASEPTIC METHOD. 127

To young physicians the decision of **the** question, **whether the** gut be alive or necrosed, may offer a good **deal** of difficulty. **The** responsibility is great, and uncertainty about a point **of** such importance extremely perplexing. Where necrosis is fairly established, the shriveled, parchment-like appearance, the yellowish-gray color, the absence of reflex motion on pinching, and the great fragility will at once characterize the condition. But where necrosis is just developing—that is, where thrombosis **of** the terminal vessels with bloody infarction has gone **so far as** to surely compromise the integrity of the gut, but the signs of necrosis are as yet unrecognizable—decision may be very difficult indeed.

The causes producing intestinal necrosis are not identical in different cases. Local, well-circumscribed necrosis, limited to the **extent of the** strangulating ring, and very often found in femoral hernia, **is due to local** anæmia produced by the pressure of the constricting band.

In other cases **the** local pressure exerted by the constricting **band** upon the neck of the hernial contents may **be** insufficient to destroy the vitality of the intestine in actual contact with the constricting **tissues**. But pressure that would be hardly sufficient to cut off arterial supply, will often compress to such an extent the veins leading *away* from the strangulated gut as to completely arrest circulation. Venous engorgement and gangrene of the convex portion of the **intestinal** knuckle are then inevitable.

The decision whether a portion of intestine, subjected to prolonged **acute** anæmia **by local** pressure, is viable or **not,** is comparatively easy. In **many of** these cases, absent circulation is often restored to the bloodless parts under **the eyes** of the surgeon. As soon as the constriction is relieved, minute red **streaks are** seen to spring up across the formerly **pale, bloodless** area ; they increase in number, and finally the parts in question assume **a rosy** hue and a normal appearance.

Sometimes, however, recovery of circulation **is tardy.** In these cases, after amply dividing the strangulating band, **a catgut** thread should be passed through the mesentery of the questionable loop of intestine, which then should be temporarily replaced in the abdominal **cavity.** The time required for restoring the circulation of the gut **is** usefully employed in attending **to** such other procedures as **may be** indicated under the circumstances. Dissection and **removal of adherent** omentum, or the **dissection** of the hernial sac, will thus occupy some time, by the end of which **the** loop of intestine can be withdrawn from the belly for examination. If the conditions be found satisfactory, the thread should be removed, and the operation finished in the usual way.

CASE IX.—Theresa Wagenglast, **cigarmaker, aged thirty**-nine, contracted, April 11, 1887, strangulation of a femoral **hernia of old standing,** situated on the left side. *April 15th.*—Admitted to German Hospital **with incessant** vomiting, induced mainly by the administration of calomel. Immediate **herniotomy.** A considerable portion of adherent omentum presented, and was tied **off in** several portions and removed. After this a very small knuckle of gut became visible, which showed an anæmic area corresponding to the locality of constriction. Recovery being tardy, a thread of catgut was passed through the mesentery, and the knuckle was replaced in the abdomen through the well-divided femoral ring. In the mean time the sac was excised. After the completion of this step, requiring about fifteen minutes, the gut was re-extracted for examination, and circulation was found fully re-established. The gut being replaced, the neck of the sac was closed with a purse-string suture, and was pushed well up in the femoral ring. Drainage and suture of the external wound. *April 15th.*—The drainage-tube was removed. *April 29th.*—Patient was discharged cured.

Where impending gangrene from venous engorgement is to be feared, the decision is generally more difficult than in the preceding class of cases. When immediate solving of the momentous question is impossible, the benefit of the doubt should always belong to the assumption that necrosis is to be expected. In these cases the neck of the hernial sac should be well divided to secure the best circulation possible, and the loop of gut should be so attached to the skin by a couple of sutures passed through the mesentery as to leave the questionable spots exposed to view. Thorough disinfection by wiping with sponges wrung out of Thiersch's solution, a light packing of iodoformized gauze around the neck of the knuckle, and a *moist aseptic dressing* (the gut being covered by a protective strip of rubber tissue) should be applied. If the gut decay, this will take place outside of the peritoneal cavity. Should it recover, the fact will be manifest within one or two hours after the operation. The gut should be then well disinfected, liberated by gentle manipulation from its newly-assumed position, and replaced in the abdominal cavity.

Case X illustrates the consequences of the replacement of the gut of doubtful vitality. It was the author's first herniotomy.

CASE X.—John Philip Iores, waiter, aged fifty-three. Very acute strangulation of twelve hours' standing of an old, right inguinal hernia. *October 27, 1878.*—Herniotomy in presence of Dr. L. Bopp, the family physician. Two knuckles of deeply-injected small intestine, aggregating to the length of ten inches, and a mass of dark-blue omentum were found in the sac. But, as the gut seemed to be turgid and viable, it was replaced. The omentum was pulled out, tied and cut off, and the stump was replaced. Septic symptoms set in immediately after the operation, with high fever and very great debility. *October 29th.*—Unmistakable signs of peritonitis, notably enormous meteorism, appeared. The restless patient disarranged the dressings during his tossing in bed, and, while vomiting, the adhesions of the wound gave way, and a large loop of intestine prolapsed. Necrosis of a portion of the prolapsed gut was evident. As much of it as was normal was replaced, the decayed part of the gut was incised, and fixed near the external wound. The patient died shortly afterward.

It must be added that, according to then prevailing notions (1878), the sac and its contents were washed with a strong solution of carbolic acid (5 : 100) before the gut was replaced. Superficial erosion of the intestinal peritonæum may have had its share in precipitating both gangrene and peritonitis.

*Necrosis of the vermiform appendix* was observed by the author once with fatal termination.

CASE XI.—Henrietta Bauland, aged forty-seven. Right femoral hernia of forty-eight hours' standing. *April 18, 1884.*—Herniotomy at the German Hospital. Vermiform appendix was found attached by its apex to the side of the sac; a knuckle of small intestine was embraced in the loop formed by the vermiform appendix, and then doubly incarcerated. Manipulation was very difficult, on account of the narrow space and the complicated state of things. The gut was slightly torn, but no intestinal contents escaped. Two Lembert's sutures being applied, the strangulation at the neck of the sac was relieved and the gut was liberated. The middle part of the *vermiform appendix was found necrosed*, and a ligature being applied above this part, the appendix was cut away. The gut was returned. The patient got on very well until April 25th, when perforative peritonitis developed. *April 27th.*—Patient died. No autopsy could be secured.

However desirable thoroughness and deliberation may be in herniotomy, undue prolongation of anæsthesia is an evil fraught with especial danger in cases of long-continued strangulation, on account of the cardiac debility present. When the patient's vitality has been much lowered by continuous vomiting, loss of sleep, and septic fever, even a brief anæsthesia may be sufficient to precipitate fatal collapse. Habitual users of alcohol and obese individuals are very poor subjects to endure anæsthesia in the presence of necrosis of the gut.

CASE XII.—Albert P., drayman, aged thirty-five, moderate but steady consumer of beer and whisky. Incarcerated right inguinal hernia of seventy-five hours' duration. The swelling was mistaken for acute orchitis, hernia being thought of by the family attendant only after fecal vomiting had set in. *March 19, 1887.*—Herniotomy at the German Hospital. Extensive gangrene of the small gut was found. Ether anæsthesia was very bad, the patient struggling all the while during the operation. If ether was crowded, respiration became irregular, the face pallid, and syncope threatening. Artificial anus was established, and the case was finished with all possible expedition, anæsthesia lasting altogether for thirty minutes. Deep collapse following, the patient did not rally in spite of copious hypodermic stimulation, and he died two hours after the completion of herniotomy.

It is plausible to assume that in similar cases herniotomy performed with the aid of local anæsthesia would offer better chances of success than if it be done in general ether or chloroform narcosis.

One of the eleven fatal cases died of acute septicæmia induced by diphtheritic enteritis of the strangulated knuckle of gut.

CASE XIII.—Charles Etzler, baker, aged thirty-five. Very acute strangulation, of fifty hours' standing, of an old right inguinal hernia. The patient had had no medical care until a few hours before his admission to the German Hospital, when Dr. H. Kudlich was called in. He was requested to stop the violent fecal vomiting caused by a very large dose of Rochelle salts taken in the morning of January 31, 1884. Herniotomy on the evening of the same day. The large scrotal hernia contained a good-sized portion of adherent omentum and a massive conglomerate of several knuckles of small gut, bound together by firm cicatricial adhesions of old date. Free external incision of the abdominal wall until the neck of the hernial sac was completely divided. The gut looked tolerably well preserved and was replaced; the omentum was freed by dissection, and, being tied off in several portions, was cut off. The stump being replaced, the sac was tied and cut off; then the abdominal wall was sutured by several tiers of strong catgut in physiological order. The outer wound was drained, sewed, and dressed as usual. February 1st passed off without any untoward symptom, the vomiting having ceased immediately after the operation. *February 2d.*—A severe chill with much belly-ache set in, but no meteorism appeared until February 4th, the thermometer indicating all the while 105° Fahr. The patient's condition grew steadily worse, with deep coma, jaundice, and petechial patches on the legs. *February 5th.*—The sutures gave way during a vomiting spell, and a loop of healthy-looking gut prolapsed. It was not replaced. Shortly after the patient died. Post-mortem examination revealed a slaty discoloration of the mentioned bunch of coherent gut, which, being incised, appeared to be covered on its mucous side with a large number of round and confluent whitish-gray adherent patches of membrane, which involved the intestinal wall to varying depths, some of them being visible through the peritoneal covering. No peritonitis.

In Case XIV such a combination of unfavorable conditions was encountered as would baffle every effort of the most careful surgeon. The coexistence of great general debility from chronic nephritis of old standing, with necrosis of a considerable part of the upper portion of the jejunum, is fortunately very rare.

CASE XIV.—Mary Henneberg, aged thirty-five. Strangulated femoral hernia of four days' standing. *March 28, 1889.*—Herniotomy at German Hospital. Necrosis of small intestine. Establishment of artificial anus. Escape of sulphur-colored fæces of acid odor. Collapsed condition of patient rendered idea of enterorrhaphy impracticable. *March 29th.*—Examination of scanty urine revealed the presence of much albumin and hyaline casts. Though the patient rallied from the anæsthesia, and no further local trouble developed, she did not pick up strength, which was mainly attributed to the high location of the intestinal leak. *April 4th.*—An attempt was made to anæsthetize the patient in order to perform enterorrhaphy, but she collapsed at the beginning of anæsthesia to such an extent that the idea of operating had to be abandoned. On April 15th she died of inanition. On post-mortem examination a far-gone degeneration of both kidneys was found.

CASE XV.—Chaic Zuckermann, aged forty-eight. Strangulated right inguinal hernia. Fecal vomiting of four days' standing. Bad general condition. Herniotomy at Mount Sinai Hospital. *January 14, 1890.*—Necrosis of gut and part of sac. Difficult liberation of gut. Artificial anus established. Patient died of collapse eleven hours after operation.

Eight of the successful operations for strangulation were done on inguinal (one preperitoneal, Case V), five on femoral herniæ.

```
Cured.................................................. 13 patients.
Died................................................... 11    "
                                                        —
    Total.............................................. 24
```

In dividing the strangulating band in femoral hernia, the incision should be directed inward toward Gimbernat's ligament. But, where the space is very narrow or the condition of the gut doubtful, free incision of the *fascia lata* parallel to the large vessels, and preparatory exposure of the femoral canal, would be more proper.

To incise the strangulating bands sufficiently to enable the surgeon to withdraw additional portions of gut for examination does not insure facile reposition by any means; and forcible crowding back of the congested and vulnerable intestine through an insufficiently wide orifice may lead to its rupture. Therefore, the dilatation must be very ample to permit easy reposition without the use of undue force.

As long as the sac is not closed, and communication is open with the peritoneal cavity, irrigation of the wound must stop, otherwise large portions of the lotion may find their way into the abdomen. The use of strong solutions of carbolic acid or mercuric bichloride on the prolapsed gut is not advisable and is unnecessary. As soon as the gut is replaced, the sac should be wiped clean with a disinfected sponge, and another small sponge, fastened to a thread of catgut, should be pushed into the inguinal canal to serve as a barrier to the influx of blood into the peritoneal cavity. If the patient is seen to bear anæsthesia well, inguinal herniotomy can be supplemented by the addition of the suture of the inguinal canal, as described under the heading of "Radical Operation of Hernia."

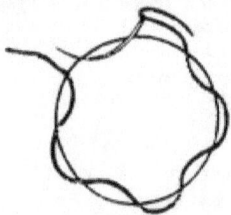

FIG. 106.—Purse-string suture, employed for occluding the neck of the hernial sac.

Should, however, collapse be present or imminent, and prolongation of anæsthesia inadvisable, a thread of strong catgut is passed through the neck of the sac (see cut) as high up as possible, assistants holding well apart the artery forceps by which the edges of the cut through the sac are secured. This suture resembles a purse-string in its workings (Fig. 106). It is tightened and knotted, and will securely occlude the peritoneal cavity. Then the external wound is well irrigated with corrosive-sublimate lotion, a drainage-tube is placed well up to the purse-string suture, and the edges of the skin are brought together with catgut stitches. The dry dressings are applied so as to cover up the scrotum and both inguinal regions, a slit being left in the middle for the penis, which should protrude from the bandages. The use of a "hip-rest" will facilitate the application of the otherwise difficult dressing. In private practice, a common hassock or footstool, wrapped in a clean towel or slipped into a clean pillow-case, will make a capital hip-rest.

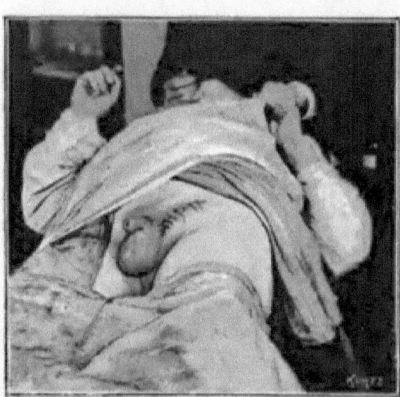

FIG. 107.—Herniotomy. Suture of external wound.

In female patients the compresses are held down by a spica bandage.

The dressings should fit snugly, especially about the edges, and should not be too scanty.

Three days after the operation the dressings should be changed, to permit withdrawal of the drainage-tube. Five or six days more will complete the essential part of the cure.

Fig. 108.—Volkmann's "hip-rest."

The patient's bowels should be moved forty-eight hours after the operation by a large enema of soap-water. Should fever set in from peritoneal irritation, a saline purge may be administered with good effect.

As long as the patient is in bed, nutrition should be simple and moderate. No patient should be permitted to go about his business before a truss can be worn with comfort. But there is no objection to his being up and about the room with a well-fitting pad and spica.

*Synopsis of successful cases hitherto not accounted for:*

CASE XVI.—Mrs. C. Reinhardt, aged fifty-four, left inguinal incarcerated hernia of three days' duration. Operation, November 15, 1882. Cured, December 11th.

CASE XVII.—Chas. Roensch, four months old, congenital incarcerated hernia. Operation in German Dispensary, January 26, 1883. Cured, February 22d.

CASE XVIII.—G. John. See history, page 24.

CASE XIX.—Fred. Hipp, mechanic, aged sixty, right external inguinal hernia. Operation at German Hospital, April 6, 1884. Cured, May 1st.

CASE XX.—Mrs. Emma T., aged forty-seven, left femoral hernia. Operation, March 25, 1887. Cured, April 10th.

CASE XXI.—Anna Brown, aged fifty, left femoral hernia. Operation at Mount Sinai Hospital in September, 1880. Discharged cured, end of October.

CASE XXII.—Martin Thorwarth, cooper, aged sixty, right inguinal hernia. Operation, February 12, 1880. Cured, March 5th.

CASE XXIII.—Adelaide K., aged forty-five, strangulated left femoral hernia. Fecal vomiting of a week's duration. Herniotomy. *June 17, 1888.*—Gut viable. Cured, June 26, 1888.

CASE XXIV.—Jacob Feldstein, errand-boy, aged fourteen, strangulated right inguinal hernia *May 10, 1889.*—Operation at Mount Sinai Hospital. Author's modification of Macewen's suture. Discharged cured, June 21, 1889.

Fig. 109.—Manner of applying dressing for wounds of scroto-inguinal region.

SPECIAL APPLICATION OF THE ASEPTIC METHOD. 133

*b.* **Radical Operation for Hernia.**—In performing herniotomy for strangulation on a patient whose general condition is good, the additional steps for radical cure may be at once carried out to great advantage. (Case XXIV.)

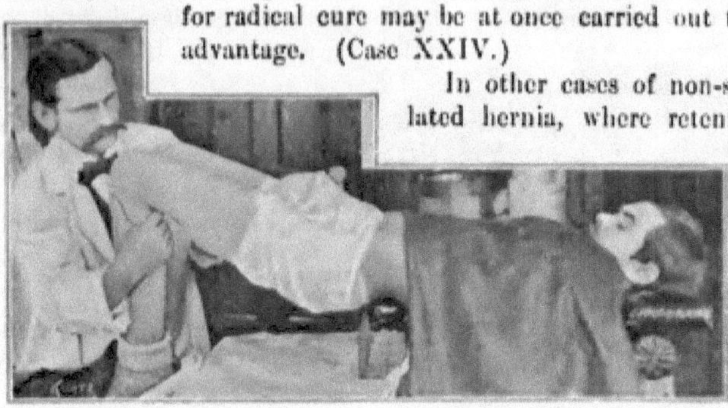

In other cases of non-strangulated hernia, where retention by

FIG. 110.—Herniotomy. Patient on "hip-rest," with completed dressing. Lateral view.

truss of a very large scrotal hernia is impracticable on account of wide distention of the inguinal canal, or where adhesions of the prolapsed gut or omentum to the sac render reduction impossible and make attempts at wearing a truss a torture to the patient, radical operation is proper and justified. Due observance of the rules of asepsis makes this operation very safe as far as the production of purulent peritonitis is concerned. Still, some danger of septic infection can never be excluded with positive certainty. Therefore, bloody radical operation should be discouraged for a hernia that can be retained by a properly constructed truss.

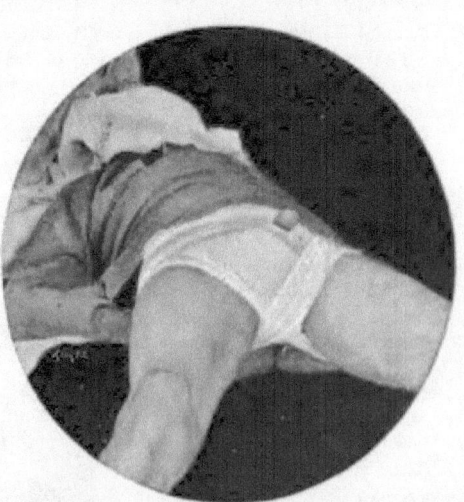

The author has, in his first twelve cases, followed Czerny's directions in performing radical operation of hernia, the several steps of which are as follows :

FIG. 111.—Completed dressing of scroto-inguinal region. Anterior view.

After due preparation by a laxative, preferably castor-oil, the patient's pubic region and scrotum, especially on the side of the rupture, are shaved, and cleansed the day before the operation with brush, soap, and hot water, and are wrapped up in a clean towel dipped in a three-per-cent solution of carbolic acid. This wet compress is again covered with a suitable piece of oiled silk or rubber tissue, and fastened on with a T-bandage. On the day of the operation the patient is placed on the table and anæsthetized, a full and good anæsthesia being especially desirable. After repeated disinfection, the hernial sac is exposed by a sufficiently long incision, in which all bleeding vessels are to be secured by ligature. The upper angle of the wound should be located well above the upper margin of the inguinal ring so as to permit easy manipulation.

The sac is incised, and its edges are taken up by a number of artery forceps, which being held apart, an excellent view of the contents of the hernia can be had. Adhesions of the omentum to the sac will be found the most common cause of the irreducibility, the gut being rarely adherent. The author has observed only one case of old hernia in which adhesions of the gut were present (case Mau). The favorite place of omental adhesions is the anterior portion of the neck of the sac.

As soon as the sac is open, the use of the irrigator has to be discontinued, to prevent entrance of large quantities of irrigating fluid into the peritoneal cavity. The lotions used for rinsing hands, sponges, and instruments ought to be very mild to prevent even superficial corrosion of the peritonæum. The author has generally used Thiersch's boro-salicylic solution.

A suitable sponge, fastened to a stout piece of silk or catgut, is pushed well up into the inguinal canal to prevent the entrance of blood into the abdomen. Care must be taken not to select a too brittle sponge, as it may happen that, on removing it, some portion of it may become detached and remain in the belly.

The sac must be split open to within a quarter of an inch of the external inguinal ring, and the adherent omentum must be detached from the sac by preparation. As soon as the distal attachments of the omentum are severed, it is withdrawn a little farther from the inguinal canal, and, being deligated in small portions with reliable catgut, it is cut away by the knife, or, preferably, the thermo-cautery. After this the sac is wiped out clean, and, the sponge being withdrawn from the inguinal canal, the stump of the omentum is replaced in the abdominal cavity.

In dissecting up adherent gut, great caution must be observed not to injure it. Where the adhesions are very close and extensive, it would be better to excise the attached portion of the sac with the gut, and replace them together in the peritonæum.

CASE I.—Henry Mau, shoemaker, aged sixty-two. Very large scrotal hernia, containing adherent gut. The inguinal ring was so dilated that the tips of three fingers could easily be slipped within the abdominal cavity. *February 23, 1886.*—Radical operation at the German Hospital. Ether anæsthesia produced violent retching and coughing, so that the irresistible escape of gut from the wound rendered operation impossible. Chloroform being administered, quiet anæsthesia was achieved. The adherent thick gut was dissected away, together with the adhering portions of the sac, and was returned to the abdominal cavity. The remnant of the sac was separated, closed at its neck with a purse-string suture, and was cut away. The wide gap of the inguinal ring was closed with eight sutures of stout catgut, and the external wound was drained and sewed up. Uninterrupted recovery. *March 25th.*—The patient was discharged cured, with instructions to wear a light truss. In November, 1886, he presented himself with a relapse. His truss had been broken, and he neglected to have it repaired. In a fit of violent coughing the rupture reappeared.

The contents of the sac being disposed of, *excision of the sac* is the next thing to be done.

In most cases this can be readily accomplished by stripping up the sac from the surrounding tissues with the fingers, the scissors being only occasionally needed to sever resisting bands, which generally contain vessels requiring ligature. In some instances, however, especially in cases of congenital hernia, the separation of the sac is not easy. The sac proper is not well defined, and in some localities consists of nothing but the bare peritonæum. Hence it is difficult to get it out uninjured and in one piece. Another difficulty is presented by the close relations of the cord and its vessels to the sac. The greatest care must be taken to properly recognize them, as otherwise they may be accidentally damaged.

CASE II.—William Litzebauer, baker, aged twenty-seven. Left inguinal irreducible hernia. *February 5, 1886.*—Radical operation at the German Hospital. Liberation of adherent omentum, which was deligated and cut away. In dissecting up the sac, the *vas deferens* was cut across. A short piece of stout catgut was introduced into the patent ends of its lumen, and the duct was united by four fine catgut sutures passed through its involucrum. The sac being removed, the external ring was closed by six stout catgut sutures. The external wound was drained and sewed. *February 7th.*—Purulent urethral discharge was noted; no fever. *February 15th.*—Change of dressings. Wound healed by adhesion, left testicle somewhat swollen and painful. Tube was removed. *February 27th.*—Urethral discharge disappeared, testicle notably decreased in size. *March 10th.*—Discharged cured, with slightly enlarged testis.

*Congenital irreducible hernia* is comparatively frequent. Four of the twelve cases operated on by the author belonged to this class. One was complicated with *undescended testicle.*

In two of these cases *castration* had to be performed along with the radical operation.

CASE III.—August B., painter, aged twenty-four. *August 23, 1883.*—Radical operation at the German Hospital. The omentum was found adherent to the left testicle, and contained near its adhesion to this organ a hard, pigmented tumor of the size of a walnut. The sac and the tunica propria of the testis were dotted with a large number of pigmented spots. Therefore the omentum, sac, and testicle were all removed. Closure of inguinal ring by catgut sutures. Treatment of external wound as usual. *September 20th.*—Discharged cured.

CASE IV.—George W., cattle-raiser, aged thirty-six. Direct inguinal hernia of left side, containing the undescended testicle. *August 24, 1885.*—Radical operation at Mount Sinai Hospital. The attached omentum was freed and removed. The atrophic testicle was also taken away. Suture as usual. *September 4th.*—Patient strained at stool, whereupon the external wound reopened, but subsequently healed by granulation. *October 2d.*—Patient was discharged cured.

In a third case of congenital hernia, in an infant, eclamptic attacks caused repeated protrusion of the intestine, that could not be reduced without the employment of anæsthetics.

CASE V.—Carl Schlichter, eight months old. *April 18, 1886.*—Prolapse of the gut during a convulsive seizure. Dr. Meltzer, the family attendant, administered chloroform, whereupon the author reduced the gut with some difficulty. The accident had occurred the fourth time in spite of a truss. Radical operation was at once performed. *May 5th.*—Patient discharged cured.

CASE VI.—Franz Faulhaber, laborer, aged twenty-two. Left congenital omental hernia. *July 28, 1885.*—Radical operation at the German Hospital. Omentum adhering to sac treated as usual. Sac was cut away below from its reflexion upon the testicle, and above close beneath the purse-string suture. Treatment of inguinal ring and external wound as usual. Uninterrupted cure. *September 1st.*—Patient was discharged cured.

The closure of the sac is to be done by the purse-string suture, depicted by Fig. 106. Rather stout catgut must be used for this, to withstand the powerful tension required for closing the circular suture. The sac is cut away below the knot, and any bleeding vessels must be separately deligated. The stump is pushed well up within the internal abdominal ring.

*In applying Czerny's suture of the inguinal ring,* the left index-finger is intruded as far as possible, its volar aspect being directed downward and inward to protect the cord, which should be kept near the inferior and inner angle of the slit of the inguinal aperture. A strongly curved needle, armed with stout catgut, is passed first through the conjoined tendon, then through Poupart's ligament, all subcutaneously, and the ends of the thread are secured in a pair of artery forceps and reflected upon the abdomen, where they are received by an assistant. This first suture should be placed as high up the inguinal ring as possible. At intervals of a third of an inch from four to seven stitches are applied in the manner indicated; then they are tied firmly by surgeon's knots in the reverse order. A small-sized drainage-tube is placed in the wound, and the integument is united by finer catgut sutures, the tube being brought out through the lower angle of the incision. An antiseptic dressing is next applied in the manner shown by Figs. 108, 109, 110, and 111.

The first change of dressings should be made on the third day, when the tube is also removed. As soon as the wound is completely closed, the patient is permitted to get up with a spica bandage or truss.

The patients should be directed to continue the use of a light truss, as this is the only reliable security against recurrence.

In one case a fibromatous node in the adherent omentum was the chief source of pain complained of by the patient.

CASE VII.—Jacob Christman, laborer, aged thirty-nine. *August 15, 1885.*—Radical operation at the German Hospital. A hard, irregular node was occupying the middle of the prolapsed and adherent omentum. It was removed with the same. Discharged cured, September 19th. The node was fibromatous in character.

In another case a subserous fibro-lipoma was located outside of, and was closely connected with, the neck of the sac.

CASE VIII.—Carl Dille, laborer, aged thirty. Subserous fibro-lipoma and left adherent omental hernia. *March 12, 1887.*—Radical operation at the German Hospital. Removal of omentum and sac, together with neoplasm. Sutures as usual. *April 9th.*—Discharged cured.

The remaining four cases presented nothing unusual, and all recovered without mishap :

CASE IX.—Charles Niemann, locksmith, aged thirty. Adherent left omental hernia. *February 19, 1887.*—Radical operation at the German Hospital. *March 12th.*—Discharged cured.

CASE X.—Martin Hussmann, baker, aged twenty-five. Adherent right omental hernia. *March 3, 1887.*—Radical operation at the German Hospital. *April 7th.*—Discharged cured.

CASE XI.—Henry Mehle, barber, aged twenty-five. Adherent right omental hernia. *January 8, 1886.*—Radical operation at the German Hospital. *February 12th.*—Discharged cured.

CASE XII.—Mr. M. D., merchant, aged thirty-nine. Very massive, growing, adherent omental hernia of the right side. *May 26, 1887.*—Radical operation at Mount Sinai Hospital. *June 16th.*—Patient discharged cured.

*Author's Modification of Macewen's Operation.*—To test its value, a modification of Macewen's method of curing hernia was employed in fifteen consecutive instances. Some of the cases were very grave, either on account of the large size of the hernia, or because of the presence of unusual complications. A serious mishap occurred only once, and consisted in the sloughing of the plug formed of the sac, resulting in the relapse of the disorder.

The initial steps of this procedure are identical with those in Czerny's method. The deviation in the technique commences after the stripping up of the sac is accomplished. In addition to the detachment of the sac proper, the parietal peritonæum is also detached in the shape of a halo just inside of the internal abdominal ring for the distance of about three quarters of an inch, forming a pocket for the reception of the plug to be directly described. The sac is not deligated and cut off as in Czerny's procedure, but a stout double catgut thread is stitched to its distal extremity, and "passed in a proximal direction several times through the sac, so that, when pulled upon, the sac becomes folded upon itself like a curtain." After this the double thread is divided, each end being separately threaded in a stout curved needle. One of these threads is passed through the abdominal wall, just above and half an inch to the inside of the internal abdominal ring, while the other is carried through Poupart's ligament below and to the outside of the same aperture. None of the stitches should include the skin. When the two threads are pulled upon, they will first fix the plug formed by the sac inside of the abdominal ring, and secondly, by being tied in a knot, will very effectually approximate the edges of the inguinal hiatus. After this the subsequent stitches are passed through the conjoint tendon and Poupart's ligament as indicated by Macewen, with the difference, however, that they are simple interrupted sutures, and less complicated than his process. When all the stitches are *in situ*, they are closed one after the other, beginning from the top. No drainage-tube was employed in ten cases; in three, drainage was dispensed with after the first change of dressings; in two cases bland suppuration necessitated the use of drainage-tubes for a longer period of time.

CASE I.—*Enormous Irreducible Inguinal Hernia* of five years' standing. Anna Finkelstein, aged thirty-four. The sac reaching down to the knee-joint. Operation January 24, 1888, at Mount Sinai Hospital. The sac contained almost all the intestines and the left ovary and tube. Two thirds of the sac were removed as unnecessary, Macewen's plug being formed of the remnant. Discharged cured, March 12, 1888.

CASE II.—*Irreducible, Very Large Inguinal Hernia; Bottom of Sac containing Abscess with Fish-bone.*—Lazar Menasse, tailor, aged forty-two. Operation November 12, 1888, at Mount Sinai Hospital. In the bottom of sac an abscess containing a fish-bone was opened, the intestine forming one of the walls of the abscess. Disinfection and replacement of gut. In dissecting sac the vas deferens was cut: *castration*. Discharged cured, December 27, 1888. ("New York Medical Journal," November 22, 1888.)

CASE III.—*Very Large Non-retainable Inguinal Hernia, relapsed after Radical Operation by another Surgeon.*—Otto Pahlmann, waiter, aged forty. Relapse of hernia nine months after first operation (probably Banks's). *Second operation* at Mount Sinai Hospital, January 16, 1890. *Castration* was done in order to do away with cord. Patient contracted severe gastro-enteritis two weeks after the operation, hence his discharge was delayed till March 22, 1890.

CASE IV.—*Very Large, Non-retainable Inguinal Hernia.*—Operation January 18, 1889. Discharged cured, February 17, 1889.

CASE V.—*Very Large, Non-retainable Inguinal Hernia of Left Side; Double Encysted Hydrocele of Cord on Right Side.*—Operation at Mount Sinai Hospital, December 9, 1889. Hydrocele sacs were incised and drained at the same time. Discharged cured, February 1, 1890.

CASE VI.—*Large, Non-retainable Inguinal Hernia.*—Franz Wahl, baker, aged thirty-four. Operation February 21, 1889, at German Hospital. Discharged cured, March 1, 1889.

CASE VII.—*Large, Non-retainable Inguinal Hernia.*—Henry Kattenhorn, grocer, aged twenty-six. Operation October 21, 1887, at German Hospital. Discharged cured, November 17, 1887.

CASE VIII.—*Irreducible Inguinal Hernia.*—Franz Bosch, tinsmith, aged twenty-three. Operation December 16, 1887, at Mount Sinai Hospital. Adherent omentum excised. Discharged cured, December 31, 1887.

CASE IX.—*Irreducible Inguinal Hernia.*—Mayer Menaffe, tailor, aged forty-two. Operation August 27, 1887, at Mount Sinai Hospital. Omentum excised. Discharged cured, October 4, 1887.

CASE X.—*Irreducible Inguinal Hernia.*—Herman Neugroesche, cigar-maker, aged twenty-three. Operation March 18, 1889, at Mount Sinai Hospital. Omentum excised. Discharged cured, April 7, 1889.

CASE XI.—*Irreducible Inguinal Hernia.*—Abraham Blum, peddler, aged seventeen. Operation November 16, 1888, at Mount Sinai Hospital. Omentum excised. Discharged cured, December 27, 1888.

CASE XII.—*Irreducible Inguinal Hernia.*—Gustave Rinknitz, laborer, aged twenty-four. Operation October 13, 1888, at German Hospital. Omentum excised. Discharged cured, November 5, 1888.

CASE XIII.—*Umbilical, Irreducible Hernia.*—Annie Smith, housewife, aged forty-three. Operation May 14, 1888, at German Hospital. Omentum excised, sac treated according to Macewen. Umbilical ring pared, then sutured with six silk-worm gut button sutures. Discharged cured, June 17, 1888.

CASE XIV.—*Reducible Inguinal Hernia of Moderate Size; Sloughing of Sac; Relapse of Hernia.*—Gustave Sprenger, waiter, aged thirty-one. Operation August 13, 1888, at Mount Sinai Hospital. Local reaction with moderate fever, wound reopened. Sloughed sac came away with moderate suppuration. Discharged cured, September 15, 1888. Relapse in May, 1889. Sloughing probably caused by too tight suturing. Had the patient worn a truss, relapse would have been prevented.

CASE XV.—*Reducible Inguinal Hernia of Recent Origin; Very Slender Sac; Relapse.*—Phil. Meagher, student, aged seventeen. Operation July 20, 1889. Sac was found to be very thin and slender, forming a rather inadequate plug. Cured, August 15, 1899. As patient wore no truss, relapse was noticed in April, 1890.

To secure the patient against the danger of a relapse, the wearing of a truss seems to be very advisable.

It has been urged, notably by McBurney, Weir, and Abbe, of New York, that, after radical operation, healing of the external wound by granulation is preferable to primary union, on account of the larger mass of cicatricial matter resulting from the granulating process. To the author this advantage seems of doubtful, certainly of only passing, value, as the massive cicatrix, first hard and resisting, must in the course of time become atrophied, soft, and yielding, and will not be able to withstand for a long time the constant impact of the intra-abdominal pressure. The analogy of this fact with the experiences gathered about the wounds resulting from laparotomy can not be gainsaid. These, when the healing of the abdominal incision was not by primary union, and the cicatrix produced by a long process of granulation is very wide and massive, regularly terminate in ventral hernia.

### 3. *Laparotomy.*

*a.* Exploratory Incision.—Although the aseptic method has very materially reduced the dangers of exploratory laparotomy, its wanton and unnecessary practice must be deprecated on several grounds. *First* of all, no surgeon is absolutely secure in his practice against accidental and unexpected, often unexplained, wound infection. *Secondly*, the dangers of anæsthesia, and of conditions indirectly caused by it, as nephritis, pneumonia, thrombosis, and embolism, are ever present, and usually surprise the surgeon when least expected.

Exploratory incision is only justified where, in the presence of a disorder threatening life, all known means for establishing a diagnosis have been exhausted without positive result, or where the extent and exact relations of a mechanical disturbance can not be estimated without ocular inspection and digital examination.

Due observance of the rules against infection will exclude suppurative peritonitis with great certainty. The detail of the procedure is treated in the chapter on abdominal tumors.

CASE I.—Fred. Kahn, aged eleven. Intestinal obstruction of seven days' duration. Fecal vomiting, very great tympanites, and threatening exhaustion. No fever. *June 27, 1882.*—Laparotomy under ether. In the right iliac fossa an immovable convolution of small gut could be felt. The incision was sufficiently extended to enable the author to inspect the locality. It was found that the tip of the vermiform appendix was attached to the parietal peritonæum. A large loop of the ileum had slipped through the hiatus thus formed, and was there incarcerated. The vermiform appendix was cut between two ligatures, and the loop of intestine became free. Reduction of the enormously distended intestines was impossible. At the suggestion of Dr. A. Seibert, an enema was administered, and it brought away a large quantity of gas, whereupon the somewhat collapsed gut could be replaced, and the abdominal incision closed. The operation lasted thirty minutes. Deep collapse followed, in which the patient died twelve hours after the operation.

Very likely an early operation would have been followed by a better result.

CASE II.—Mary Block, aged twenty-seven. Symptoms of subacute intestinal obstruction, with fever and vomiting of all ingesta. Moderate distention of abdomen. Tumor in right loin, which is painless on pressure. Trouble of a week's standing. *January 8, 1889.*—Laparotomy reveals a convolution of much reddened, distended, and œdematous gut in the right iliac fossa, surrounded by pale and normal-looking empty intestine. The hyperæmic portion of gut proved to consist of about two feet of the lowest part of the ileum, twisted on its mesentery, the vessels of which showed marked venous stagnation. The twisted coils of gut were easily restored to their normal position, as no adhesions had as yet formed. The tumor in the right loin was found to be a somewhat enlarged movable kidney. Closure of wound. Uninterrupted recovery. Patient discharged cured, February 26th.

CASE III.—Philippine Pahler, aged thirty-five. Pyloric cancer of stomach. *February 18, 1886.*—Probatory abdominal incision at the German Hospital, with a view to possible resection of the pylorus. The extension of the disease to the retro-peritoneal glands, the pancreas, and omentum put the contemplated step out of question, wherefore the incision was closed. *March 11th.*—Patient discharged with firmly healed wound.

CASE IV.—Albert Schroeder, painter, aged thirty. Large retro-peritoneal tumor located behind hepatic flexure of colon, causing intestinal stenosis. *August 8, 1882.*—Probatory incision at the German Hospital established the fact of the inoperability of the swelling—a sarcoma of the mesocolic glands. Closure of wound. *August 9th.*—Patient died in collapse.

*b.* **Abdominal Tumors:**

(*a*) GENERAL REMARKS.—To avoid infection from without, it is first necessary to carefully shave the belly and pubic region of the patient, then to scrub it well with soap and brush, and finally to rub it off with a 1 : 1,000 solution of corrosive sublimate. The navel ought to be very thoroughly cleansed of deposits of dirt. The scrupulous cleansing and disinfection of hands, instruments, sponges, and other utensils should render unnecessary the application to the peritoneal cavity of disinfectant lotions, which, by their corrosive properties, may produce mischief.

The usual measures adopted for protecting the body of the patient against wetting and undue cooling off, as the wrapping up of the extremities in flannels, and the spreading of rubber cloths over the trunk and lower limbs, leaving exposed nothing but the abdomen, demand special care and attention. *Excessive loss of body heat is a great factor in determining collapse, and should be guarded against most sedulously.*

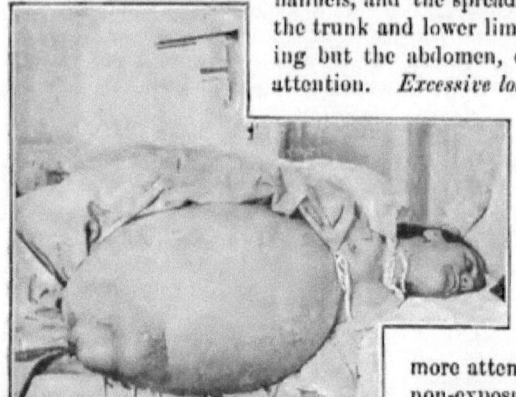

FIG. 112.—Ascites and ovarian tumor. Patient ready for operation in the lateral posture. Case of Dr. W. L. Estes, of Bethlehem, Pa.

The principle of *non-exposure* applies equally to the contents of the abdominal cavity. The greater the incision, the more attention must be paid to the non-exposure of the intestines. *Hot, flat sponges or towels should hide from view everything except the very spot subjected to surgical manipulation.*

The use of the *spray apparatus* during abdominal operations is harmless, but unnecessary. Certainly it forms a very objectionable feature of the original Listerian method, and has been abandoned in general as well as abdominal surgery by most operators. The author has not used the spray apparatus since 1881.

*The control of hæmorrhage* is of the utmost importance to the success of abdominal operations. This and the former requirements can be best fulfilled by an intelligent observance of the rules laid down in the paragraphs on the technique of surgical dissection and the removal of tumors. The principles

FIG. 113.—Protection of the intestines by flat sponges arranged about the tumor.

there explained remain unchanged, their application to abdominal tumors only being somewhat modified by the peculiarities of the locality.

*An ample incision is the first condition of the safe removal of an abdominal tumor.* When a unilocular, non-adherent cyst is to be exsected, a small incision will be ample, because the cyst, however large, can be emptied by tapping, and is thus reduced to the elongated proportions of a flat band, which can be extracted through the small incision without much force until the pedicle comes in view.

Multilocular cysts that can not be emptied readily, or solid tumors, or growths with many adhesions, must be freely exposed, to enable the sur-

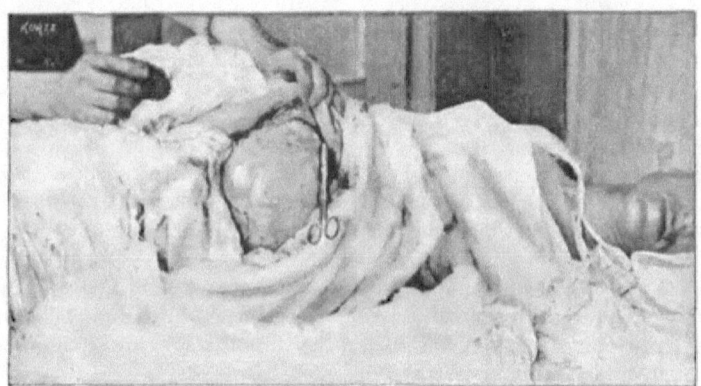

FIG. 114.—Protection of the intestines in ovariotomy by hot towels.

geon *to see* what is to be done. Accidental laceration of the gut, bladder, or large veins will not easily occur while the adhesions binding the tumor to these organs are exposed to view.

Disregard of this plain and rational rule is the cause of many an accident and mishap that might be easily avoided otherwise.

NOTE.—However important the incision and final suture of the abdominal walls may be, it must not be forgotten that they do not represent the critical part of most abdominal operations. The abdominal incision, being a preliminary measure, should not occupy too much time. Of course, it must be done *lege artis*, BUT WITH EXPEDITION. Bleeding vessels need not be tied here, as the pressure of the hemostatic forceps, exerted for ten or fifteen minutes, will effectually arrest hæmorrhage. Here, as elsewhere, cutting between two forceps will be more expeditious and safer, than the use of the grooved director.

The skillful and unstinted use of mass ligatures by means of Thiersch's spindle apparatus will render the dissection even of extensively adherent abdominal tumors remarkably bloodless and safe. Strong catgut is preferable to silk, as the latter is known to have been the cause of suppuration in a good many cases, although the silk was prepared in a seemingly proper fashion. Extensive masses of tissue, especially if their shape approaches that of a membrane, should not be included in a single ligature, as they are very apt to slip at the edges. It is safer to divide them into a number of smaller portions which should be separately tied. This rule applies to the omentum especially.

Adhesions or pedicles of a more cylindrical shape can be safely tied in one mass without risking the slipping of the ligature. Every mass should be included in two ligatures, between which it can be severed with the knife or, better, the thermo-cautery.

Transfixion of pedicles with a sharp Peaslee's needle is not advisable, as large veins passing into the mass may thus be cut open and cause troublesome hæmorrhage from a point not included in the ligature. It is better to use a blunt instrument, such as Thiersch's spindle, or a dressing or artery forceps, which will pass through any pedicle easily without injuring the vessels.

Where the adhesion or pedicle is too short, and the tumor too large, to admit of easy manipulation under the guidance of the eye, the use of a temporary elastic ligature, with or without preliminary transfixion to prevent slipping, will be found a welcome expedient. To this, a rather stout, solid band of (*not rotten*) pure gum-elastic, and one or more round probe-pointed steel needles are necessary. The pedicle is first transfixed singly or crucially, then the rubber band is thrown around the needles beyond the place of transfixion. The ends of the tightened rubber are crossed and secured at the crossing by a stout pedicle-clamp. After this the tumor can be cut away, and the pedicle, becoming more accessible, can be divided and tied off with catgut in several portions. As soon as this is done the clamp is loosened, the rubber is removed, and the tied-off masses are trimmed and seared with the actual cautery.

*Close adhesions of the gut require special care.* Recent adhesions are easily separated by blunt preparation, but cause a good deal of oozing. Much wiping and sponging of the oozing points is apt to prolong hæmorrhage, for reasons explained elsewhere. It is better to cover these points with a flat sponge, and to let them alone till hæmorrhage ceases spontaneously. The blood that found its way into the abdomen must be sponged out at the final toilet. *Old adhesions of the intestine* are very dense, and efforts at their blunt separation may easily lead to injury of the gut. Dissection by the scalpel, the line of section being well away from the intestine, will be found the most expeditious mode of proceeding. Spurting vessels must be tied, and as soon as the adhesion becomes less close and the formation of masses by blunt separation possible, mass ligatures should be applied.

Forcible blunt preparation in the vicinity of large veins, more especially of the large plexus regularly encountered in the bottom of the small pelvis near the uterus and its adnexa, is hazardous, on account of the hæmorrhage often caused by laceration of the delicate walls of these vessels. Careful isolation and double deligation, with subsequent cutting between the ligatures, are the best safeguard against dangerous hæmorrhage.

Blunt dissection, preferably by the tips of the fingers, is, however, eminently proper where the peritonæum is to be stripped up from underlying tissues. It is, in fact, the only safe way of separating tumors that are located between the folds of the broad ligament, in the mesentery, or in any portion of the retro-peritoneal space.

Exploratory puncture and aspiration of exposed abdominal cysts of unknown contents with a fine, hollow needle is very advisable, as the exact knowledge of the nature of the cystic contents may materially modify subsequent steps of the operation.

If the cystic fluid be bland, its escape into the peritoneal cavity does not signify much, provided that careful cleansing be employed before the closure of the wound. But when the cyst contains purulent or fetid serum, accidental soiling of the peritonæum by it may effectually destroy all chances of recovery.

Whenever puncture of an exposed tumor is determined on, whether by a small or large-sized instrument, good care must be taken to prevent, during and after the act, the escape of cystic fluid through the puncture-hole into the abdominal cavity. To do this it is necessary to surround the needle or trocar with a number of flat sponges laid on the tumor. As soon as the piston is withdrawn the nature of the fluids appearing in the barrel of the syringe will become manifest. If it be clear and limpid, no further precaution need be taken. Should the fluid appear to be turbid, or manifestly purulent, the barrel should be emptied and refilled and emptied again, until the tension of the sac becomes so far reduced, that its transfixed portion may be raised in a fold and secured by a large clamp. The sponges used for this step of the operation should be at once discarded.

To prevent laceration of the sac or capsule, the utmost gentleness and care should be practiced in handling the tumor. The use of sharp retractors and vulsellum forceps, or forcible traction with or without blunt force of any kind, are extremely ill-advised. Not only may the sac be torn, but large veins spread out over the surface of the tumor may be injured, and give rise to uncontrollable hæmorrhage. The aperture of a torn vein can not be easily occluded by any kinds of artery-clamp, first, because of its irregular shape and extension, and principally because the tension of the capsule of a solid tumor precludes the formation of a fold that could be conveniently grasped.

NOTE.—The author recalls an instance witnessed by him where, during the removal of a large uterine growth through an inadequate incision, sharp retractors were used in forcibly developing the mass from the abdominal cavity. Several large veins being torn, profuse hæmorrhage set in. The incision was somewhat, but still insufficiently, enlarged, and, more force being applied, the tumor was finally brought out of the abdomen. But very soon it became evident that, in consequence of the forcible manipulation, the transverse colon, which was closely adherent to the posterior aspect of the tumor, had been extensively torn. Enterorrhaphy did not save the patient's life, which was forfeited by the injudicious management induced by superstitious fear of a "large" abdominal incision.

The tenet of making small incisions for the removal of abdominal tumors had its origin in the justified disinclination to expose a large peritoneal surface to the contaminating and refrigerating effect of the atmospheric air. And unnecessarily long incisions are certainly to be avoided. But the surgeon's discretion must decide the question of the size of the incision, the principle of *safe dissection under the guidance of the eye* being herein of the first importance.

Undue cooling off of the peritonæum is a very undesirable thing, on account of the collapse it may induce; therefore, all portions of the abdominal organs that are not actually under dissection should be carefully covered up by large flat sponges or clean towels wrung out of hot Thiersch's solution.

NOTE.—To always have a sufficient supply of warm sponges and towels, the following arrangement will be found convenient: A tin pan or basin, containing the sponges or towels immersed in Thiersch's solution, is rested on the tops of two clean bricks stood on edge. A blazing alcohol-lamp is placed between the bricks and underneath the vessel, which, being covered with another pan, will preserve unchanged the temperature of its contents. For larger operations, three or four similarly prepared pans can be conveniently arranged on a separate table.

Whenever a stout adhesion or a pedicle is deligated and cut through, it should be dropped back into its natural position, where it should be inspected for a short while to see whether hæmorrhage is thoroughly controlled by the ligature. Oozing points should be touched with the thermocautery, but care must be taken not to go too near the ligature, for fear of burning it.

Oozing points located on the gut should never be touched with the thermo-cautery.

It is best not to tap at all dermoid cysts or tumors containing clearly septic fluid, as the integrity of the cyst-wall is the only guarantee of preventing contamination of the abdominal cavity by cystic fluids. Rather increase the external incision, and remove the tumor intact.

The relations of the bladder to the tumor should be carefully considered. *Greig Smith advises not to empty the bladder before operation,* and it is undeniable that a full bladder can not be well overlooked or injured. Injury to an empty and collapsed bladder, on the other hand, has repeatedly occurred in the presence of abnormal adhesions of the organ to the tumor. To further ascertain the extent of adhesions of the bladder, the introduction and manipulation of a solid male urethral sound will be found very useful.

NOTE.—Catheterism should be done, if possible, by a person not employed about the wound, or, if this be not feasible, careful cleansing and disinfection of the hands should follow it.

After the removal of the tumor, the toilet or cleansing of the abdominal cavity has to be attended to. Sponges attached to long handles are very convenient for this purpose. With them first the lumbar, then the vesico-uterine recesses, finally the utero-rectal or Douglas's pouch, are to be thoroughly cleansed and dried.

In the presence of large denuded surfaces lacking peritoneal investment, a glass or hard-rubber drainage-tube is to be inserted into the bottom of the small pelvis. It can be brought out through a counter-opening made into the vagina from Douglas's pouch, or through the lower angle of the abdominal incision.

In the former case, the external end of the tube projecting into the vagina or in the vulva must be wrapped in a packing of iodoformized gauze, which ought to be changed whenever it gets saturated. When the tube is brought out through the abdominal incision, its outer end must be so dressed as to be easily accessible. Every hour the serum collecting in its bottom should be exhausted with a pad of absorbent borated cotton fixed to a handle, or with a long-nozzled syringe. In the intervals the tube should be loosely filled with a strip of iodoform gauze. As the serum diminishes, this process is gone through with at longer intervals. As soon as the tube remains dry for several hours, generally about the third day, it can be withdrawn.

NOTE.—Miculicz has successfully substituted for the drainage-tube a loose packing and fillet of iodoformized gauze, brought out through an angle of the wound. The exsiccation of the secretions by this arrangement is certainly very effective. The fillet should be removed on the third or fourth day.

*The closure of the abdominal wound should be done as rapidly* as thoroughness will permit, *simplicity and solidity of the suture* being the main desiderata.

A Peaslee's needle is thrust on one side through the entire thickness of the abdominal wall, including the peritonæum, and is brought out in a similar manner on the other. The points of entrance and emergence should be at least two inches from the edges of the wound. A piece of well-disinfected silver wire or stout silk-worm gut, armed with a quill, or a leaden button and shot, is threaded through the eye of the needle. This is then withdrawn, bringing out the end of the thread from one side of the

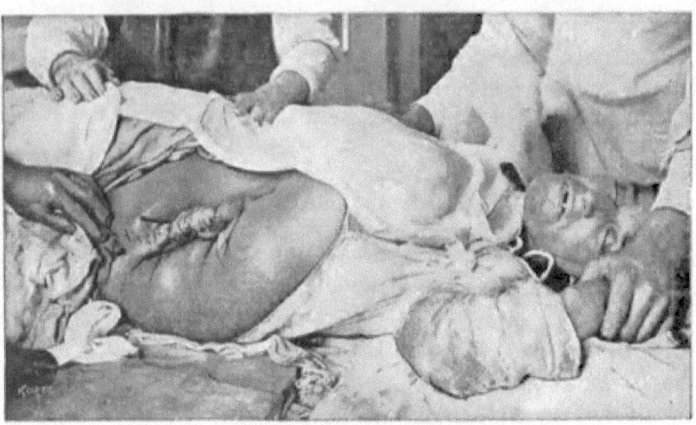

FIG. 115.—Completed quilled suture of abdominal incision.

wound to the other, where it is temporarily secured by an artery forceps. Three, four, or more retentive sutures of this kind are passed at intervals of about an inch, until the entire length of the wound is covered by them.

NOTE.—While the stitches are being passed, a flat sponge should be kept spread over the intestines to receive the blood escaping from the stitch-holes.

If the patient's condition be good, the peritonæum may be separately united by a row of catgut sutures placed between the silver or silk-worm gut stitches. But this is not essential.

After the withdrawal of the flat sponge, and a final cleansing of the peritonæum by sponges fixed to long handles, a quill is applied to the unarmed end of the thread, and is tightened until the edges of the incision are raised in the shape of a low ridge. Or, if lead buttons are to be used, one of these is slipped on the thread with a perforated shot, the thread is tightened, and the shot is pinched. After this, a sufficient number of exact "sutures of coaptation," made of fine catgut, secure the edges of the incision. (Figs. 115 and 116).

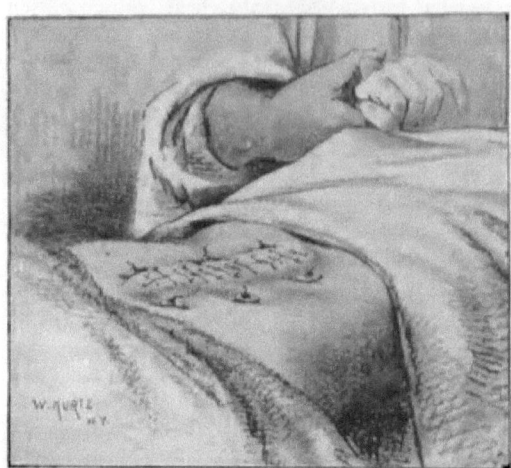

FIG. 116.—Completed plate and shot suture of abdominal wounds.

The dressings consist of a few strips of iodoform-gauze, and an ample compress of sublimated gauze over it, all snugly fastened by several strips of adhesive plaster and a broad flannel or gauze bandage.

On from the eighth to the tenth day the dressings are changed, and the retentive sutures are removed; but the bandage must be worn for some time to serve as a support to the fresh cicatrix.

(*b*) SPECIAL OBSERVATIONS:

*a. Ovarian Tumors.* — Probatory puncture of an abdominal tumor through the walls of the belly is not an indifferent matter. If the tumor be cystic, and its wall very tense, escape of a limited quantity of cystic contents is unavoidable. Bland and very thin contents may escape in large quantities without causing irritation. A large number of cases are on record in which probatory puncture of cysts of the broad ligament was followed by cure.

CASE.—Mrs. Francisca N., liquor-dealer's wife, aged thirty-four, was tapped, August 31, 1877, for a large abdominal cyst. About a gallon of fluid, characteristic of a cyst of the broad ligament, was removed, but a considerable quantity was left behind. In a short time the flabby, fluctuating swelling disappeared entirely, and the woman remained free from any further trouble.

Escape of minute portions of purulent cyst-fluid is apt to cause circumscribed peritonitis, resulting in more or less extensive adhesions. Larger quantities of septic matter, that find their way into the peritoneal cavity, may produce fatal purulent peritonitis.

The preparations, with a view to the aseptic performance of exploratory or evacuating puncture, must be very thorough, as the use of an unclean needle or trocar may be the source of peritonitis or suppuration of the sac. The hollow needle or trocar to be used must be sterilized either by boiling for an hour, or by incandescence in the alcohol flame.

When an exposed cyst is to be tapped or emptied by incision, the patient should be turned over on her side. An assistant should prevent the escape of gut; another one should surround the place of tapping with a circle of sponges to receive fluid that may escape alongside of the instrument. Tait's trocar is, on account of its simplicity, the best one of all instruments devised for evacuating cysts.

As soon as the cyst begins to collapse, its folds should be taken up with large clamps. The empty cyst is then withdrawn to the pedicle, which is tied in one or more portions and cut off.

CASE I.—Mrs. Dorothy Grunewald, aged sixty-one, multipara. Unilocular cyst of the left ovary. *December 19, 1882.*—Ovariotomy. External incision four inches long. Cyst presenting, patient was brought in lateral position. Tapping, evacuation, and extraction. Rather stout pedicle transfixed with thumb-forceps, and tied in four portions, then cut off and dropped back into the abdomen. Uninterrupted recovery. *January 4, 1883.*—Discharged cured.

Multilocular cysts can be best emptied by making a free incision through their presenting part, through which the hand can be carried within the tumor to break up intervening septa. All this should be done extra-abdominally if possible.

When a cyst is found extensively adherent, its contents should be carefully mopped out with a sponge, and the interior of the sac should be disinfected while the patient is in the lateral posture. After this a large sponge is thrust into and left within the cavity until the cyst is dissected out.

CASE II.—Miss Lucretia Bernard, aged seventy-two, virgin. Very large multilocular ovarian cyst of the right side, causing intense dyspnœa. *August 8, 1881.*—Puncture and partial evacuation at Mount Sinai Hospital, resulting in marked relief of the dyspnœa. *August 10th.*—Fever set in, with some abdominal tenderness, and suppuration of the cyst was apprehended. *August 13th.*—Ovariotomy. Incision twelve inches long. Broad, recent adhesion of the sac to the anterior abdominal wall severed by blunt preparation. Patient being brought into the side position, the cyst was first tapped, then incised, and its volume was much reduced by breaking down septa by the hand. Some hæmorrhage occurring, a large sponge was thrust into the sac, and the patient was returned to the supine position. A number of adhesions to the right side of the parietal peritonæum and ascending colon were divided between several double mass ligatures of silk. Short pedicle was similarly secured. Toilet of peritonæum; closure of incision. Moderate elevations of the temperature. Uninterrupted healing of wound. *November 15th.*—Abscess of right groin was incised. Three silk ligatures were discharged. *August 11, 1882.*—Patient died of an intercurrent disease not connected with ovariotomy.

CASE III.—Mrs. Lena Dochtermann, aged thirty-nine, multipara. Very large multilocular cyst of right ovary. General condition very poor; chronic bronchial catarrh and chronic enteritis, with diarrhœa, ascites, and anasarca. *April 19, 1886.*— Ovariotomy. Extensive adhesions of cyst to anterior and lateral parietes; to transverse

## SPECIAL APPLICATION OF THE ASEPTIC METHOD. 149

colon, omentum, and the bladder. A large number of mass ligatures were made. Hæmorrhage insignificant. Duration of operation two hours and a half. Patient died in collapse seven hours after the completion of the operation, temperature remaining subnormal to the last.

*Cysts of the broad ligament* generally present great difficulties on account of their situation between the peritoneal folds of the ligament. If they extend low down into the small pelvis, their dissection is occasionally impracticable, and always very difficult. The utmost circumspection and care must be exercised not to provoke hæmorrhage by injuring large veins in the bottom of the wound, and all adhesions, not yielding to gentle blunt dissection with the fingers, must be fashioned into suitable masses, doubly tied with Thiersch's spindles, and then divided. In cases baffling the skill or enterprise of the surgeon, the sac should be properly trimmed and stitched to the skin, so as to convert it, if possible, into an extra-peritoneal recess. Drainage of the sac is indispensable.

CASE IV.—Mrs. Ethel D., aged twenty-one, nullipara. Rather immovable cyst of the right broad ligament of the size of a child's head. *April 6, 1887.*—Ovariotomy. Incision five inches long. The cyst had dissected its way out from between the folds of the broad ligament, and had pushed away the parietal peritonæum of the anterior abdominal wall on the right side to such an extent as to remain entirely extra-peritoneal. The sac was tapped and emptied, then it was easily separated from its attachments by blunt preparation. About one fourth of a square foot of peritonæum was detached. Finally, the pedicle was reached, secured in three ligatures carried through by means of Thiersch's spindles, tied, and cut off. The cavity was mopped out with corrosive-sublimate lotion, drained by two ordinary rubber tubes, and the external wound united and dressed in the usual manner. *April 7th.*—Nothing alarming had occurred, the temperature ranging about 99° Fahr. *April 8th.*—Temperature 101·5° Fahr., with a good deal of tympanites and dyspnœa. Pulse of varying intensity and rhythm, about 125 beats per minute, and rather weak. The outer bandage had to be loosened, and energetic stimulation by hourly enemata, consisting of one ounce of brandy and two ounces of warm water, were administered, till the pulse became decidedly fuller and more regular. *April 10th.*—Some flatus passed spontaneously, the meteorism diminished markedly, and the temperature fell to the normal standard. *April 11th.*—Patient consumed a few oysters and a little champagne, her nourishment having consisted until then of milk and lime-water. On the same date slight uterine and vesical hæmorrhage was noted. The former may have been dependent upon subinvolution remaining behind after a recent miscarriage; the vesical hæmorrhage seems to have been due to detachment of the superior and lateral vesical wall during dissection. *April 13th.*—A saline laxative was administered, causing some nausea and vomiting with a good deal of griping, but resulting in three copi-

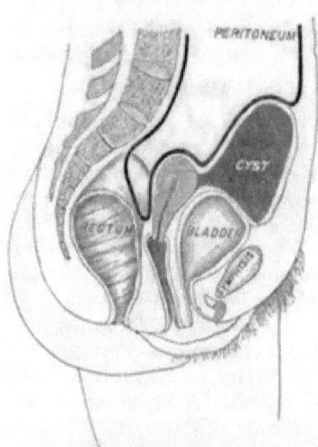

FIG. 117.—Diagram of cyst of the broad ligament. (Case IV.)

ous stools. The same day the drainage-tubes were shortened. The wound was found healed by adhesion except where the tubes lay. Three of the plate and shot sutures were also removed, and two were left behind. The catgut sutures had been all absorbed. *April 18th.*—The tubes were entirely withdrawn and remaining sutures removed. *April 20th.*—The patient left the bed the first time. *April 25th.*—The wound was entirely healed. (Fig. 117.)

It seems that the extensive detachment of the peritonæum from its nutrient vessels led to a grave disturbance of its circulation, and perhaps to partial (*aseptic*) necrosis. An adhesive peritonitis of the intestinal investment apposed to the denuded parietal peritonæum was set up, causing paralysis of the muscular layer of the gut with meteorism. As soon as the devitalized parts of the peritonæum were enveloped by fresh exudations, the irritation ceased.

β. *Removal of Uterine Appendages.*—The extension upward of septic conditions from the external female genital organs often leads to the establishment of acute or chronic inflammatory processes in and around the uterine appendages, causing a train of febrile and painful disturbances, which are so little influenced by general and topical treatment that their cure requires eradication by laparotomy. Though in most cases the diagnosis of salpingeal or ovarian disturbances will meet with no serious difficulty, occasionally nothing short of an exploratory section can shed light upon the nature of the affection. A comparatively short incision will usually be found adequate for the removal of the uterine appendages, which, as is now well known, can be easily and safely shelled out of their adhesions by the tips of the fingers (Tait). Occasionally a tube or a parovarian or ovarian abscess will be ruptured, and its contents will escape into the peritoneal cavity. In this case, the peritonæum has to be thoroughly flushed with hot Thiersch's solution or boiled hot water. A long, stiff drainage-tube of large caliber is connected with a funnel, and is placed successively well down into Douglas's pouch and the lumbar recesses. While the hot solution is gradually poured in, the index and middle fingers are gently moved about among the coils of intestine, so that all the peritoneal surfaces should receive the benefit of the cleansing irrigation. When the fluid is seen returning in a limpid state, irrigation may be stopped. The pedicle of the appendages is transfixed and tied off in the usual manner, the peritonæum is mopped out dry, and the wound is closed.

Case.—Mrs. Hannah M., aged twenty-three, unipara. Repeated attacks of severe local peritonitis in left parametrium, with high fever and rigors, appearing after childbirth in 1888. *May 2, 1889.*—A dead fœtus of three months was expelled. *December 10, 1889.*—During a sharp attack of local peritonitis, the temperature rising to 104° Fahr., a painful immovable tumor was made out to the left side of the uterus. Laparotomy, December 19, 1889, at Mount Sinai Hospital. The left enlarged ovary, containing an abscess of new formation, and the left tube much thickened and distended by pus, were enucleated from a mass of adhesions, tied off and removed. The appendages of the right side were found in a normal state. Cessation of febrile and painful symptoms. Intramural abscess, delaying the discharge of the patient till January 25, 1890.

In view of the great index of mortality following supra-vaginal hysterectomy for uterine fibroma, the removal of the uterine appendages, to induce menopause and subsequent shrinkage of the enlarged uterus, seems to deserve attention and faithful trial. The operation is often rendered difficult by the presence of enormously dilated veins surrounding the appendages, the management of which demands great care and circumspection. Being raised out of the pelvis by the growth, the ovaries and tubes usually occupy a high position, hence are found to be rather accessible. But when the fibromatous uterus is very large, a long incision will nevertheless be needed, to enable the surgeon to evert first one then the other side of the organ, in order to reach the annexa.

CASE.—Mrs. Susan M., multipara, aged thirty-nine, noticed abdominal enlargement since four years. Normal menstruation. Good general condition, her only complaint being the size of the belly and increasing dyspnœa. *March 13, 1889.*—Girth thirty-eight and a half inches, the apex of the solid, smooth, and freely movable tumor reaching about two inches above the navel. *January 23, 1890.*—Girth forty-one inches, the apex of the tumor reaching nearly to the ensiform cartilage. In view of the absence of hæmorrhages, there being no vital indications present, ablation of the uterine appendages was recommended in preference to supra-vaginal hysterectomy. Lungs, kidneys, and heart were found normal. *March 1, 1890.*—Removal of uterine appendages at Mount Sinai Hospital. Incision of eleven inches in length, to enable the operator to expose first one then the other side of the uterus. The intestines were not seen at all. Difficult isolation of appendages on account of enormously distended veins. Rather high temperatures followed the operation, accompanied by a very conspicuous shrinking of the tumor, and by the appearance of a copious, dark, sanguinolent discharge from the uterus. The vagina was regularly irrigated and kept plugged with iodoform gauze. A week after the operation an intramural abscess was opened. The bowels were moved on the third day. The fever persisted for nearly four weeks, then gradually diminished, to disappear entirely by the beginning of April. On the 24th of this month patient was discharged convalescent, with good appetite, sound sleep, and no pain, the granulating wound still open, but healing rapidly. The tumor had shrunk to about one half of its former size.

γ. *Supra-vaginal hysterectomy* for large myo-fibroma of the uterus may be indicated either by profuse loss of blood at the menstrual epoch, or by other causes, rendering the patient's life unendurable. An operation should be determined on only after a faithful trial of less incisive remedies known to induce involution of uterine fibromata has plainly failed to give relief.

The preparations for the operation are to made with all possible care directed to the avoidance of septic infection. Hæmorrhage is to be prevented by the application of single or double mass ligatures to the uterine adnexa on both sides of the uterus, and a stout elastic cord to the cervix. Under favorable conditions (that is, when the cervix forms a slender pedicle to the otherwise movable womb), the application of double ligatures can be obviated by cutting off the blood-supply of the organ from all sides by two continuous lines of mass ligatures converging from the free margin of the adnexa toward the cervix. A suitable-sized mass is first formed at the margin of the broad ligament by means of Thiersch's spindle, and is tied off with strong catgut or silk. A second mass adjoining the first one is now isolated, and the thread, being carried around it and back through the aperture made for the application of the first ligature, is firmly knotted.

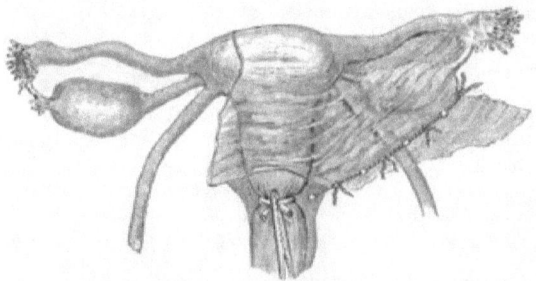

Fig. 118.—Diagram showing the arrangement of mass ligatures in supra-vaginal hysterectomy.

A third mass is isolated by Thiersch's spindle, and the thread is carried back through the hole made for the isolation of the adjacent mass, and the application of the preceding ligature. Thus the cervix will be soon reached. While an assistant raises the tumor well above the pelvis, an elastic ligature is thrown around the elongated cervix; being tightened, it is secured by a stout pedicle-clamp. This step will have completed the isolation of the uterus, which can be now exsected without loss of blood, the line of section being carried just outside of the chain of ligatures. (Fig. 118.)

With increasing experience the intraperitoneal treatment of the uterine stump is more and more abandoned in favor of the extraperitoneal method, which is undoubtedly much safer. The stump is crucially transfixed with two long shawl-pins on the distal side of the elastic ligature, and as much of its mass as can be safely removed is gradually pared away. The upper portion of the abdominal wound is closed in the usual manner. Then, the stump being suitably adjusted in the lower angle of the wound, its peritoneal covering, situated on the proximal side of the ligature, is attached to the parietal peritonæum by a number of interrupted catgut stitches all around its circumference, thus shutting off the peritoneal cavity from further contamination. The raw surface of the stump is either seared with the thermo-cautery and dusted with iodoform, or painted with some perchloride of iron solution and then dressed in the usual manner. The stump will shrink considerably within two or three days, and the ligature can be removed on the fifth day. The slough will come away in about a week or ten days, and the remaining granulating surface will heal in from four to six weeks.

In the presence of adhesions, or a broad implantation of the myoma into the deeper parts of the pelvis, the same rules of dissection are to be heeded that have been elucidated in a former paragraph relating to abdominal tumors.

The author's only case of supra-vaginal hysterectomy ended fatally by septicæmia. The sources of infection were presumably the sponges, managed by two raw nurses at Mount Sinai Hospital.

## SPECIAL APPLICATION OF THE ASEPTIC METHOD. 153

CASE.—Mrs. S. Levy, aged thirty-three, multipara. Very large **fibro-myoma of the corpus uteri.** Severe metrorrhagia at each menstruation, with increasing anæmia and great helplessness from the size of the **tumor.** *June 7, 1883.*—Hysterectomy at Mount Sinai Hospital. Incision six inches long. Easy deligation **of** adnexa in **two rows** of mass ligatures; elastic ligature of cervix; ablation of the tumor and adnexa. **Searing of the** surface of **the** small stump by thermo-cautery. The smallness of the stump induced the author to treat it like an ovarian pedicle, and it was replaced in the abdominal cavity after securing of the elastic ligature by a knot of strong silk. Hardly any blood was lost, and a smooth course of healing was expected. But all hopes were shattered by the development of septic symptoms in the night following the operation. *June 8th.*—High fever, retching, and sharp abdominal pain were present, but no signs of peritonitis could be made out. Twenty-nine hours after the operation the patient died in **coma.** Post-mortem examination revealed an abscess of the abdominal wall in the line of suture, and a grayish discoloration of the peritonæum near the elastic ligature. A few drachms of turbid, bloody serum were found in Douglas's **pouch.** No sign of peritonitis.

**Investigation** showed that **during the operation** the management of **the sponges by** the nurses had been **a careless one**; that a too large number of persons were intrusted with **the care of the** sponges. The practical outcome of this experience was the **order that** the sponges should be attended to by one person only, and that this person should always be the most experienced and reponsible one of the available number.

The preceding case shows that fatal septicæmia may be induced by infection of the peritonæum, and yet purulent peritonitis may be absent. Perhaps there was not enough time for the development of peritonitis.

*Many rapidly fatal cases, classed by various surgeons under the heading of "shock," or "exhaustion," would, on closer inquiry, turn out to be cases of acute septicæmia.*

δ. *Nephrectomy by abdominal section* is justified in cases of degenerated floating kidney when the urine gives sufficient evidence of chronic pyonephrosis with or without stone.

CASE.—Mrs. S. Weissenstein, **aged forty-six.** Noticed fourteen years ago a movable painless lump in her right hypochondrium. Since about nine months very acute symptoms of cystic trouble set in, **and** the lump became larger and painful. Constant desire to urinate, continuous **fever,** with occasional rigors, and large quantities of pus **in the** urine brought her to a **very low state.** A smooth, hard, kidney-shaped movable tumor of the size of a large man's fist could be felt in the right hypochondriac region. *January 11, 1887.*—Examination under chloroform. *The left kidney could not be made out distinctly.* **The** urine was scanty and acid, amounting **to** about twenty ounces per day, of the consistency of cream, **and** contained very large quantities of pus. *January 15th.*—Abdominal nephrectomy **at the** German Hospital. The tumor being exposed, the hand was slipped into the left lumbar part of **the** peritoneal cavity, when *the left kidney could be distinctly felt.* **After this** the peritonæum and its capsule were split along the whole anterior aspect of **the** enlarged kidney, **and the** organ was easily peeled out. A pedicle was formed **of the** ureter and vessels, and was tied **off in** two masses. After the removal of the tumor, the large retro-peritoneal cavity was carefully mopped out and loosely packed with strips of iodoformed gauze. These were brought out near the upper angle of the abdominal wound. The edges of the incision through the posterior lamella of the peritonæum and the renal capsule were stitched to the peritoneal lining of the anterior abdominal wall. The outer wound was united in the usual way. The patient lost very little blood, but during the operation threatening heart-weakness necessitated the subcutaneous exhibition of camphor and whisky. She rallied pretty well, and passed some perfectly clear urine shortly after the operation. *January 16th.*—Temperature, 100° Fahr. Patient cheerful, and suffering very little pain. Urine continues clear and very concentrated. In the night several fainting-spells. The night nurse did not pay sufficient attention to the patient, who died in a fit of syncope early in the morning of January 17th. Post-mortem examination failed to show any morbid change aside from the abdominal wound, which was found dry, and just as fresh as at the time of the operation. With more untiring stimulation, the patient might have survived. The enlarged right kidney had lost its textural character, and was converted into an irregular sinuous bag, containing six uratic stones of various sizes, surrounded by a quantity of pus.

*c.* **Gastrostomy.**—*Impassable cicatricial stenosis of the œsophagus* is a very strong indication for the establishment of a gastric fistula. Threatening starvation will be thus averted, and an opportunity will at the same time be created for attempting *retrograde catheterism* of the œsophagus, which may succeed.

CASE.—Hedwig Meyer, aged twenty-four. Cicatricial impassable stricture of the œsophagus twelve inches from incisors, caused by swallowing pure carbolic acid. Liquids only could be swallowed, with frequent regurgitations. Extreme emaciation. *April 17, 1886.*—Gastrostomy at the German Hospital. Immediately below and parallel with the left costal arch, an incision of two and a half inches exposed the peritonæum. After stanching the slight hæmorrhage, the peritonæum was incised, and the edges of the peritoneal incision were taken up by four artery forceps. The left lobe of the liver was found presenting. This being pushed aside, the anterior wall of the empty stomach came in view, and was withdrawn from the wound with a pair of thumb-forceps. The cardiac portion of the organ was drawn well into the wound, and was transfixed with a Peaslee's needle to prevent its slipping back. The peritoneal covering of the stomach was stitched to the everted edges of the parietal peritonæum by two tiers of interrupted silk sutures. The artery forceps were of very great service in securing the apposition of broad peritoneal surfaces. The external wound was packed with iodoformized gauze, and dressed antiseptically. No reaction following, the packing was removed on April 20th, and the Peaslee's needle was withdrawn. After this an incision one half inch long was made into the stomach, and a short piece of stout drainage-tube snugly fitting into the aperture was placed in the stomach, and was secured from slipping in by a large safety-pin. Its opening was closed by a cork stopper. Previous to this the lips of the mucous membrane were stitched to the outer skin. From this date on daily attempts were made to pass the stricture with a sound, introduced into the œsophagus from below, through the gastric wound. *May 13th.*—Dr. Bachmann, the house-surgeon, succeeded in passing from below an elastic catheter armed with a mandrel through the stricture. Milk injected into the catheter made its appearance in the fauces. *May 14th.*—A small-sized sound was passed from above. Alimentation was carried on both artificially through the drainage-tube placed in the stomach, and by the mouth. Gradually, as the ability to swallow solids returned, more and more food was taken by the mouth, and the drainage-tube was withdrawn from the stomach. The gastric fistula closed spontaneously by the end of June. *August 26th.*—Patient was discharged, with directions to continue the use of the œsophageal bougie.

*In cases of cancer of the œsophagus*, gastrostomy does not yield favorable results. Of eight cases, mostly men past middle age, and all presenting the picture of more or less extreme emaciation, five died in a few (all within twelve) hours after the operation. The slight depression of the heart's action by anæsthesia was sufficient to induce fatal collapse. The sixth case survived the operation for thirty-two days, but was losing ground steadily in spite of artificial feeding by the tube placed in the stomach. A great deal of difficulty was experienced in this case on account of the considerable leakage that was taking place alongside of the tube. Apparently the incision had been made too large, and gastric juice was escaping in varying quantities into the dressings. The gradual emaciation and final dissolution were in a great measure due to this constant loss of albuminoid substances.

In two cases the operation brought about a very marked improvement in the patient's condition. Both gained in weight and strength, the weight of one increasing twelve pounds in four weeks after gastrostomy. One survived the operation for seven, the other for five months.

The outer dressings of a gastrostomy wound are arranged in the following manner: A split compress of iodoformized gauze, similar to that used in tracheotomy dressings, is slipped in under the safety-pin holding the drainage-tube, and is arranged around the same. A piece of rubber tissue, or sheet rubber, somewhat larger than the gauze compress, is provided with a not too large slit in its middle, which then is also slipped on the end of the tube by being passed first over one, then over the other end of the pin. The rubber should fit snugly to the tube. Over this is laid a succession of two or more sublimate-gauze compresses of increasing size, each provided with a slit for the passage of the corked-up end of the rubber tube. The safety-pin, which was underpadded by the iodoformed gauze and rubber sheet, is covered up by the subsequent compresses, which are snugly bandaged to the trunk. Over the outer bandage another apron of rubber tissue is pinned, the rubber tube projecting from a slit in its middle. The object of this is to protect the bandage from soiling by regurgitant food.

Feeding is to be done at first in short intervals; later on, larger quantities of food can be introduced in four daily doses.

*d.* **Colotomy.**—Rectal obstruction, most commonly by syphilis or cancer, is an accepted indication for the establishment of an artificial anus, either in the groin or in the loin. Lumbar and inguinal colotomy each has special advantages and drawbacks, the consideration of which must determine the choice of the method preferable in a given case. While lumbar section is extra-peritoneal, nevertheless injury to the peritonæum is very apt to occur; finding of the colon is not easy; sometimes it is impossible without opening the peritonæum, notably when there is a well-developed mesocolon. The shape of the artificial anus after the lumbar operation is mostly excellent on account of the ample mass of tissues traversed by the fistula; but the situation of the aperture is unhandy, the patients generally requiring the aid of a second person for cleaning and dressing the artificial anus.

Inguinal colotomy is a short and easy operation, and provides for an opening located accessibly for the manipulations of the patient in cleaning and dressing the aperture. Its drawbacks are the necessity of incising the peritonæum—a circumstance which has lost most of its terrors since the introduction of the aseptic method—and the tendency to troublesome prolapse of the intestinal mucous membrane. The latter difficulty can be overcome by proper management.

(a) *Lumbar colotomy.*—Finding of the posterior aspect of the colon is very much facilitated by insufflation of the thick gut. This can be done either by a bellows attached to a soft catheter passed in beyond the stricture, or by the similar employment of a siphon bottle filled with mineral water charged with carbonic acid. The mouth of the siphon is connected with the catheter by a piece of rubber tubing, then the siphon is inverted and the valve is opened. The carbonic-acid gas, collecting about the end of the glass tube reaching to the bottom of the bottle, escapes into the gut, and produces a visible bulging of the colon.

When the stricture is impassable and inflation not practicable, recognition of the colon may offer great difficulty. The landmarks are the kidney above, and the reflexion of the peritonæum externally, but occasionally they are of little practical use.

CASE I.—Mrs. C. O., aged fifty-six. Very extensive far-gone cancer of the rectum with involvement of the uterus. The stricture was very long and impassable. *June 25, 1882.*—Lumbar colotomy was attempted. Though the kidney and the reflexion of the peritonæum were clearly discerned, the incision opened the peritonæum, and the protruding gut turned out to be small intestine. The poor condition of the patient made further prolongation of anæsthesia undesirable, therefore the gut was attached to the skin and incised. The wound healed promptly, giving much relief, but the patient died four weeks after the operation from emaciation, due in part to insufficient nutrition caused by the high position of the intestinal aperture. Post-mortem examination showed that the intestinal fistula was midway between the stomach and cæcum.

CASE II.—Mrs. Mary Brunner, aged forty-three. *August 23, 1885.*—Lumbar colotomy at Mount Sinai Hospital under ether. *August 24th, 25th.*—Acute lobar pneumonia of the entire right lung, to which the patient succumbed. The colotomy wound had closed by primary adhesion. Presumably the pneumonia was caused by the entrance of foul oral secretions into the right bronchus during the operation.

(b) *Inguinal colotomy.*—A vertical incision is preferable to one parallel with Poupart's ligament. With the former, the fibers of the oblique muscles will be cut across their course and will retract, giving ample space for a clear insight and free manipulation. Asepticism has to be maintained as in all abdominal operations mainly by scrupulous cleanliness.

The peritonæum is sufficiently incised to grasp the presenting colon with the fingers for withdrawal, and its edges are secured with four artery-forceps. The gut will be known by its tæniæ and the epiploic appendices. A loop about four inches in length, having neither a too long nor too short mesocolon, is withdrawn, and its mesial and distal halves are stitched to each other in front and in the rear so as to cause the formation of a spur (A B, Fig. 120). The sutures are made with an ordinary straight sewing-needle, the suturing material being catgut No. 3. The stitches should include only the peritoneal covering of the intestine. The loop is then dropped back into the peritoneal incision, and its sides are stitched to the parietal peritonæum all round with two tiers of catgut sutures. In doing this the parietal peritonæum can be well everted by the artery-forceps attached to it, and a broad surface of contact between it and the gut can be thus secured. Finally, the gut is transversely incised and the intestinal mucous membrane is sewed to the outer skin. To prevent prolapse of the mucous membrane, a loop of gut is to be selected that has a mesocolon of not greater length than just to permit its easy approximation to the parietes. The formation of the spur as suggested by Verneuil has this advantage, that fecal matter will not find its way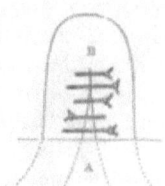

Fig. 120.—Formation of spur in inguinal colotomy

into the lowest part of the rectum situated below the artificial anus, and thus painful and otherwise disagreeable regurgitation of fæces will be avoided. At the same time, secretions forming in the distal section of the rectum will not be retained, but can escape through the fistula.

The proposition of completely dividing the loop of extracted colon, sewing the upper end into the wound, and closing by suture and dropping back the distal end, is feasible, but is met by a serious objection. The stricture may lead to complete occlusion, and the secretions of an ulcerated cancer may so distend the closed gut as to lead to rupture of the sutured part and to fatal peritonitis.

CASE I.—Mary Steiger, aged fifty-nine. Extensive rectal cancer with a number of periproctitic abscesses causing profuse purulent discharge through the anus. Emaciating hectic fever and distressing fecal retention. *August 13, 1885.*—Inguinal colotomy at the German Hospital. The thick gut was withdrawn, and was closed with two ligatures of stout silk carried through the mesocolon by the point of a thumb-forceps. The peritoneal incision was covered with two flat sponges and the gut was cut through between the ligatures. A little fecal matter escaped and was caught by the sponges, whereupon they were changed. The open lumen of the gut was mopped out cleanly, and well irrigated with Thiersch's solution. After this the distal end of the gut was closed by two tiers of Lembert sutures made with catgut, and was returned to the abdominal cavity. The peritoneal layer of the mesial end was stitched to the parietal peritonæum and the mucous membrane to the outer skin. The patient rallied well from the operation, but the high fever and profuse discharge from the anus continued. *August 18th.*—The patient died under septic symptoms. On autopsy, the wound was found healed by the first intention, likewise the sutured distal end of the gut. The peritonæum was normal, but a very large retro-peritoneal abscess, communicating with the rectal pouch above the cancer, extended high up along the front of the sacrum, and contained a large quantity of extremely fetid pus.

CASE II.—Stephen Y., government official, aged sixty-one. Far-gone rectal cancer, with involvement of the prostate and old strictures of the pendulous part of the urethra. *November 15, 1886.*—Inguinal colotomy with formation of spur at Mount Sinai Hospital under ether. *November 16th.*—Lobular pneumonia, probably caused by aspiration of mucus during the anæsthesia. By November 25th the acute febrile symptoms had subsided, but profuse purulent sputa were continually expectorated. The bladder also caused much trouble, although the tight strictures had been well dilated. The urine contained much pus, later on blood, coming from the ulcerated portion of the cancer occupying the neck of the bladder. The colotomy wound healed kindly, and a satisfactory artificial anus had been secured. The chronic bronchial catarrh, fetid cystitis, and later pyelo-nephritis, however, hastened the death of the patient, which occurred on December 23d.

Aside from Case II, inguinal colotomy with the formation of a spur was successfully performed by the author for rectal cancer altogether six times at Mount Sinai Hospital. The histories do not present sufficient interest to warrant their detailed enumeration. All the patients were materially improved, and survived the operation for periods of from six to eighteen months.

(c) *Excision and Suture of Gut (Enterorrhaphy).*—The object of enterorrhaphy is either the repair of injuries, morbid or traumatic, or the establishment of a direct artificial communication between more or less distant sections of the intestinal tract. Looked at from the standpoint of antisepticism, the cases requiring enterorrhaphy can be roughly classed in two divisions, which also fairly represent a corresponding gradation of the gravity of the procedure.

First come chronic conditions, requiring enterorrhaphy alone, as for instance fecal fistulæ due to a moderate amount of loss of substance of the wall of the gut. In many of these, where the loss does not involve more than one third of the circumference of the intestine, liberation of the adherent edges, followed by longitudinal suture, will be found sufficient. Longitudinal suture in a case where more than one third of the circumference of the gut was lost would result in stricture. Hence in these instances suture has to be preceded by transverse excision. It may be said that in these cases (excepting those of fecal fistula in the jejunum), the patient's general condition being good or fair, the time and circumstances of the operation can be selected and arranged so as to render it comparatively safe. This statement is borne out by a large proportion of recorded recoveries.

Next we have to mention as belonging to the same division, tumors and cicatrices requiring excision and subsequent enterorrhaphy, or the establishment of an artificial interosculation.

The second class of cases, much more difficult to deal with, consists of acute disturbances, in which enterotomy, enterectomy, and the various forms of enterorrhaphy have to be done under an urgent and peremptory vital indication.

It would lead too far to enter here into a detailed consideration of all the conditions belonging to this group, and we must content ourselves with a compendious enumeration. Persistent intestinal hæmorrhage, the removal of a gall-stone or foreign body impacted in the intestine, acute perforation of the gut by an ulcer, as for instance in typhoid fever; then rupture and stab and gunshot injuries of the stomach, small or large intestine; finally, necrosis from hernial strangulation, especially if the site of the lesion is in the jejunum. These may require enterorrhaphy with or without preceding enterotomy or enterectomy.

It stands to reason that in this class of cases the ratio of recoveries can not be as favorable as in the preceding one. But, with an increasing tendency to resort to early operative measures, and with the improvement of the technique, multiplying successes offer decided encouragement to surgical enterprise.

As enterorrhaphy is, either before or after the establishment of fecal fistula, most commonly employed for the repair of lesions due to intestinal necrosis from hernial strangulation, the detailed consideration of the procedure as adapted to this subject will be selected as an example.

The liberation and withdrawal of the involved intestinal loop is the first problem to be solved. Where there is no perforation, and adhesions are absent, the gut can be easily brought out as soon as the strangulating band is divided; but, where a fecal fistula is present, this step will demand care and attention to prevent infection of the peritonæum. If the circumstances of the case were such as to permit before the operation a thorough preparation of the gut by evacuation and disinfection, the adherent edges of the fecal fistula can be directly dissected out until the gut is free to follow gentle traction. But where preparation was impossible, it is safer to open the peritoneal cavity above and near the inguinal canal, exposing the involved knuckle, which will enable the surgeon to apply a clamp each to the ascending and descending part of the loop on both sides of the lesion, thus excluding the possibility of accidental escape of fecal matter. As soon as the knuckle of gut is liberated, it is freely withdrawn from the peritoneal cavity, so as to render the subsequent steps of the operation practically extraperitoneal. The extracted intestine is placed on a disinfected towel and the abdominal wound is packed around the emerging and returning portions of the knuckle with strips of gauze, to shut off the peritoneal cavity. After this two temporary ligatures consisting of pieces of well-disinfected tape are passed close to the gut through its omentum at a safe distance from, and one above the other below the fistula. Being safely knotted, they will prevent the escape of intestinal contents. The clamps are now removed, and the interior of the tied-off portion of the gut is carefully cleansed by irrigation and thorough wiping out with a small sponge.

In excising the damaged part of the viscus, the sections are made with straight scissors at a right angle to the axis of the gut, good care being taken to carry them through unmistakably sound tissues. Formerly a corresponding wedge-shaped portion of the mesentery was also excised with the intestinal cylinder, but this detail has been abandoned as unnecessary, and the mesentery is spared as far as it appears healthy. The bleeding points being deligated, the gap in the mesentery is stitched up with catgut, care being taken to bring about a precise apposition of the mesenteric attachments near the intestine.

In observing the transverse section of the gut, it will be seen that the peritoneal and muscular layers having retracted, the mucous membrane projects for about an eighth of an inch. A stitch passed through this mucous margin is called Czerny's suture. The stitch which without penetrating the entire thickness of the intestinal wall brings into apposition two serous surfaces is known as Lembert's suture. The best suturing material is fine, well-disinfected China bead silk, threaded at each end through a slender, round sewing-needle.

The intestinal suture should commence by a Czerny stitch corresponding to the mesenteric attachment. The stitch is closed so as to leave the knot on the mucous surface. After this a Lembert suture is applied close

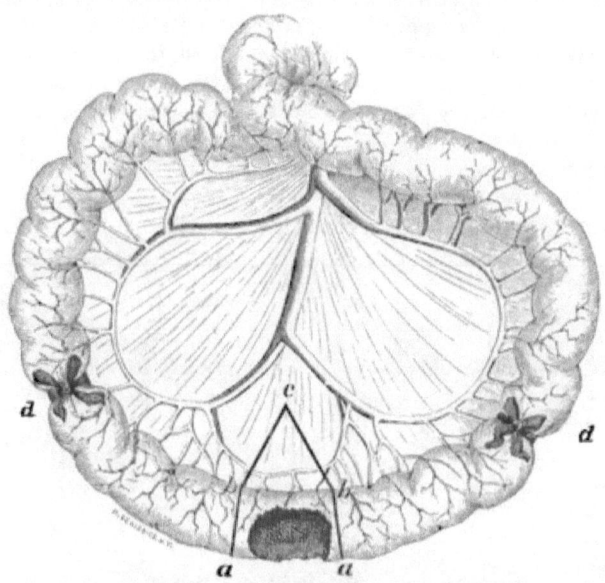

FIG. 120 A.—Loop of small intestine. *a b*, Lines of section through the gut, removing the gangrenous portion. *b c*, Same through the mesentery. *a a*, Gangrenous portion of ileum. *d d*, Occlusion of the afferent and efferent tubes by tape ligatures. (Wyeth.)

to one side of the mesenteric attachment; the needle, entering the peritoneal and muscular layers about one eighth of an inch from one margin of the cut, and passing between the peritonæum and mucous membrane for about

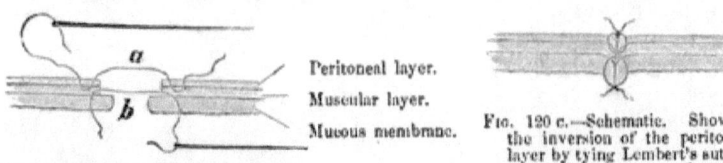

FIG. 120 B.—Schematic. *a*, Lembert's, and *b*, Czerny's sutures. (Wyeth.)

FIG. 120 C.—Schematic. Showing the inversion of the peritoneal layer by tying Lembert's suture, and of the mucous membrane by Czerny's suture. (Wyeth.)

one third of an inch, is made to emerge again through the peritonæum. The other end of the same thread is similarly passed through the other margin of the gut at a point exactly corresponding to the first stitch, and this suture is knotted. Thus the wall of the intestine will be inverted. The intervals between the Lembert sutures should not exceed one fourth, those between Czerny's sutures one third of an inch. The stitching should be done first on one then on the other side of the mesenteric attachment, Czerny's stitch always preceding two or three Lembert sutures, until about three quarters of the circumference of the gut are united. The last two Czerny stitches can not be knotted within the lumen of the gut, the knots remaining imbedded between the mucous and muscular layers. To give additional security against leakage, another continuous suture, comprising the two external coats, is rapidly applied just outside of Lembert's line of sutures. When the work of stitching is completed, the intestine is cleansed with Thiersch's solution, and, the packing and tape ligatures being removed, is dropped back into the peritoneal cavity. As an additional safeguard, a pledget of iodoform gauze is placed well up against the line of sutures, and is brought out at the lower angle of the external wound, which is closed in the usual fashion, except where the gauze emerges. An aseptic dressing completes the procedure. An opiate is administered and food is withheld for twenty-four hours, after which time liquid nourishment can be sparingly given. Rectal alimentation will husband the patient's strength, and will in a measure control thirst. Under favorable circumstances the iodoform packing can be withdrawn on the third or fourth day. Should leakage occur, it will find its way out along the track of the pledget of gauze.

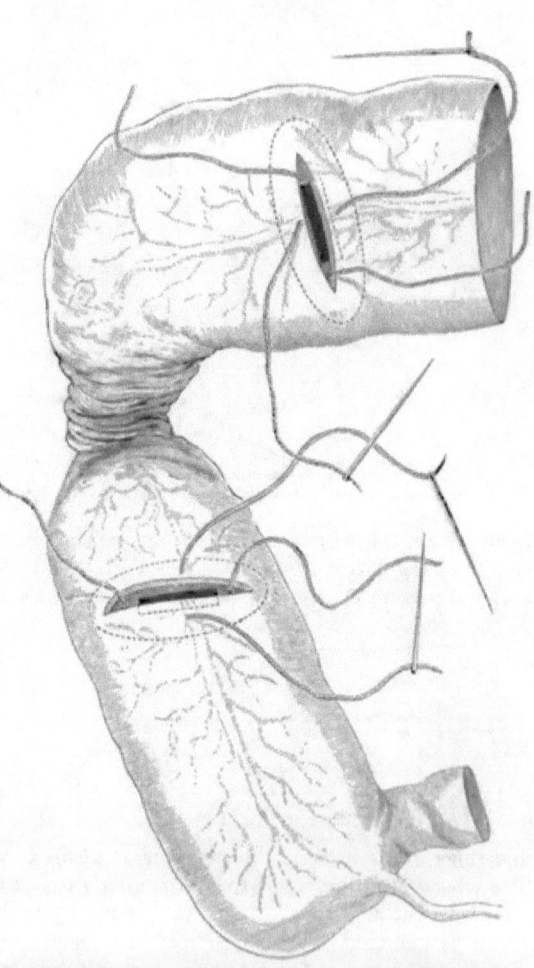

FIG. 120 D.—Senn's bone plates applied to colo-colostomy before tying together. (Abbe.)

The only serious drawback to this operation is the long time consumed in applying the stitches. Senn, of Milwaukee, was the first to approach successfully the question of abbreviating the procedure. He suggested, instead of the large number of interrupted sutures, the use of two elliptic decalcified bone disks, each armed with four silk threads passed through so many needles, by means of which two portions of gut could be made to anastomose. Originally, Senn's device was intended merely to establish lateral interosculation of two distant portions of gut in certain cases, as for example in intestinal obstruction dependent on cicatricial or neoplastic character, and thus save time by eliminating the necessity for precedent excision and subsequent enterorrhaphy. The places for the establishment of the new communication having been selected, a transverse incision of from one to two inches' length is made into each section of gut (see Fig. 120 D). A bone plate is slipped into one of these newly-made apertures, and is attached to its edges by the transfixion of the entire thickness of the gut with the four needles which belong to the bone plate. After the second aperture is provided for in a similar manner, the corresponding threads are tied firmly. Thus the two apertures are brought to correspond exactly, and the peritoneal surfaces in the vicinity of the openings are retained in close and secure contact. For the sake of greater security, a few Lembert stitches are applied outside of the line of contact. Senn also advises that the surfaces to be brought into apposition should be scratched with the point of the needle, to hasten speedy agglutination.

Care must be taken that the newly-united sections of intestine are joined together so as to permit the unimpeded progress of intestinal contents in a homonymous direction. (This was not observed in the illustration (Fig. 120 D), taken from Abbe, where the nature of the anastomosis would compel the intestinal current to change its direction to the extent of 180°.) As far as expenditure of time is concerned, the rapidity of this simple process places it far above the Lembert-Czerny sutures.

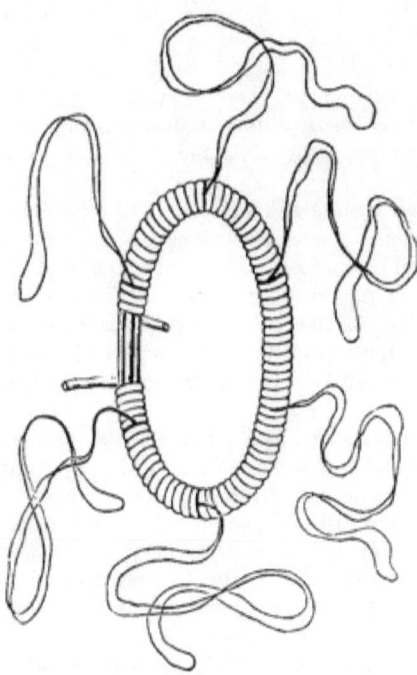

FIG. 120 E.—Apposition ring of catgut. (Abbe.)

Abbe has still further improved the method by substituting easily procurable catgut rings for Senn's bone plates,* the preparation of which is somewhat circumstantial (see Fig. 120 E).

Strangely, the development of the new operation was brought to its logical culmination by a homœopathic practitioner of Philadelphia,† who, after excising twelve inches of intestine for gangrene, successfully united the gut, end to end, by means of rings made of ordinary rubber drainage tubing. Though found in curious company, the paper containing the report of the case bears internal evidence of credibility and knowledge.

* Robert Abbe, Complete Obstruction of the Colon successfully relieved, etc. "New York Medical Journal," March 23, 1889.

† Van Lennep, "Hahnemannian Monthly," vol. xxiv, No. 10, October, 1889.

### XII. HYDROCELE, VARICOCELE, AND CASTRATION.

1. *Hydrops of the tunica vaginalis of the testis* is either an essential disorder *per se*, or is symptomatic of some acute or chronic affection of the testicle. If it be produced by acute epididymitis and orchitis, it is transient; but if its cause is tuberculosis, or cancer, or syphilis of the testicle, it assumes the character of a chronic complaint. For the sake of a correct prognosis the recognition of secondary hydrocele is important, as it is improbable that, brought on by these affections of the testicle, hydrocele can be cured by either tapping and injection or the radical operation.

If the hydrocele is very tense, preliminary tapping is advisable, in order to afford an opportunity for estimating the condition of the testicle. Should this be found rugged, swollen, and hard, it is very doubtful that measures directed to the cure of the effusion will be successful, unless the condition of the testicle be improved by appropriate treatment. Gummy swellings will usually disappear under antisyphilitic medication, and with them the hydrocele. Tuberculosis and cancer, on the other hand, will require castration.

*The cure of simple hydrocele by tapping and subsequent injection* with tincture of iodine or pure carbolic acid is safe, and is generally followed by cure. The only caution to be taken is a proper disinfection of the trocar or cannula to be used, by either boiling in carbolized lotion (five per cent), or by heating the instrument in an alcohol-flame. Care must also be exercised not to leave behind in the sac too large a quantity of the tincture of iodine, as there is on record a case of acute iodine-poisoning brought on by that circumstance.

*Volkmann's radical operation* is also safe, and offers the best chances of a permanent cure; but it necessitates longer confinement of the patient than the preceding method. The author has performed this operation successfully forty-eight times on thirty-one patients, and no serious disturbance was ever observed during the course of healing. In each case cure was complete in from two to three weeks, and was permanent. Lately the operation was done with the aid of local anæsthesia by cocaine.

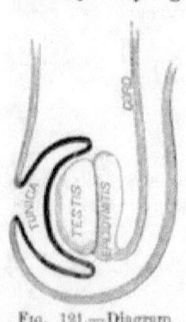

Fig. 121.—Diagram illustrating Volkmann's operation for hydrocele.

The procedure is as follows: The penis and scrotum are shaved, scrubbed off, and disinfected. A rubber band or drainage-tube is tied about the root of the penis and scrotum, and about twenty minims of a five-per-cent solution of cocaine are injected along the prospective line of incision. The skin and dartos are incised for about two inches, and the exposed tunica is opened. A grooved director is slipped into the sac, which is then slit open, this incision being somewhat shorter than the cutaneous one. The sac is mopped out with a sponge dipped in a five-per-cent solution of carbolic acid. After this the tunica is stitched to the skin by a continuous suture of fine catgut. A small drainage-tube is inserted and secured from slipping in by transfixion with a safety-pin. The constricting rubber band is removed, and the scrotum is held compressed between two sponges for a few minutes to stanch any possible hæmorrhage. A small strip of disinfected rubber tissue is laid on the wound, which is enveloped, together with the entire scrotum, in a dry dressing, held down by a roller bandage applied in the manner described in the paragraph on herniotomy. (Fig. 121.) The dressings are changed on the third day after the operation. On the second day the movement of the bowels is attended to by enema or laxative. On changing the dressings the patient can be permitted to get up and to exercise moderately. The wound is dressed with a strip of iodoformed gauze until it is healed.

2. *Varicocele* of a moderate degree is best treated according to Keyes's plan, which consists of *subcutaneous ligature of the distended veins with catgut*. The scrotum being cocainized, the cord is separated from the varicose veins, and is held in the grasp of the thumb and index-finger of the left hand, the patient standing during the procedure. A straight Peaslee's needle, armed with a loop of silk, is thrust through the scrotum from in front until its eye appears behind the scrotum. The left hand releasing its grasp is used for placing the ends of a medium-sized thread of catgut into the loop of silk, which is then pulled through forward and out of the anterior puncture-hole, and the catgut is released from the silken loop. Now the left hand grasps again the scrotum, and the needle is reinserted exactly into the anterior puncture-hole, and carried around the varices externally to them, and close to the scrotal integument backward, until it emerges precisely from the posterior puncture. The other end of the catgut thread is then taken up by the loop of silk, and is brought out through the anterior aperture by withdrawing the needle. Both ends of the ligature are now seen emerging from the anterior puncture-hole. They are tightly knotted, cut off short, and disappear in the scrotum as soon as released. A slight amount of hard swelling will appear around the place of ligature the next day, but this will not prevent the patient from attending to his vocation.

The author has employed this method with the best success in ten cases.

*Extensive varicocele* can be cured only by free exposure, double ligature, and excision of the dilated veins. Under aseptic precautions this measure is free from danger.

CASE.—E. Luhning, baker, aged twenty-one. Large varicocele of the left side, extending down to the middle of the inner aspect of the thigh. *April 25, 1882.*—At the German Hospital the scrotal varices were exposed by incision, and a large plexus was separated and tied above and below. The intervening veins were exsected. Another incision of eight inches in length exposed the varicose veins extending down the thigh, and they were also exsected after being secured by double ligature. A rather wide strip of attenuated skin had to be removed along with the veins, preventing entire closure of the femoral wound by suture. Uninterrupted cure of the scrotal wound by primary union of the femoral one by granulation. *June 22d.*—Patient was discharged cured.

Four more somewhat less extensive cases were treated in a similar manner, and all healed by the first intention.

Care must be taken not to remove all the veins of the pampiniform plexus. In the author's sixth case necrosis of the testicle was caused by too extensive excision of the dilated veins.

CASE.—Joseph Stern, baker, aged twenty-two. Extensive varicocele of the left side. *March 17, 1886.*—Excision of varices at the German Hospital. *March 27th.*—Necrosis of testicle was noted. A few of the stitches had given way, and the yellowish, discolored testis was distinctly visible. *April 8th.*—The testicle came away with very moderate sero-purulent secretion. *April 26th.*—Patient was discharged cured.

3. *Castration* is indicated by neoplasms, tuberculosis, or syphilis of the testicle, in the latter case, however, only when the disease is not amenable to systemic treatment, and is a source of much suffering.

*The author's procedure for castration* is as follows: The patient's genital region is shaved, scrubbed with soap and hot water, and disinfected with corrosive-sublimate lotion, or, if any open ulcer or fistula be present, these are finally syringed or touched up with an eight-per-cent solution of chloride of zinc. *First, the seminal cord is exposed* well above the diseased testicle, and, being separated, is taken up by the index of the left hand. The vessels composing it are successively grasped by separate artery-forceps, while the vas deferens remains intact. As soon as all the vessels are thus secured, they are nipped off one after the other with the scissors in front of the artery-forceps, and are at once tied. The vas deferens is cut through. Before being released, the mesial end of the severed cord is somewhat relaxed and carefully inspected, to see whether all bleeding be stanched or not.

By making the division of the cord the first step of the operation, the subsequent parts of the procedure are made decidedly less bloody. Dissection of the testicle proper is much easier and more rapid than if the reverse order is observed, and the stump of the cord serving as a convenient handle, contact of the surgeon's fingers with ulcerating surfaces or fistulæ can altogether be avoided. A few more ligatures will be generally needed along the bottom of the scrotum.

A drainage-tube is inserted, extending from the inguinal ring down to the lower angle of the cutaneous incision, and then the wound is united by interrupted catgut sutures, the edges of the cut being held pinched up by the fingers in passing the stitches. A dressing similar to that used after herniotomy is applied and left on generally for three or four days. The tube is removed with the first dressing. Tying of the cord in mass saves a little time in operating, but the stump generally necroses, and cure is very much delayed by the slow process of its detachment.

Castration was performed by the author twenty-nine times; in twenty-four cases for tuberculosis. One of these cases died of croupous pneumonia, probably induced by ether anæsthesia.

CASE.—Moses H., merchant, aged sixty. *January 24, 1887.*—Castration for tuberculosis of right testicle at Mount Sinai Hospital under ether. The operation did not present anything unusual, and the patient did well after it until two clock on the afternoon of January 26th, when suddenly high fever with dyspnœa appeared, and developed into coma within a few hours. At 6 P. M. the thermometer indicated 106·7° Fahr. in the rectum; at 9.55 P. M. the patient died. Dullness at the base of the right lung, made out a few hours before death, corresponded to an area of fresh lobar pneumonia found at the autopsy. The wound, peritoneal cavity, and kidneys were normal.

Twenty-three cases castrated for tuberculosis all recovered.

In one case castration was done for syphilitic gumma of the left testicle of five years' standing, which had remained uninfluenced by various kinds of constitutional treatment.

CASE.—John W. G., brewer, aged thirty-eight. Large hydrocele caused by chronic specific disease of the testicle. *March 4, 1887.*—The hydrocele was incised, and the testicle was found very much enlarged; the rugged and hard epididymis was occupied by a solid fibrous mass extending well into the glandular tissue of the testicle. Castration was at once done. *March 15th.*—Patient discharged nearly cured, the place of exit for the drainage-tube presenting a small spot of granulations.

In two cases ablation of the testicle had to be done for malignant neoplasm. They recovered.

CASE I.—Jacob Praeger, tailor, aged seventy-two. Very large giant-cell sarcoma of right testis. *December 4, 1879.*—Castration. Preparation of the bowels by laxatives was insufficient, and on the third day after the operation violent colic developed, which could not be controlled by opiates. In the night a large stool escaped into the bed, the dressings and the wound were soiled, and in a few hours fever set in. The wound was injected with an eight-per-cent solution of chloride of zinc, which checked the fever. Much sloughing tissue came away, but patient recovered, and was discharged cured about five weeks after the operation.

The author's experience in this case taught him the valuable lesson of *never trusting the patients' statement regarding the action of their bowels*, and never leaving the manner of preparation of the intestine to their judgment. In this case the patient assured the author that citrate of magnesia acted on him like a charm. Citrate of magnesia was taken, with the result reported above. Had a good dose of oil or calomel raked out the flaccid and coprostatic gut of the old man before the operation, his life would not have been endangered by subsequent fecal infection of the wound.

CASE II.—Siegmund Hertz, clerk, aged thirty-two. *August 24, 1885.*—Castration of right testicle for myxosarcoma at Mount Sinai Hospital. Primary union. *September 15th.*—Patient discharged cured.

*Twice castration was done for spontaneous gangrene of the testicle.* Both cases recovered. The record of one was lost; that of the other is as follows:

CASE.—George Otto, butcher, aged thirty-nine, admitted, February 2, 1880, to German Hospital with an enormous emphysematous swelling of the left testicle. The organ had nearly the size of a man's head, was dusky red and hot, showed crepitus, and gave tympanitic percussion-sound. The patient, a powerfully built man, showed symptoms of most acute septic intoxication. He stated, on being shaken out of his stupor, that the swelling had come on three days ago suddenly with much pain after a probatory puncture. Immediate ablation of the organ was done. The skin was preserved, and the very large wound cavity was filled with a packing of carbolized gauze. An almost immediate improvement of the patient's general condition followed. The

SPECIAL APPLICATION OF THE ASEPTIC METHOD. 167

wound healed rather rapidly by granulation. *February 26th.*—Patient was discharged cured. Examination of the specimen showed bloody infarction of the testis and epididymis, with far-gone disintegration and softening of the tissues. The tunica and subcutaneous connective tissue were in a state of emphysematous gangrene.

### XIII. ASEPTIC OPERATIONS ON THE RECTUM.

1. **General Observations.**—The aseptic performance of rectal operations done for hæmorrhoidal or other tumors requires a careful preparation of the gut. It consists, first, of the administration of a cathartic like castor-oil or calomel several days, in elderly subjects a week before the operation, followed up by the daily exhibition of a saline laxative, to be given on an empty stomach. Four hours before the time of the operation a large enema of soap-water is administered, and, as soon as it has acted, a full dose of opium is given by mouth, or is introduced into the rectum in the shape of a suppository.

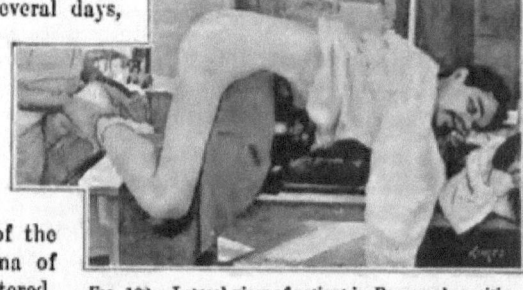

FIG. 122.—Lateral view of patient in Bozeman's position.

When the anæsthetized patient is laid on the operating-table, a good-sized sponge attached to a stout silken thread is thrust well up the rectum, and, the sphincter being thoroughly stretched by manual force, the anus and rectal pouch are flushed with a stream of corrosive-sublimate lotion (1 : 1,000) thrown from an irrigator.

During the progress of the operation irrigation has to be kept up constantly at short intervals. When the peritonæum is approached, or has to be invaded by the surgeon, Thiersch's solution is substituted for the mercuric lotion as an irrigating fluid.

2. **Hæmorrhoids.**— A varicose condition of the hæmorrhoidal veins of recent origin, caused by some dis-

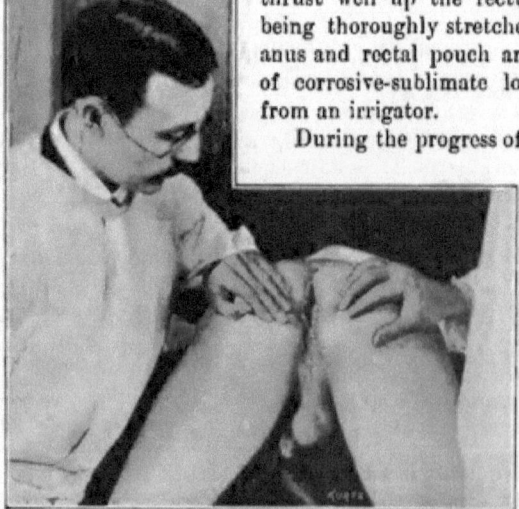

FIG. 123.—Posterior view of patient in Bozeman's position.

turbance of the portal circulation, is often amenable to general treatment by fulfilling the causal indication. Removing a fecal retention, or regulating the portal circulation with a dose of calomel, followed up by a course of Carlsbad salts, will often do away with the hæmorrhoids caused by these conditions. Or regulation of the heart's action by digitalis in valvular lesions will be followed by marked improvement. When the hæmorrhoidal nodes are in a state of acute phlebitis, marked by painful hot swelling and fever, topical applications of cold in the shape of enemata of ice-water or iced compresses will give much relief.

Aggravated cases, however, especially when there is a state of prolapse of the mucous membrane of the anus, can be cured only by operative measures.

Of all operations for the cure of hæmorrhoids, that by ligature commends itself as one of the simplest. This statement is based on an experience gathered from several hundred cases operated by the author according to various methods.

The manner of procedure is as follows: The anæsthetized patient is brought either in the lithotomy position, with a hard cushion under his buttocks, or he is arranged in Bozeman's manner for the operation of vesico-vaginal fistula (Figs. 122 and 123). This latter position is especially useful where the assistance needed for holding the patient in the lithotomy position can not be procured. In both cases the feet and legs of the patient should be protected from exposure by a wrapping of rubber sheets. These should be covered over with clean towels wrung out of mercuric lotion for the protection of the assistants' hands from contamination.

Selecting the lithotomy position, the patient's palms should be brought in contact with his soles, and this relation should be secured by tight bandaging. The operator, well protected by a rubber apron, takes a seat in front of the patient, and proceeds to vigorously stretch the sphincter ani muscle with his thumbs inserted in the anus. As soon as the sphincter is paralyzed by stretching, the hæmorrhoidal nodes, external and internal, will spontaneously protrude. A sponge secured with a thread of silk is thrust into the rectum, and the field of operation is cleansed by irrigation. The lowest node is grasped with an artery forceps, and, being well drawn out, is circumscribed by a shallow incision made with a pair of curved scissors. A curved needle is taken, armed with a double thread of stout disinfected silk, and with it the base of the tumor is transfixed from without inward. The silk is cut near the needle, and, the threads being separated, the base of the node is tied in two portions. The node is cut off below the ligatures, and then the remaining nodes are attended to in a similar manner. When the operation is finished, some iodoform powder is rubbed into the nodal stumps, and, after a final irrigation, the sponge is withdrawn from the rectum, which is mopped out dry with another sponge attached to a long stick or sponge-holder. (Fig. 124, A and C.)

A hollow tampon is next prepared by wrapping a few layers of iodoformized gauze around a piece of stout rubber tubing three inches long. This is introduced into the rectum well beyond the sphincter, and its protruding end is transfixed with a large-sized safety-pin. (Fig. 125.)

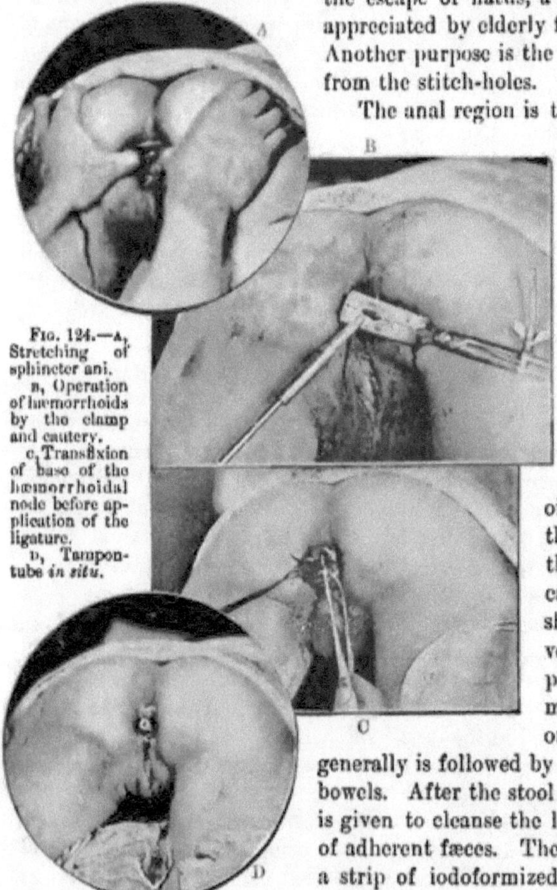

The object of this tampon is twofold. Its main object is to facilitate the escape of flatus, a circumstance highly appreciated by elderly flatulent individuals. Another purpose is the prevention of oozing from the stitch-holes.

The anal region is thickly anointed with vaseline, and, the safety-pin being under-padded with a few strips of iodoformized gauze, a large pad of corrosive-sublimate gauze is held down to the anus by a T-bandage. (Fig. 126.)

Forty-eight hours after the operation four ounces of sweet oil are injected into the rectum through the rubber tube, which can be withdrawn a short while after with very little pain to the patient. A large enema of soap-water is at once administered, and generally is followed by an evacuation of the bowels. After the stool another small enema is given to cleanse the hæmorrhoidal stumps of adherent fæces. The anus is dressed with a strip of iodoformized gauze and a pad as before.

FIG. 124.—A, Stretching of sphincter ani. B, Operation of hæmorrhoids by the clamp and cautery. C, Transfixion of base of the hæmorrhoidal node before application of the ligature. D, Tampon-tube *in situ.*

The next morning a dose of salts is given, and, stool following, the rectum is again washed out afterward. This practice may have to be repeated once or twice within the next few days.

The patient may be permitted to get up about ten days after the operation, but must remain at home till after the detachment of the ligatures.

*Cauterization with fuming nitric acid* was formerly also much employed by the author; but in one case almost fatal hæmorrhage occurred from a small artery just within the sphincter on the detachment of the eschar. Since then the author has abandoned this practice.

CASE.—Mr. M. P., gilder, aged thirty-one. *February 24, 1882.*—Cauterization of external and internal hæmorrhoids with nitric acid. *March 10th.*—At 2 A. M. the author was hastily summoned to the bed-side of the patient, and found him in a collapsed condition. He reported that shortly after supper he felt a desire to stool, and had a copious evacuation. Evacuations followed since then about every hour, but, the closet being dark, he could not say whether the stools were bloody. At 1 A. M., on coming back to bed from the water-closet, the patient fainted. Being brought to bed, another stool followed, consisting of a large clot and some liquid blood. The patient was at once anæsthetized, and, a speculum being inserted, a rather large-sized artery was seen spurting from where an eschar had been detached just inside of the sphincter. The vessel was seized and tied, and the patient made a good recovery.

*Langenbeck's clamp and actual cautery method* is very good and safe, its only drawback being the necessity for a cautery apparatus. Care must be taken not to grasp with the clamp the nodes too near their base, as the resulting eschar is apt to be very large, and anal stricture may follow. The hollow tampon is very useful in this method also, and its use can be warmly recommended (Fig. 124, B).

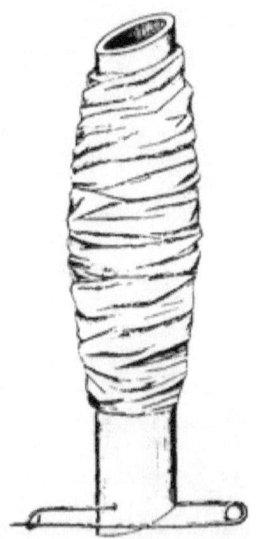

FIG. 125.—Tampon-tube.

*Whitehead's Method by Excision and Suture.*—Whenever an aggravated case of hæmorrhoids is found associated with more or less considerable prolapse of the rectal mucous membrane, excision of the irreducible nodes with subsequent suture of the circular wound is advisable. The author has tested the value of this operation in forty-seven unselected cases. In the hands of an expert surgeon the operation gives excellent results, and can be done rapidly and with moderate loss of blood. If the wound heals by the first intention, the time required for a cure is shorter

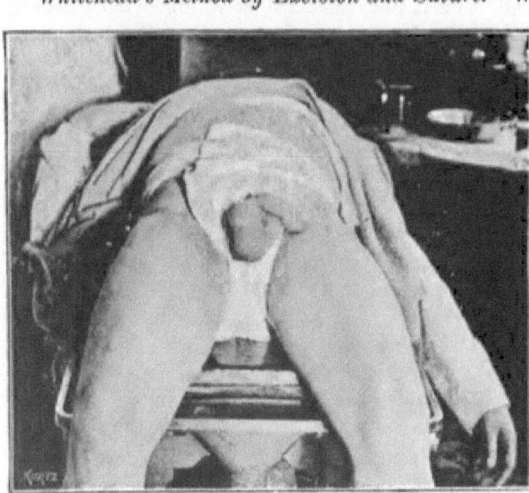

FIG. 126.—T-bandage *in situ.*

than with any other effective method, and the course of the after-treatment is remarkably free from untoward complications. The drawbacks of this process are, that few general practitioners have the requisite expertness to perform it well and rapidly. Failure of primary union in the entire circumference of the gut entails unavoidable stricture, requiring methodical dilatation, or even renewed excision and suture. The *modus procedendi* is as follows :

After stretching the sphincter, a sponge is placed well up in the rectal pouch, and the field of operation is thoroughly disinfected. A circular incision, including the entire ring of hæmorrhoids, is carried around the anal aperture, close to the mucous membrane. About one half of the muco-cutaneous margin should be preserved intact. Fleshy tabs belonging to the outer skin should not be included in this circle, and no skin should ever be removed, as this is sure to result in ectopia of the normal mucous membrane. The incision is deepened until the fibers of the sphincter are exposed, when the mucous membrane containing the turgid hæmorrhoids can easily be stripped up all around from the underlying tissues to a line about an inch beyond the nodosities. Cut vessels are at once deligated. The detached cylinder is longitudinally divided on one side. From the end of this longitudinal cut a circular section is carried through the detached gut, severing it from the rectum. This circular section should be done *gradatim*—that is, after cutting off an inch or so, the mucous membrane should be stitched to the outer skin as soon as it is divided, and so on, until the entire circumference of the hæmorrhoidal mass is cut off and the margins stitched. If tension is great, the process will be much facilitated by the insertion of a buried catgut stitch carried through the submucous layer on one, and through the subcutaneous tissue on the other side. This will render the apposition of mucous membrane and cutaneous margin easy, and the sutures will not cut through prematurely. From ten to fifteen stitches will be needed to unite the cut edges. After the removal of the sponge from the rectal pouch, a tampon tube is inserted and the case is treated in the usual fashion.

To secure perfect asepsis, the wound should be frequently irrigated during the progress of the operation, but especially well before the closure of the sutures. The patient's bowels are to be moved not before forty-eight and not later than sixty hours after the operation. In successful cases the cure will require from two to three weeks.

3. Rectal Tumors.—Since the publication of Volkmann's remarkable results achieved by extirpation of the rectum for cancer, the operation, formerly condemned, has met with frequent imitation. The author's melancholy record of six deaths out of nine operations has nothing to inspire great confidence. It must be said, however, that most of these operations were performed under very unfavorable conditions. All the patients presented instances of very extensive involvement of the gut, requiring in each case the removal of more than three inches—in one case, nine inches—of intestine. Almost all of them were performed during

the first years of the author's independent surgical activity, when his mastery of the difficult technique, both of the aseptics and hemostasis of the region in question, was imperfect. Much unnecessary hæmorrhage was incurred, and several of the most important cautelæ against infection remained unemployed. Accordingly, two patients died shortly after the operation, of collapse due to acute anæmia; two died of purulent peritonitis, caused by infection of the incised peritonæum; one died of septicæmia, induced by the presence of a large retroperitoneal abscess, extending far up in front of the vertebral column. One patient, a very fat, flabby woman, died of lobar pneumonia at a time when the wound was nearly healed.

Three cases of very extensive removal of the rectum made a remarkably short and easy recovery.

CASE I.—Ed. Turner, mechanic, aged twenty-nine. Extensive soft adenoid cancer of the rectum, of rapid growth. The involved part of the gut was freely movable, although its upper limit could not be reached by the tip of the index-finger. *November 18, 1884.*—Extirpation of the rectum at Mount Sinai Hospital. As the growth did not extend downward to within an inch of the sphincter, this muscle was preserved. The coccyx was exposed by a posterior median incision, and was exsected. The mucous membrane of the lower end of the gut was dissected up in the shape of a cylinder, and was closed by a ligature to prevent the escape of rectal contents during the operation. Every vessel was immediately secured and tied, either at being cut or before division, if it could be previously recognized. The levator ani muscle was detached by dissection from the intestine. All resisting bands of tissue, mostly containing vessels, were secured by double mass ligatures before being divided. Most difficulty was met with in freeing the gut from its attachments to the deep pelvic fascia, but by dint of mass ligatures this was also overcome. As soon as the pelvic fascia was passed, the intestine readily yielded to traction, and was withdrawn until the upper limit of the tumor was distinctly felt through the walls of the gut. The peritonæum was detached anteriorly by blunt separation, but it had to be incised on the posterior aspect of the rectum to permit complete removal of the growth. The gut was grasped with a large clamp-forceps about an inch above the tumor, and was severed. The patent orifice of the rectum was carefully cleansed and disinfected, and, the clamp being removed, a number of vessels of the rectal wall were secured and tied. During the whole operation the wound was almost constantly irrigated with corrosive-sublimate lotion (1 : 2,500). The peritoneal incision being closed by catgut suture, the wound was loosely packed with iodoformized gauze after the insertion of two drainage-tubes into its bottom, and the gut was attached to the skin by two silk sutures. The ends of the drainage-tubes were left projecting from the dressings, and the wound was flushed through them at regular intervals of an hour. The temperature remained normal except on the sixth day, when it rose to 103° Fahr. The patient complained of colicky pains, and a saline purge was administered. A stool following, the fever disappeared. The wound was carefully cleansed by irrigation after each stool, and healed in spite of its great extent in six weeks. The removed portion of the gut measured, when laid upon the table, just five inches.

The resulting incontinence of the widely patent gut was remedied by a proctoplasty performed February 28, 1885, at the German Hospital. The divided ends of the preserved sphincter muscle were dissected out, and were united by a row of catgut stitches placed in the median line. In April, 1890, the patient was free from relapse.

CASE II.—Eugene Haffner, waiter, aged twenty-four. Relapsing cancer of rectum after extirpation done by Dr. F. Lange. *February 24, 1887.*—Extirpation of additional two inches of the gut at the German Hospital. Peritonæum was found descended to within half an inch from the skin. It had to be freely incised, and was subsequently closed by five catgut sutures. Uninterrupted recovery. *April 2d.*—Patient was discharged cured.

In the third case Kraske's procedure was employed. It consists in the preliminary removal of the coccyx and a portion of the lower half of the sacrum, whereby the removal of the rectum is very much facilitated.

CASE III.—Koppel Barscheinik, tailor, aged forty-six. Circular and massive cancer of anus and rectum, extending five inches beyond the anal aperture. General condition fair. *March 1, 1889.*—Excision of rectum by Kraske's method at Mount Sinai Hospital. Coccyx and sacrum were exposed by a median incision. The former being removed, the sacrum was denuded, and by means of the chisel and mallet its lower half was divided in the median line up to the height of the third sacral foramen, where its left lower segment was entirely severed by a transverse cut. After this the anal end of the gut, together with the sphincter, were dissected out and tied off in mass to prevent the escape of rectal contents. Thus the liberation of the diseased part of the rectum became a remarkably easy task. The hæmorrhage was readily controlled, the gut severed an inch and a half above the limits of the disease, the opened peritonæum closed with a few catgut stitches, and the stump of the gut loosely sutured to the upper angle of the wound. The open parts of the wound were packed with iodoform gauze. Duration of the operation, forty-five minutes. The removed portion of the rectum measured six inches. The patient rallied well. On the fifth day fever was observed, and an abscess, located between the rectal stump and sacrum, was found and opened, after which the temperature fell to the normal standard. By July 2d the wound had healed and about three inches of the rectal mucous membrane had gradually prolapsed. The gut was separated from its attachments, the cicatrix was divided down to the original site of the anus, and the intestine was attached to its original habitat. A triangular segment was excised from the lower end of the intestine to somewhat narrow its aperture. The stitches partially gave way, and a third smaller operation had to be done September 30th, to secure the desired result. This stretching out of the rectal stump did away with the tendency to prolapse; the patient's solid fæces were very well retained by a sort of sphincter about three inches above the anal opening. His general condition had improved remarkably, and he is now (May, 1890) attending to his business.

The main source of infection is the interior of the gut. To exclude this danger, the lower end of the rectum must be closed by a circular ligature. When the gut is divided above, care must be taken to prevent soiling of the wound by escaping intestinal contents.

### XIV. ASEPTICS OF THE BLADDER.

1. **Catheterism.**—Infectious processes rarely originate in the bladder itself. Their most common way of entrance is by the urethra from without; next to this come the modes of infection from within—that is, by descent from the kidneys or by extension of contiguous septic processes from the organs located in the vicinity of the bladder, as for instance from peritoneal or retro-peritoneal suppurations.

As before indicated, the most common source of infection of the bladder is an unclean catheter. *The ordinary methods of cleansing metallic catheters by flushing with hot or cold water, and subsequent rubbing off with a clean towel, are altogether inadequate.* In order to secure their absolute cleanliness, the same processes of sterilization must be employed that were recommended for cleansing other hollow tubes—notably, aspirating needles and trocars. Boiling for an hour in water, or passing the instrument through an alcohol flame until all organic matter contained in its lumen is volatilized by burning, is meant thereby. Only after smoke and steam have ceased to escape from the catheter can it be declared to be surgically clean.

Before use, the cleansed catheter should be placed in a tray or flat pan filled with tepid salt water (6 : 1,000, or one heaped teaspoonful to a quart of boiled water); the surgeon's hands should be previously well washed with soap and hot water, and the instrument should be anointed with iodoformized vaseline of the strength of 1 : 50 (fifteen grains to two ounces).

NOTE.—The ordinary solutions of corrosive sublimate or carbolic acid corrode the mucous membrane of the urethra and bladder, often causing intense pain and reflex symptoms. The resulting denudations of the epithelial layer all may serve as portals of subsequent infection, manifesting itself in the form of urethral fever, urethritis, cystitis, and, in extreme cases, metastatic processes. None of these very active germicides should be introduced into the healthy urethra or bladder: first, because they are unnecessary; and, secondly, because they may do harm. Simple immersion of a filthy catheter into these germicidal lotions will not disinfect it sufficiently, and, if some of the strong solution be carried into the urinary passages along with a filthy catheter, the chances of infection will only be increased by the combination. Catheters that were immersed in strong disinfectant solutions should be freed from them before being used.

In passing the instrument into the bladder for exploration or evacuation, the utmost gentleness should be exercised, not only for the sake of the patient's comfort, but also because it is of importance not to injure the urethral mucous membrane. Certain parts of the normal male urethra will often raise obstacles to the passage of the instruments which should never be overcome by force, but only by patient and gentle manipulation.

The first obstacle is usually met at the suspensory or triangular ligament. Holding the shank of the catheter parallel with the abdominal wall while gently extending the penis upward in the same direction, thus pulling the latter over the former like a glove-finger over a finger, will easily guide the beak of the catheter around the promontory formed by the inferior margin of the symphysis pubis.

The second obstacle will be occasionally found in the sinus of the bulbous portion. This pitfall must be avoided by exerting digital pressure upon the perinæum, and indirectly upon the beak of the catheter while gently depressing its handle. In sensitive urethræ, the compressor urethræ, or "cut-off" muscle, will offer by reflex contraction considerable resistance to the progress of the operation, especially if an instrument of small caliber be employed. It is injudicious to force this obstacle. A better plan is to abide the moment when the muscle will relax, the instrument being held against the resisting band by gentle pressure. As soon as relaxation begins, the point of the catheter will be felt slipping through the contracted part of the urethra.

The enlarged prostate is the last and most difficult, because deepest, impediment that may retard the operator. A long-beaked instrument will penetrate to the bladder easier than any other one. The handle of the catheter must be deeply depressed between the thighs of the patient, and, if this be insufficient, the tip of the left index-finger introduced in the rectum must aid the entrance of the beak by gentle upward pressure.

*Properly performed catheterism of a healthy urethra and bladder should not be followed by hæmorrhage.*

**Soft** *catheters* made of gum **elastic or** webbing impregnated **with resin-**ous matter are never safe unless their history is known to the **operator.** They should be new, or, at least, such should never be employed **that had** been previously used on a septic case, or were not carefully cleansed, **disin-**fected, and preserved in a proper manner after use.

Soft gum-elastic or Nélaton catheters are cheap, and need not be preserved after having been used **in a septic case.** Before employing a soft catheter, it must be soaked **for ten minutes in** hot soap-water and flushed out with it; then it is disinfected **with** a strong germicide lotion, preferably corrosive sublimate, from which **it must be** freed again by another flushing with salt water before it is anointed with iodoformized vaseline for introduction. After use, the catheter should be again flushed out thoroughly with carbolic or mercurial lotion, dried, and put away in **a tight box or** wide-mouthed bottle. If needed frequently, the catheter **should be** kept immersed in **a** two-per-cent carbolic lotion. Before use, however, **the** adherent carbolic lotion must be always removed by washing in salt water. The author saw a considerable number of cases in which catheterism had to **be done for some time after** rectal operations, and in which troublesome urethritis **developed on** account of the corrosion caused by frequent contact of the urethral mucous membrane with **the** carbolic acid adherent to the elastic catheter.

**Searching a non-dilated bladder** for stone, **tumors,** or foreign bodies **would lead to superficial injury of** the mucous membrane; therefore, **dilatation, by injecting three or four** ounces of salt water, should precede every exploration. **After completion of the** search, clots **should be** removed by irrigation with the saline solution.

These remarks **refer to bladders** only that discharge normal urine.

Whenever examination of **the** urine gives evidence **of** a catarrhal **or septic** condition, every intravesical manipulation must be preceded by disinfection of the bladder by Thiersch's solution, or a lotion consisting of one part of permanganate of potash to **five** thousand parts of tepid water. The operation should be completed by another disinfecting irrigation of the organ.

2. **Litholapaxy.**—The rapid and complete evacuation of **the** bladder in one session, of all fragments produced **by** crushing **concrements with** a lithotrite, forms a most valuable improvement of the technique of lithotripsy. Bigelow's evacuator enables the **surgeon** to free the bladder at once of all sharp-edged fragments of stone. **This** circumstance justifies the prolonga-

tion of the operation to an extent formerly considered unsafe, as subsequent irritation caused by the presence of sharp fragments is thus done away with.

Before introducing the lithotrite, strictures ought to be cut or divulsed, and the bladder ought to be thoroughly washed out with tepid permanganate-of-potash or boro-salicylic solution. After this the bladder is filled with from three to four ounces of tepid boro-salicylic lotion, and the lithotrite is introduced well anointed with iodoformized vaseline. The penis is tightly deligated with a piece of rubber tubing, and the stone, being grasped, is crushed first into a number of larger, and subsequently into as many small fragments as possible. The crushing instrument is removed and is replaced by the evacuating catheter, which is connected with the evacuating bulb, that was previously filled with boro-salicylic lotion. All small fragments are next sucked out of the bladder by the apparatus. Should a peculiar click indicate the fact that one or more fragments, too large to pass the catheter, are still remaining, the lithotrite must be introduced anew to complete their reduction to a proper size, after which complete evacuation will meet no difficulty. The bladder is washed out again until the irrigating fluid returns free from blood, and the patient is brought to bed.

Small stones, especially of the softer varieties, are eminently suited for this treatment, which has the great advantage of a short convalescence; but its disadvantage of a possible relapse from failure to remove all fragments can not be denied.

CASE I.—M. Witzkal, peddler, aged fifty. *April 5, 1884.*—Litholapaxy at the German Hospital. Uratic stone with phosphatic shell weighing four drachms fifty-five grains. Duration of operation thirty-five minutes. Discharged April 28th. In June, patient was readmitted for stone, which was removed by Dr. Adler by median lithotomy.

CASE II.—Mr. E. B., clerk, aged twenty-one, renal colic followed by symptoms of stone in the bladder, which was diagnosticated by sounding. In March, 1887, lithotrity and evacuation. The bladder symptoms continued until June, when Dr. Schede, of Hamburg, removed another small calculus.

The author performed litholapaxy in four more cases.

CASE III.—Edward Mink, baker, aged twenty-one. *January 26, 1881.*—Rapid lithotrity for a phosphatic calculus weighing two hundred and fifty grains. *March 5th.*—Patient discharged cured.

CASE IV.—Henry Bowitz, agent, aged forty. *April 24, 1884.*—Litholapaxy for uratic calculus, weighing three drachms and ten grains, at Mount Sinai Hospital. *May 10th.*—Patient discharged cured.

CASE V.—Francis Johnson, druggist, aged forty-seven. Phosphatic calculus, ammoniacal urine. *October 6, 1883.*—Rapid lithotrity at Mount Sinai Hospital. Weight of stone, forty seven grains. Duration, fifty-five minutes. Discharged cured, October 27th.

CASE VI.—Philip Prinz, shoemaker, aged fifty-nine. Rapid lithotrity for small uratic calculus, done January 25, 1887, at German Hospital. On the day following the operation all the symptoms of stone disappeared, but the patient sustained a burn of the legs requiring surgical treatment. This delayed his discharge until March 17th.

Intense forms of cystitis caused by the presence of calculi require after lithotrity continued treatment of the bladder by irrigation.

3. **Cystotomy.**—In perineal as well as in suprapubic cystotomy, the condition of the urine should serve as a guide in determining whether aseptic or antiseptic measures have to be observed during the operation. When the normal condition of the urine indicates that the vesical mucous membrane is in a healthy state, strong disinfecting solutions should not be used within the bladder, and the surgeon's chief attention should be directed to the careful cleansing of his instruments, in order to avoid the introduction of filth into the bladder. For purposes of filling and cleansing, a saline or Thiersch's solution will be all sufficient.

In cases characterized by pyuria, with or without ammoniacal odor, or with outright fetidity of the urine, disinfection of the bladder must precede and follow each operation.

The rules of asepticism referring to the treatment of the external wound must also be scrupulously observed. During the after-treatment, drainage of the bladder may be required, especially in cases where a septic condition of the organ would render retention of fetid urine undesirable or risky. A rather stout rubber drainage-tube inserted in the bladder will answer every practical purpose.

(*a*) PERINEAL SECTION :

CASE I.—Fred. Kurtz, aged fifty-five. Phosphatic stone, ammoniacal urine. *February 1, 1881.*—Lateral lithotomy at the German Hospital. Weight of stone, three drachms and forty grains. No reaction or fever. Continued washings of bladder with salicylic-acid solutions. *April 10th.*—Discharged cured.

CASE II.—Hugo Liedtke, aged three and a half. Small uratic stone. *March 19, 1881.*—Lateral lithotomy with the assistance of the family attendant, Dr. Hassloch. Weight of stone, eighteen grains. *April 15th.*—Discharged cured.

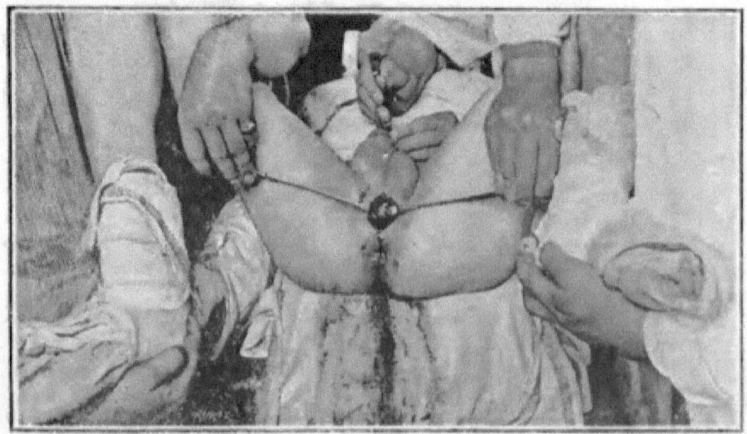

FIG. 127.—Arrangement of patient for perineal cystotomy. Feet wrapped up in disinfected towels.

(*b*) SUPRAPUBIC SECTION.—Tumors, a very large prostate, encysted or very large stones, oxalic concrements, or rebellious cystic hæmorrhage from dilated veins of the neck of the bladder, indicate the selection of the high operation. Petersen and Garson's proposition to distend both bladder and rectum before cutting, marks a most valuable improvement of the method, as injury to the anterior reflection of the peritonæum can be thus avoided. A soft rubber bag, or "colpeurynter," similar to Barnes's dilator, is introduced into the rectum, and is filled with from fifteen to eighteen ounces of water. Escape of the water is prevented by attaching an artery forceps to the end of the tube.

Seven or eight ounces of tepid salt water or boro-salicylic lotion are injected into the bladder, and the penis is tied with a piece of rubber tubing. The patient's shaved suprapubic region is carefully disinfected, and a median incision is made, commencing about three inches above, and extending to the symphysis. The recti muscles are separated, and the prevesical fat is incised. *Care must be taken not to injure the reflexion of the peritonæum, which may be looked for in the upper angle of the wound.* In many cases the peritonæum will not come in view at all. Should distention of the rectum and bladder not suffice to push up and out of the way the peritoneal fold, this must be separated from the bladder by blunt dissection, to be done preferably by the tips of the fingers. Vessels crossing the prevesical space should be divided between double ligatures.

The bladder is transfixed on each side of the median line with curved needles, carrying fillets of silk. The vesical incision is made between these hold-fasts with a sharp-pointed bistoury. In cases of doubt, the presenting organ may be first punctured with a hypodermic needle. While the silken threads keep the vesical wound patulous, the surgeon's finger explores the interior of the bladder. Stones are then extracted with forceps, or the scoop, or even with the fingers, tumors are inspected and excised under the guidance of the eye, and bleeding varices of the neck of the bladder are grasped and tied off or touched with the thermo-cautery.

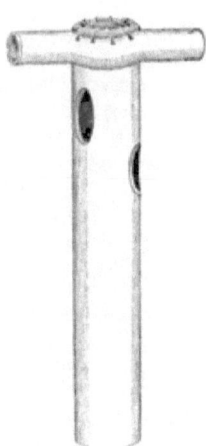

FIG. 128.—T-shaped drainage-tube for suprapubic cystotomy. (Trendelenburg.)

After thorough irrigation, a T-shaped drainage-tube (Fig. 128) is inserted in the bladder, and the external wound is loosely packed with iodoformized gauze. A split compress of the same material is arranged about the projecting end of the tube, and is covered with a number of compresses consisting of corrosive-sublimate gauze. The skin all around the wound is profusely anointed with iodoformized vaseline, and the dressings are held down by a few turns of a roller-bandage. The patient is brought to bed, and is laid on his side upon a circular air-cushion, his back being supported by a number of cushions held up by the backs of several chairs, or by boards stuck into the side of the bed. As the lateral position has to be maintained for three days at least, sides should be changed every two or three hours. The drainage-tube projecting from the dressings is connected with a longer tube, that is led into a urinal placed alongside the patient in or out of bed. As soon as the urine ceases to be bloody, and its reaction becomes acid, the patient may be allowed to assume the supine posture. The drainage-tube can be removed on the fifth day, when the wound will be usually found in a state of healthy granulation. The packing of iodoformized gauze has to be continued as long as urine escapes through the wound. As soon as urination *per vias naturales* is re-established, the wound should be dressed as any other superficial wound.

CASE I.—Martin Gyr, laborer, aged fifty. Large oxalic calculi of ten years' standing, with undilatable bladder. Wretched general condition. *April 12, 1886.*—Suprapubic lithotomy at the German Hospital under chloroform, which was preferred to ether on account of the presence of casts in the urine. Two immovable stones were found occupying the contracted bladder. They were grasped, freed by rotation, and extracted one after the other. They showed on extraction two freshly broken surfaces, corresponding to as many pedicle-like projections, branching into two diverticles, each containing a separate calculus. One of these calculi was extracted, the other and smaller one was left behind, as the patient's poor condition verging on collapse did not justify continuation of the operation. The patient did not rally from the collapse, and died three hours after the completion of the lithotomy.

The suprapubic incision gave free access to the bladder, and enabled the author to conduct the search and extraction of the calculi under the guidance of the eye. Removal or even the finding of the encysted calculi would have been utterly impossible from a perineal wound. Weight of calculi, one ounce, five drachms, and twenty grains.

CASE II.—Mr. Adolph W., plumber, aged fifty-six. Vesical trouble of three years' standing. Urine slightly acid, turbid, containing much pus, but no casts. *March 30, 1887.*—Exploration of the very irritable bladder with the stone-searcher yielded no positive result. *April 18, 1887.*—On exploration in ether anæsthesia, stone was found. A Thompson lithotrite being introduced, a large stone was grasped, and on rotation was felt to grind against another calculus. Suprapubic lithotomy. Extraction of three stones, each weighing about forty-three grammes, their aggregate weight being four ounces and three grains Troy weight. *April 20th.*—Temperature, 100·5° Fahr.; urine clear, acid, containing no blood; its daily quantity eighty ounces. *April 23d.*—Patient was allowed to occupy the supine position. *April 25th.*—The drainage-tube was withdrawn and the packing removed. A soft catheter was introduced by the urethra, and the bladder was irrigated through it. The catheter was left in the bladder; the external wound was repacked. Temperature, 98·5° Fahr. *May 1st.*—Thrombosis of right femoral vein, apparently due to defective circulation caused by confinement. The right lower extremity enormously increased in size. Treatment: Elevated posture; later on, moist packing, and elastic compression by Martin's bandage. *May 25th.* —Lithotomy wound nearly closed; passed some water through urethra. *June 4th.*— Lithotomy wound closed; urination normal. Patient up and about most of the time; œdema of thigh fast diminishing. *June 20th.*—Swelling of thigh almost gone; patient discharged cured. *July 25th.*—General condition excellent. Patient entirely recovered.

CASE III.—Mr. Meyer B., liveryman, aged thirty-nine. Symptoms of very acute cystic catarrh of four months' duration, causing the loss of fifty pounds of flesh. Almost constant desire of and very painful micturition, the acid urine containing blood, pus, some mucus, uric acid, and oxalate-of-lime crystals. The prostate was very painful on touch, but not appreciably enlarged. The patient had become morphiophagous, and was thoroughly demoralized. Stone was searched for unsuccessfully by a surgeon. *June 17, 1886.*—Suprapubic cystotomy at Mount Sinai Hospital. No stone was found, but the mucous membrane of the bladder presented a most marked state of hyperæmia and thickening, profusely bleeding at the slightest touch. The inflammation was most pronounced about the trigonum and the neck of the bladder, where the reddening and tendency to hæmorrhage were most intense. Trendelenburg's T-shaped drainage-tube was inserted, and the case was treated in the lateral position. The cystic irritation ceased at once, the blood and pus in the urine diminished, and morphine was discontinued. *July 17th.*—The patient was removed to his home, where he made a rapid and perfect recovery. In March, 1887, a slight degree of catarrh of the neck of the bladder was cured by irrigation with permanganate-of-potash lotion. The patient remained well ever since then.

CASE IV.—Joseph Goldstein, aged sixty-six, bladder trouble of old standing. Calculus is diagnosticated by the sound. Fair general condition. *August 8, 1888.*—Epicystotomy at Mount Sinai Hospital. Removal of two large uratic calculi, weighing together 1,190 grains. Tardy closure of wound. Patient discharged cured, October 5, 1888. A sinus leading down into the bladder reopened three times, but ultimately healed in the spring of 1889.

CASE V.—Mr. George L., musician, aged fifty-seven. In July, 1888, vesical calculus was diagnosticated. *August 15, 1888.*—Epicystotomy. Removal of uratic stone weighing 201 grains. Slight iodoform intoxication. Cured, September 16, 1888.

CASE VI.—Samuel Bader, tanner, aged twenty-seven. *July 6, 1888.*—Removal of sarcoma of trigomus by transverse incision at Mount Sinai Hospital. Died August 15, 1888, of septicæmia. Autopsy revealed left multiple pyonephrosis.

CASE VII.—Julius Basch, actor, aged thirty-five. *January 31, 1890.*—Removal of diffuse papilloma of bladder at Mount Sinai Hospital. In spite of double pyonephrosis, wound was healed May 28, 1890. Long-continued drainage seems to have somewhat abated the kidney trouble, as there was no fever since March, 1890.

CASE VIII.—Solomon Loewenthal, janitor, aged fifty-four. *October 28, 1887.*—Upper cystotomy at Mount Sinai Hospital for chronic prostatic ulcer and extreme irritability of bladder under ether anæsthesia. Died of acute lobar pneumonia (autopsy) November 6, 1887.

CASE IX.—Solomon Posner, tailor, aged thirty-seven, suprapubic cystotomy. *November 2, 1888.*—At Mount Sinai Hospital for tubercular cystitis. Died February 5, 1889, after nephrectomy (see history, page 282).

CASE X.—Linche Kester, tailor, aged twenty-seven, chronic cystitis with irritable bladder. *August 2, 1888.*—*Perineal* cystotomy at Mount Sinai Hospital. Discharged with closed wound and much improved condition of bladder, September 17, 1888.

# PART II.

# ANTISEPSIS.

## CHAPTER VI.

*NATURAL HISTORY OF IDIOPATHIC SUPPURATION. TREATMENT OF SUPPURATION.*

### I. THE CAUSE OF SUPPURATION OR PHLEGMON.

It would far transcend the limits of these essays to enter into a detailed presentation of all vegetable organisms known to lead a parasitic existence in the living human body. But a few glimpses into this new world of beings, more or less hostile to human health and life, may be welcome to the busy practitioner, who lacks time or opportunity for independent research.

Rosenbach's classical investigations have revealed the fact that the most common source of suppuration is the implantation and thriving in the living human tissues of a minute globular fungus or micrococcus, called from the

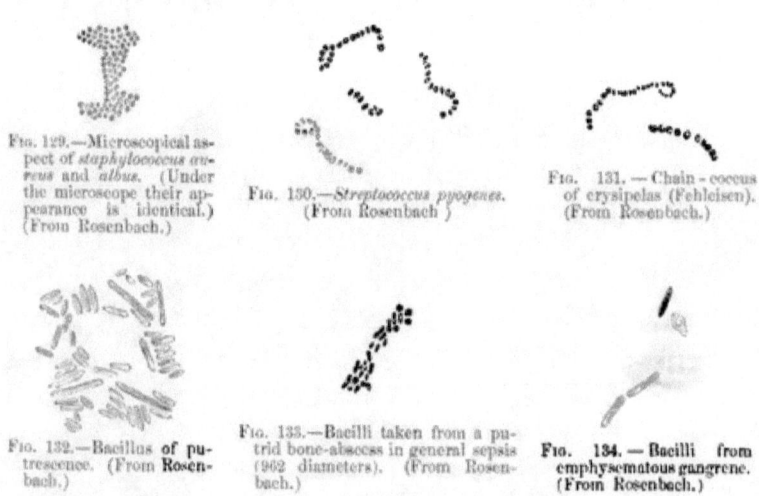

Fig. 129.—Microscopical aspect of *staphylococcus aureus* and *albus*. (Under the microscope their appearance is identical.) (From Rosenbach.)

Fig. 130.—*Streptococcus pyogenes*. (From Rosenbach.)

Fig. 131.—Chain-coccus of erysipelas (Fehleisen). (From Rosenbach.)

Fig. 132.—Bacillus of putrescence. (From Rosenbach.)

Fig. 133.—Bacilli taken from a putrid bone-abscess in general sepsis (962 diameters). (From Rosenbach.)

Fig. 134.—Bacilli from emphysematous gangrene. (From Rosenbach.)

golden yellow color of the mold it forms on a peptonized meat-agar culture-soil, "*Staphylococcus pyogenes aureus*," or the *golden grape-coccus*. It is called grape-coccus (*staphyle*, grape) on account of the agminated or bunched arrangement of the single cocci that compose a colony. (Fig. 129.)

This coccus is found in almost all forms of acute suppuration—in phlegmon, glandular abscesses, and in acute, infectious osteomyelitis. By certain methods of manipulation, a pure or unmixed culture of this fungus can be raised upon glass plates covered with a film consisting of a mixture of peptonized meat-jelly and agar agar, a vegetable form of gelatin. This mold resembles in structure the common form of mold dreaded by housekeepers, only it has a deep orange color. It has the peculiarity of thriving upon the living human tissues, causing their inflammation and ultimate death. (Plate I, Fig. 1.)

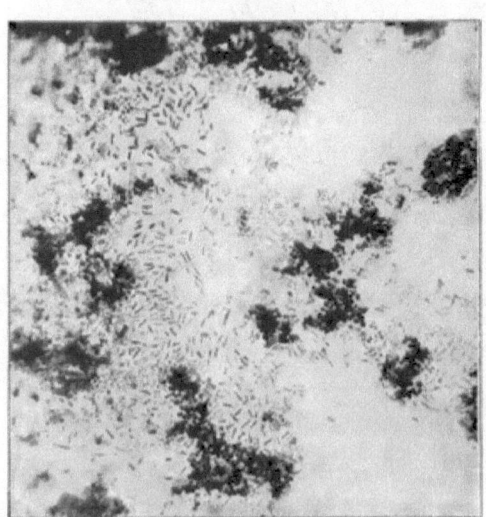

Fig. 135.—Bacilli of putrefaction and diverse forms of cocci in putrid blood. (Koch.)

Another form of grape-coccus, not so common as the preceding one, and appearing either alone or associated with the golden grape-coccus, is Rosenbach's "*Staphylococcus pyogenes albus.*" It can not be distinguished from the yellow coccus under the microscope, but the mold produced by pure culture is easily recognized by its pearly white color. (Plate I, Fig. 2.)

Both forms of grape-coccus have the clinical peculiarity of causing well-localized foci of phlegmon. All tissues within a certain area become uniformly permeated by the grape-coccus. They coagulate, then emulsify, and the result is a distinct abscess.

Another form of micro-organism—Rosenbach's "*Streptococcus pyogenes*," or *pus-generating chain-coccus*—is so called on account of the arrangement of the single globular cocci in more or less elongated chains. (Fig. 130.) Its peculiarity is to rapidly extend along the lymph-spaces and lymphatic vessels. Its emulsifying property is not as pronounced as that of the grape-coccus, but it may become very destructive to the tissues by rapid infiltration along the lymphatics, causing progressive gangrene. The peculiarity of extending along the course of the lymph-vessels, as well as its microscopical appearance, testify to its close morphological relation with the *streptococcus*, or *chain-coccus of erysipelas*, discovered by Fehleisen. (Plate I, Fig. 3, and Plate II, Fig. 4; then Fig. 131.)

Pure cultures of the pus-generating streptococcus and the coccus of erysipelas differ very distinctly in several important points (see Plate II, Figs. 4 and 5), but microscopically they can not be distinguished.

Plate I.

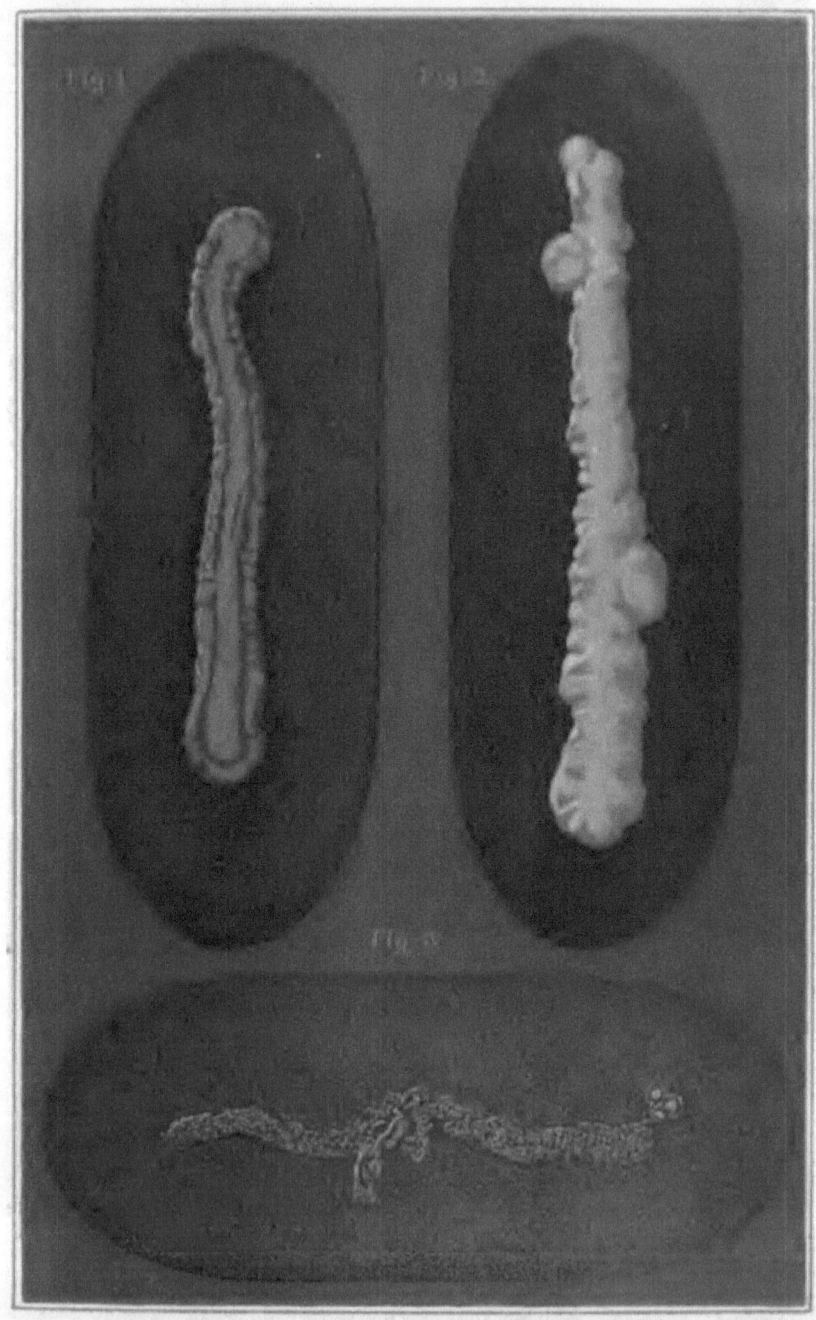

Fig. 1.—Pure culture of gold-colored grape-coccus of suppuration from a furuncle of the lip, on meat-peptone-agar, seen by reflected light.
Fig. 2.—White grape-coccus by reflected light.
Fig. 3.—Chain-coccus of pyæmia by reflected light. (From Rosenbach.)

None of the pus-generating cocci cause what is commonly called putrescence. *Decomposition of tissues, accompanied by the production of foul odors, is always due to the fermentative action of diverse forms of elongated bodies, called bacilli or bacteria.* Plate III, Fig. 8, shows a pure culture of the "*Bacillus saprogenes*," or bacterium of putrescence. Fig. 9 is a pure culture gained from an osteal focus in putrid compound fracture with fatal septicæmia. (Figs. 132 and 133.)

The accompanying chromolithographs were carefully copied from Rosenbach's monograph, and give a very life-like image of the several molds or cultures.

On account of their excellence and truthfulness, a number of Koch's renowned microphotographs, illustrating various forms of microbial growth, have been here reproduced.

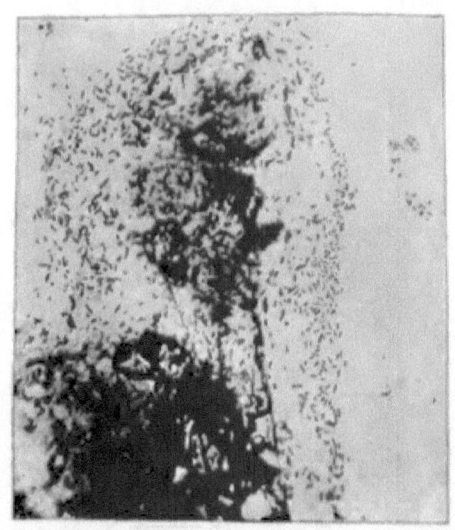

FIG. 136.—Bacteria of blue pus (700 diameters). (Koch.)

## II. PORTALS OF INFECTION.

It is safe to assume that, without exception, all forms of suppuration owe their origin to infection from without. The portals through which the pyogenic organisms known as cocci and bacteria enter the system are, on one side, the lesions of the outer integument; on the other, lesions of the mucous lining of the digestory, respiratory, and urogenital apparatus. The infection of larger accidental or surgical wounds has been treated of in the preceding chapters. Infection through minimal lesions of the skin or mucous membranes and its sequelæ will now receive attention.

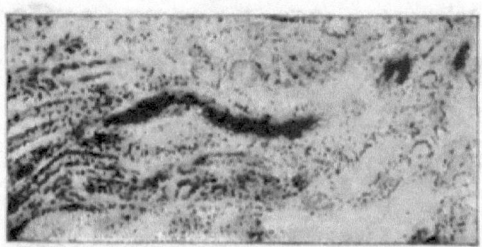

FIG. 137.—Human kidney in pyelo-nephritis. In the center, urinary canal filled with cocci (700 diameters). (Koch.)

1. **Infection through Lesions of the Skin.**—The popular tenet that a wound that bleeds well heals well, is based on correct observation. Sharp

hæmorrhage is very apt to dislodge and carry off particles of filth deposited in the wound from without at the time of the injury; and, further, it signifies an abundant blood supply, good nutrition, hence prompt union. Another point of importance is, that wounds that bleed profusely generally come under the care of a physician, and will receive at once proper attention and protection from further injury.

Small abrasions, lacerations, or punctured wounds that bleed very little, or not at all, have deservedly a bad reputation. If the injuring instrument or object does not inoculate the wound with filth, and subsequent infection is prevented by proper measures, healing will proceed without interruption.

But, as a rule, these wounds are neglected from the outset, because there is scanty or no hæmorrhage. The sharp-edged tool of the mechanic, or the pointed object handled in the daily vocation of the laboring man, is very rarely clean. In certain occupations, as that of the butcher, anatomist, or cook, the hands are frequently injured while in contact with foul organic substances, and the injuring force will at the same time inoculate filth. No hæmorrhage following, and the pain being insignificant, the matter is lightly passed over, and work proceeds without interruption. The cleansing effected by hæmorrhage is absent, the small orifice of the skin is soon filled by lymph and obliterated, and we have to deal with a hermetically sealed focus containing filth, leavened by a certain number of microorganisms, that at once must and do begin to develop and multiply, causing a destructive purulent inflammation.

Not all of these small injuries are infected from the beginning. They may and, as their frequent spontaneous healing proves, are often enough aseptic.

As a matter of fact, they do well at first, and as long as the patient takes care of them. But if, as often happens, the protecting scab is reinjured, and infection by contact with foul matter follows, the consequence is suppuration.

NOTE.—Inflammatory lesions of the skin are fruitful sources of infection, among them eczema the foremost. The intense itching leads irresistibly to scratching, and the small excoriations thus produced are often the portals of infection.

2. **Infection through Lesions of the Mucous Membranes.**—Less numerous than the lesions of the skin, yet productive of frequent mischief, are the traumatic and inflammatory lesions of the mucous membranes. Slight injuries to the lips, tongue, buccal and faucial mucous membrane are very common. In most cases a profuse flow of saliva is instantly produced by a painful injury, and, if hæmorrhage be also present, infection rarely takes place. Healthy oral cavities and their adnexa are especially exempt from infectious processes following injuries. Even gunshot wounds of these parts can heal without suppuration under favorable circumstances:

CASE.—E. L., aged eighteen, admitted to Mount Sinai Hospital, December 7, 1884, with suicidal fresh pistol-shot wound of the tongue, extending from the tip backward to the left side of the base, dividing the organ in two unequal parts. Gunshot perfora-

Plate II

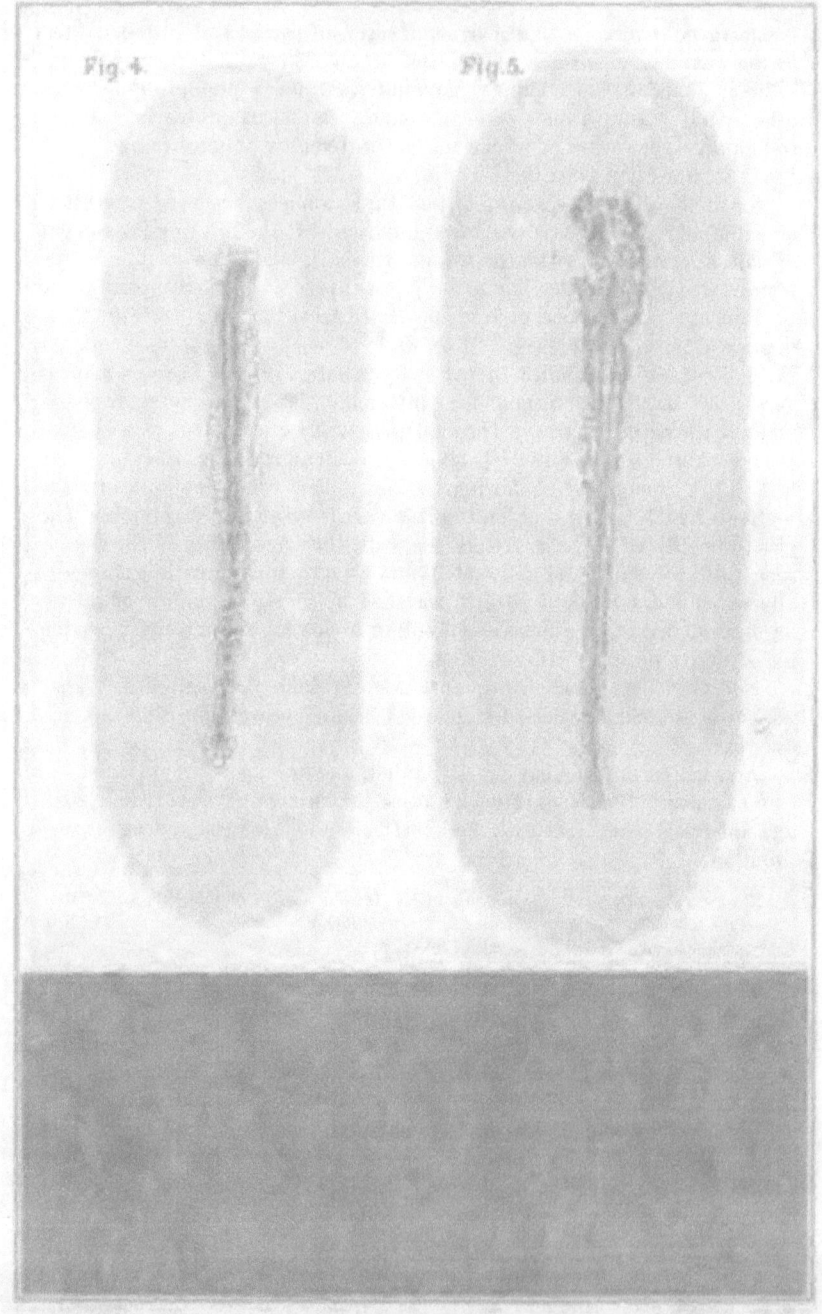

Fig. 4.—Culture of chain-coccus from a case of acute progressive gangrene. Transmitted light.
Fig. 5.—Chain-coccus of erysipelas (Fehleisen). Transmitted light.
Fig. 6.—Chain-coccus of erysipelas by reflected light. (From Rosenbach.)

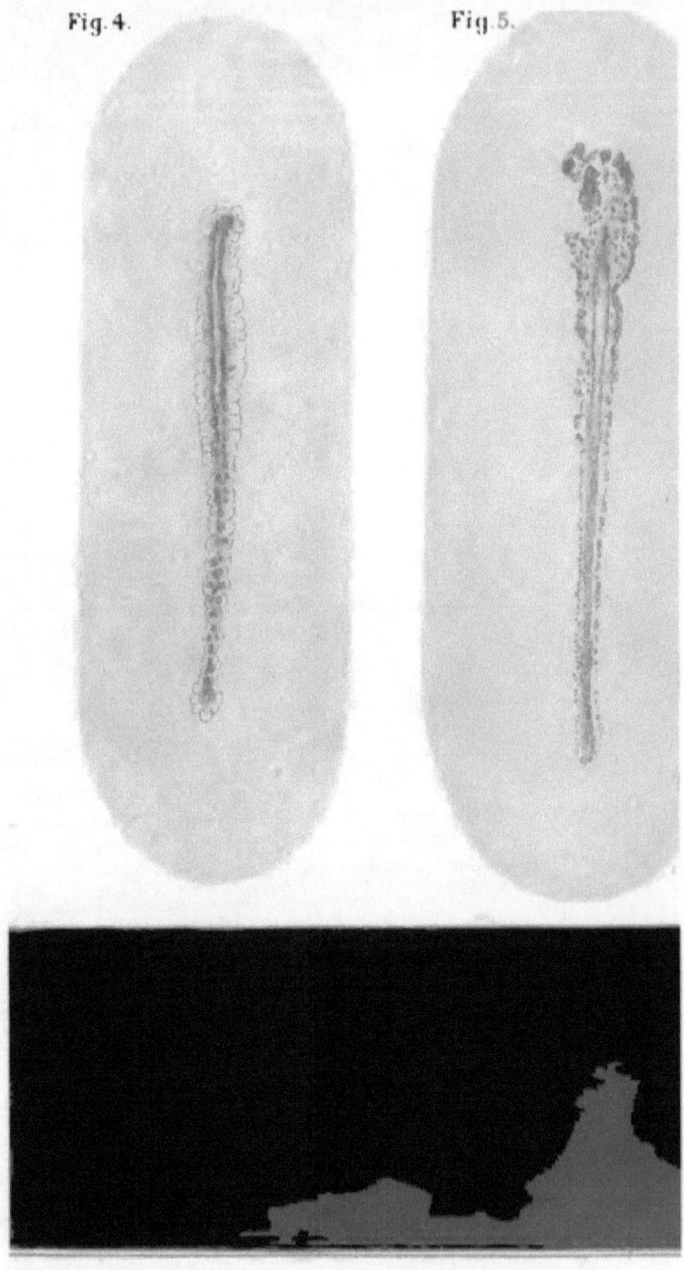

FIG. 4.—Culture of chain-coccus from a case of acute progressive gangrene. light.
FIG. 5.—Chain-coccus of erysipelas (Fehleisen). Transmitted light.
FIG. 6.—Chain-coccus of erysipelas by reflected light. (From Rosenbach.)

## NATURAL HISTORY OF IDIOPATHIC SUPPURATION.

tion of the pillars of the fauces of the left side; gunshot wound of the posterior pharyngeal wall, the point of entrance situated just back of the faucial pillars of the left side, about an inch and a quarter from the median line, all of these injuries being produced by a bullet of 22 mm. caliber. A second non-penetrating gunshot wound on the forehead without a point of exit. Free hæmorrhage from the tongue, and also a stream of arterial blood from the pharyngeal wound. The latter being in close vicinity to the left internal carotid artery, the left common carotid was tied **at** once as a preventive measure, mainly with a view to the possibility of subsequent suppuration and secondary hæmorrhage. The perfect condition of the teeth and oral mucous membrane was noted. **The** lingual wound was lightly rubbed over with **a small** sponge dipped in iodoform-powder; *the pharyngeal wound was not probed*, and hourly irrigation of the oral cavity with weak salt water was practiced. Profuse sweating, perhaps due to reflex vasomotor disturbance, set in, and persisted for **about** forty-eight hours. The febrile movement was very slight, **and** both the operation wound and the gunshot wound on the forehead, being redressed on December 15th, **were** found healed and dry under their iodoform dressings. **The** lesion of the tongue was found granulating and contracting, the perforation of the pillars of the fauces nearly closed, the point of **entrance** in the posterior pharyngeal wall firmly occluded **by a** fresh-looking blood-**clot.** Breath odorless. *December 21st.*—The flattened ball **removed** by small incision **from the top of the head, where it could** be felt beneath the skin. The entire track **of this projectile had literally healed without suppuration.** The pharyngeal wound found also cicatrized over, the ball being **imbedded** near and below the left transverse **process of the atlas, in close proximity to** the vertebral and internal carotid **arteries.** **The head was held inclined to the right** side, erection of the spine and **its** flexion to **the left** being **impossible** on account **of the** intense pain caused by the attempt. This functional disturbance diminished to such an extent within **a few** months that the contemplated extraction of **the** small projectile was abandoned.

Had the patient's oral cavity been foul from putrid **processes** accompanying an acute or chronic oral catarrh, due to dental caries or other causes, suppuration of the pharyngeal wound would have been very probable. **The** danger would have been very much **graver on** account of the possibility of extension of the suppuration and the likelihood of uncontrollable secondary hæmorrhage. *A probing **of** similar wounds **without** a clear and necessary object in view is always a **dangerous** and invariably useless step, and should be refrained from under almost all circumstances.* We may use a clean probe, and the probe **may not be the** carrier of infection; **but** its introduction will break down **the blood-clot, the** natural barrier provided by the organism itself **against infection,** and the probe will leave behind an open channel for the **entrance of possibly** fetid oral mucus into **the** narrow wound.

Next **in frequency to the inflammations** in and about the oral cavity and its **adnexa are** those **due** to **injuries and** other lesions about the anal and uro-genital orifices.

### III. ENTRANCE, PROGRESS, AND LOCALIZATION OF THE INFECTION.

As long as the integrity of the epidermis is preserved, no infection from without will take place. The integrity of the epithelial covering of the mucous membranes **does** not **seem to have the** same **protective power as the** epidermis. This may be explained by the fact that slight injuries of the mucous lining are produced much more easily than those of the skin, and are not readily ascertained on account of the normally moist condition of the parts.

As formerly stated, the slightest denudation, not deep enough to cause hæmorrhage, and just productive of a slight exudation of serum, offers a favorable point of entrance to the virus in the patulous orifices of the lymphatic vessels or lymph-spaces, thus exposed by the injury.

In lacerations or punctured wounds the infective agents are very often deeply inoculated with the point of the injuring article—that is, they are at once deposited in close vicinity to deep-seated lymph-vessels.

In the more superficial forms of injury, the implantation of the virus occurs only in the neighborhood of more superficial lymphatics, and its transmission to the deeper lymph-vessels is accomplished by forces which govern the flow of lymph from the periphery to the center. Aside from the normal current setting toward the thoracic duct, external forces and the play of the voluntary muscles have an important part in hastening the flow of lymph. So, for instance, the pressure exerted upon the lymphatics of the palm by the frequent and vigorous grasping of a tool wielded for a long time with great force, will undoubtedly help to propel the contents of the peripheral lymphatics toward the larger, more deeply situated lymphatic trunks. Or the vigorous contractions of the muscles during mastication will undoubtedly empty the adjacent lymphatics centerward, their action being aptly comparable to that of a force-pump.

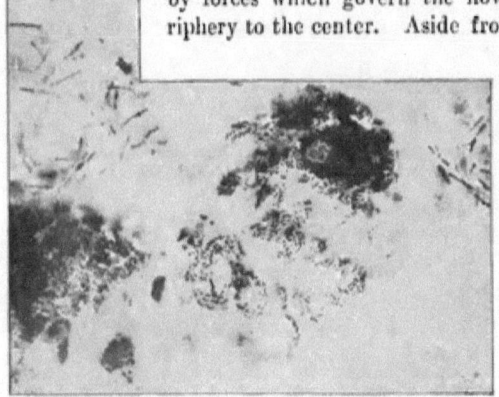

Fig. 138.—Bacilli of anthrax and streptococcus (700 diameters). (Koch.)

What was formerly denoted as external *mechanical irritation* is nothing but this *forcing of pus-generating substances into the open lymphatics* by friction or other pressure due to exercise.

The direction and extent of the spread of the infection by the lymphatics are prescribed by the anatomical arrangement of the lymph-vessels of the region concerned. Thus, on the palmar aspect of a finger, the poisoning will rapidly extend to the periosteum, as the lymphatics all tend that way. In the vicinity of lymph-glands, the infection will promptly extend to them, an intervening lymphangitic streak often clearly denoting the route by which it traveled.

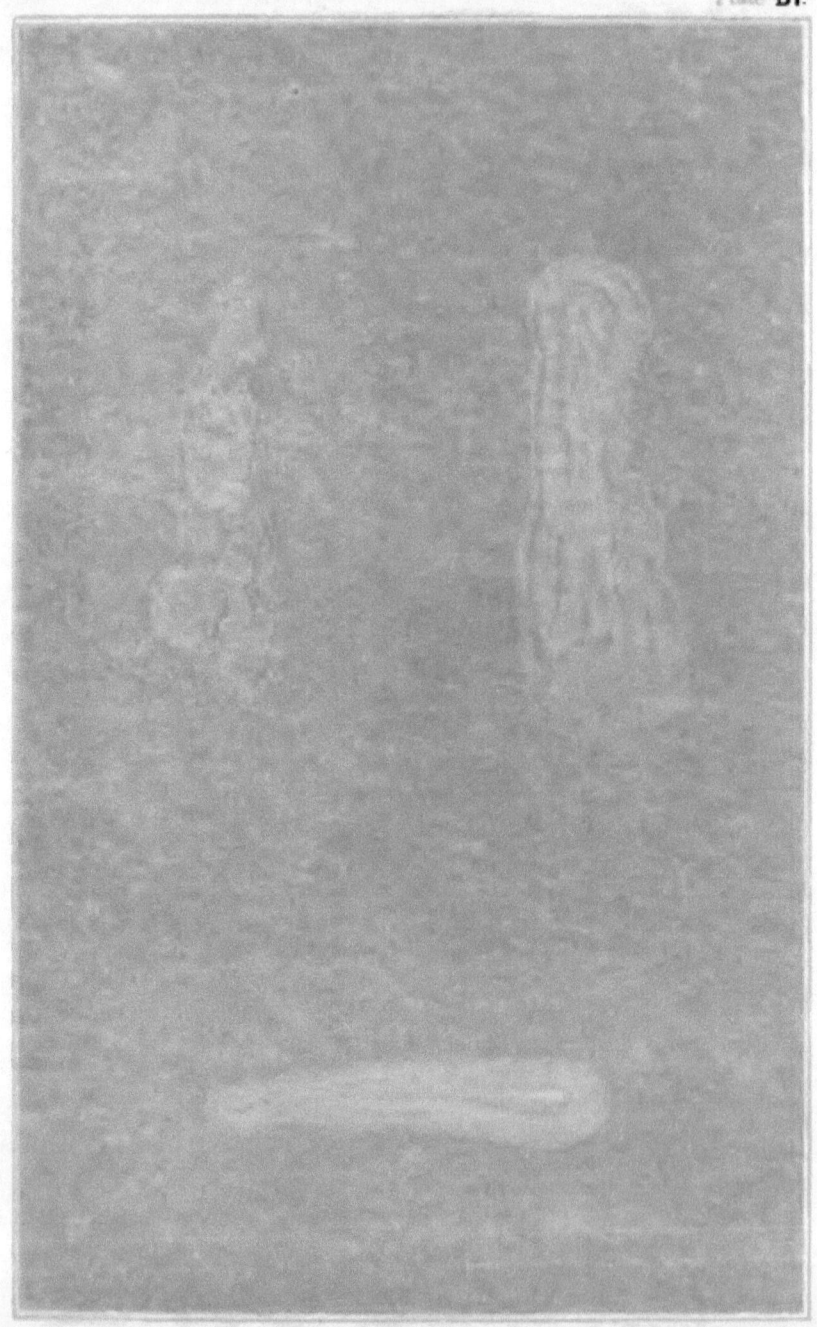

Fig. 1 – Mixed culture of golden and lemon colored and of white grape-coccus from a

# 188 RULES OF ASEPTIC AND ANTISEPTIC SURGERY.

epidermis. This may be explained by the fact that slight injuries of the mucous lining are produced much more easily than those of the skin, and are not readily ascertained on account of the normally moist condition of the parts.

As formerly stated, the slightest denudation, not deep enough to cause hæmorrhage, and just productive of a slight exudation of serum, offers a favorable point of entrance to the virus in the patulous orifices of the lymphatic vessels or lymph-spaces, thus exposed by the injury.

In incersions or punctured wounds the infective agents are very often deeply inoculated with the point of the injuring article—that is, they are at once deposited in close vicinity to deep-seated lymph-vessels.

In the more superficial forms of injury, the implantation of the virus occurs only in the neighborhood of more superficial lymphatics, and its transmission to the deeper lymph-vessels is accomplished by forces which govern the flow of lymph from the periphery to the center. Aside from the normal current setting toward the thoracic duct, external forces and the play of the voluntary muscles have an important part in hastening the flow of lymph. So, for instance, the pressure exerted upon the lymphatics of the palm by the frequent and vigorous grasping of a tool wielded for a long time with great force, will undoubtedly help to propel the con-

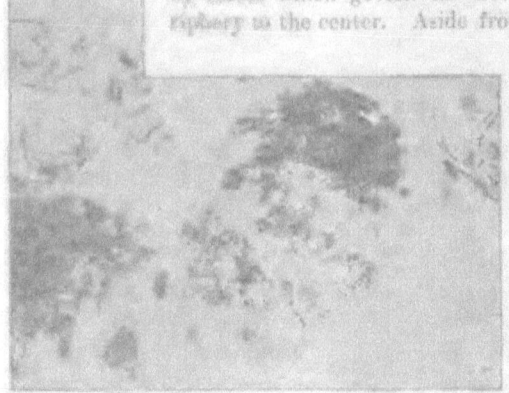

Fig. 114.—Bacilli of anthrax and streptococcus (700 diameters). (Koch.)

tents of the peripheral lymphatics toward the larger, more deeply situated lymphatic trunks. Or the vigorous contractions of the muscles during mastication will undoubtedly empty the adjacent lymphatics centerward, their action being aptly comparable to that of a force-pump.

What was formerly denoted as external *mechanical irritation* is nothing but this *forcing of pus-generating substances into the open lymphatics* by friction or other pressure due to exercise.

The direction and extent of the spread of the infection by the lymphatics are prescribed by the anatomical arrangement of the lymph-vessels of the region concerned. Thus, on the palmar aspect of a finger, the poisoning will rapidly extend to the periosteum, as the lymphatics all tend that way. In the vicinity of lymph-glands, the infection will promptly extend to them, an intervening lymphangitic streak often clearly denoting the route by which it traveled.

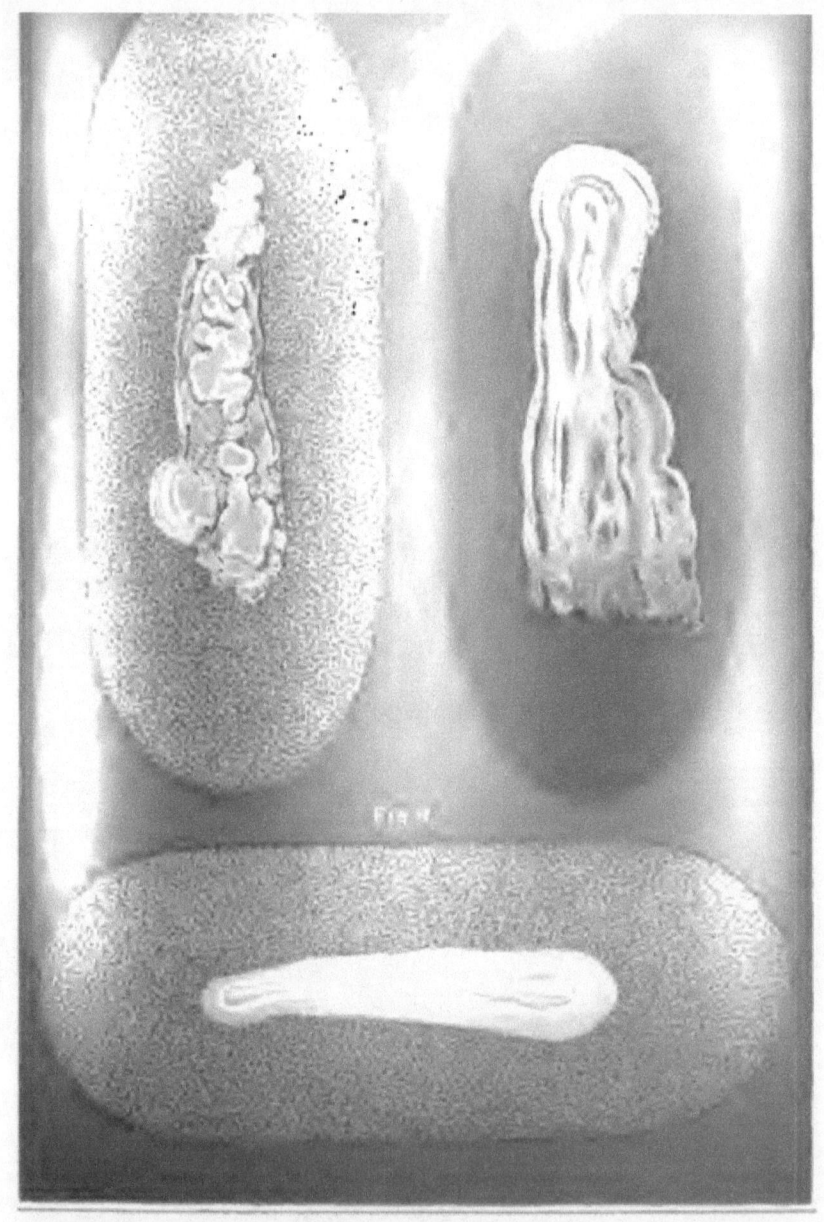

Fig. 7.—Mixed culture of golden and lemon colored and of white grape-coccus from a case of empyema. Reflected light.
Fig. 8.—Common organism of putrescence. Bacillus saprogenes. Reflected light.
Fig. 9.—Bacillus saprogenes from a focus of septic compound fracture. Septicæmia. Reflected light. (From Rosenbach.)

The varying intensity of the infection, dependent on hitherto unknown and varying fermentative qualities of different cultures of micro-organisms, will also greatly influence the rapidity and virulence of the inflammatory process. So much is well established that the intensity of the infection depends, *first*, on the virulence of the invading culture of bacteria ; *secondly*, on the quantity of fungi absorbed ; and, *thirdly*, on the power of resistance—that is, the state of health of the invaded organism.

**Mechanical Irritation.**— *Mechanical irritation by foreign substances* imbedded in tissues, such as bullets, splinters of glass, or a broken-off point of a knife-blade, is also a myth in the old meaning of the phrase. *They never cause suppuration unless infectious substances—that is, microbial filth—be adherent to them* at the time of their being deposited in the tissues. They may cause pain by pressure upon nerves, or may interfere with the play of a joint or a muscle, but, as a rule, never will cause inflammation or suppuration. Well-disinfected steel nails, driven by mallet through femur and tibia after exsection of the knee-joint, are unhesitatingly left imbedded for thirty or more days, never causing any irritation (see Exsection of Knee-Joint, page 319.)

CASE.—In 1882 a young blacksmith presented himself in the surgical division of the German Dispensary. An angular foreign body could be distinctly felt under the skin on the palmar aspect of the right forearm, midway between elbow and wrist, causing pain by impinging. The body had appeared only since a few weeks. Near the carpus a transverse cicatrix was to be seen, and the patient explained that he was cut there during a drunken brawl two years ago, and that a surgeon had tied an artery and sewed up the wound, which had healed without suppuration. Ever since then he had worked at his trade without any inconvenience until within a few days. From the incision made over the projecting body, a blackened knife-blade, four inches long and five eighths of an inch wide, was extracted, to the greatest astonishment of the patient. The small wound closed promptly.

Here we saw a massive, sharp-edged foreign body lie imbedded for two years between the muscles of the forearm without any inconvenience to the patient, until the angular base of the blade had worked out under the skin. Why did it not cause suppuration ? Apparently the blade must have been newly ground, or at any rate very clean, when it broke off in the arm of our blacksmith. Had a considerable amount of infection been carried along with it at the time of the injury, its presence would not have been overlooked so long.

*Dead organic substances*, as, for instance, blood, or cubes of animal tissues, such as muscle, tendon, or portions of liver or bone, were taken from a freshly killed animal, and introduced into the abdominal cavity of a number of other rabbits under strict antiseptic precautions. In a very large proportion of cases no reaction whatever followed. The animals being killed, it was found that blood was absorbed outright ; that muscle, liver, tendon, and bone were encapsulated ; and that their structure was gradually invaded by granulation tissue—disintegration and final absorption following after a while, proportionate to the density of the implanted bodies. In cases where the ordinary aseptic measures had been omitted, septic purulent peritonitis followed as a rule.

NOTE.—The most remarkable of Dr. H. Tillmann's experiments (Virchow's "Archiv," Bd. lxxviii, 1879) is that concerning a rabbit, in the abdomen of which an entire rabbit's kidney was deposited without causing any harm whatever. The animal being killed forty-seven days after the operation, the implanted kidney was sought for in vain, as it had disappeared by absorption, the only vestige of its former presence being a spot of tough cicatricial tissue, denoting the locality where the foreign body was attached by exudations.

This experimental observation is fully borne out by the experience gained in numberless ovariotomies, where massive pedicles, dead through stoppage of their circulation by ligature, are dropped back harmlessly in the peritonæum, to be finally absorbed—that is, they will do no harm if a culture of bacteria is not deposited on them by the operator.

**Chemical and Caloric Irritation.**—The common experience that certain acutely irritating substances, as, for instance, croton-oil, oil of cantharides, turpentine, concentrated solutions of corrosive sublimate, and others, brought in contact with living tissues, always would produce suppuration, represented a serious gap in the theory of the microbial origin of suppuration. If invariably proved, it would be more than a defect, as it would positively contradict the thesis that suppuration is exclusively and *always* the result of the development of micro-organisms. The experiments of Councilman,* who introduced under the skin of animals small glass globes filled with sundry irritating substances, and then crushed them, all led to suppuration. Scheuerlen † and Klemperer,‡ however, in going over Councilman's experiments, showed that his procedure was faulty, inasmuch as sufficient precautions had not been taken to exclude the introduction of microbes along with the croton-oil, etc. They moreover positively demonstrated by a very large number of successful experiments that, whenever thorough aseptic cautelæ were observed, suppuration never followed the introduction of even very considerable quantities of the mentioned substances. Small quantities caused some exudation of plasm, and then were absorbed outright. Afterward the fragments of the glass receptacle were found imbedded in a film of new-formed connective tissue. Larger quantities of croton-oil, for instance, caused a coagulation necrosis of a limited mass of tissue, which was found dense, bloodless, and of a yellow color. These nodes of necrosed tissue were gradually absorbed, *suppuration never following the experiment*. This fact is in full accord with other incontestable facts of the same character, as, for instance, the absorption of necrosed ovarian stumps in the abdominal cavity if there be no microbial infection present.

*Caloric irritation*, or even an outright destruction of tissues by excessive heat, presents a similar state of things. As long as microbial infection is successfully kept away from the exudations in burns of a milder charac-

---

\* Virchow's "Archiv," 1883, vol. xcii, p. 217.
† "Archiv für klin. Chirurgie," vol. xxxii, p. 500.
‡ Prize essay, Berlin University, "Zeitschr. für klin. Med.," 1885, vol. x, p. 158.

ter, and from the eschar and exudations in severer forms, no suppuration will follow. The modern use of the thermo-cautery in the peritoneal cavity, in joints, and, as a matter of fact, in wounds of the most various character and of all anatomical regions, is followed by uninterrupted union in all cases where, at the same time, adequate aseptic measures are employed. An eschar or a mass of dead tissue, whether produced by ligature, or chemical corrosion, or red heat, will never assume the irritating character of a "foreign body," in the meaning of the term as presented by the tenets of an older pathology, if the decomposing action of the presence of microorganisms is excluded by proper measures.

The behavior of *superficial burns of the skin* is fully in accord with the facts just presented.

If a bleb be raised, and is left unbroken and dry, its contents will be absorbed, and the epidermis will settle back into its normal relation to the cutis. It will turn into a dry scale, and will peel off within ten to twelve days, exposing the tender new epidermis.

How different is the course of a burn if the epidermis is torn off by accident or intentionally, and the exudations are thus exposed to the invasion of micrococci! If the surgeon do not employ timely disinfection and the application of a protective dressing, suppuration of the exposed cutis, with all its accompaniment of pain, long-continued granulation, and a very tardy healing, will follow.

## IV. DEVELOPMENT OF PHLEGMON.

From the moment that a sufficient quantity of active fungi have established themselves within the living tissues, remarkable local and general phenomena develop, known under the name of *inflammation and septic fever*.

Our object is not research into, but rather a lucid explanation of, the essence of inflammation, as understood and accepted by contemporary authorities. Hence a brief sketch of the leading features of the process is deemed sufficient.

Micrococci find a most favorable pabulum in dead or devitalized organic substances. The living tissues offer a decided resistance to the ravages of the micro-organism. The spontaneous limitation and occasional unaided cure of some forms of suppurative inflammation prove this assertion.

Bacteria can not thrive on the products of decomposition: they need for their sustenance dead but undecomposed albuminoid substances. As soon as the supply of dead animal tissue is exhausted, the micro-organisms starve and perish. Their *spores* or seeds are left behind dormant, but will become active if fresh pabulum is offered under favorable circumstances.

This explains the fact that *fresh cadavers or animal substances in the recent stages of putrescence are much more infectious than those that are in a progressed state of decomposition*. The varying intensity of different cases of infection seems to depend in a great measure upon the varying degrees of vitality of different microbial cultures. It seems to admit little doubt that the great majority of dangerous wound infections are brought about by the importation of considerable masses of very active, rapidly proliferating micro-organisms in the shape of "lumps of dirt," as Lister graphically puts it, taken from various sources of recent putrescence, so abundant in all human surroundings. The dry spores floating in the air will be easily taken care of by the living tissues, if pollution of the wound by *gross dirt*—that is, masses of organic matter in active decomposition—is avoided.

Every injury causing a wound destroys the vitality of those cells that lie in the direct path of the cutting or lacerating object. The blood and lymph exuded from the vessels coagulate, and also represent dead matter.

If a number of active micrococci are implanted into the bottom of the wound, they will at once multiply, using the blood-clot and its extensions into the blood-vessels, together with the adjacent dead or devitalized tissues, as a welcome soil for their development. This fermentative decomposition produces from its very beginning certain alkaloids or chemical, extremely poisonous substances, *the ptomaïnes*, that are very diffusible. By dint of this diffusibility, the adjacent vasomotor nerves at once come under their toxic influence, as the result of which their strong dilatation ensues, which becomes manifest in the shape of an *active hyperæmia*, "*rubor*."

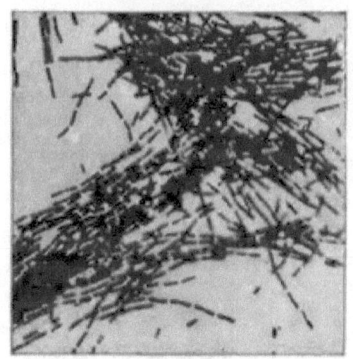
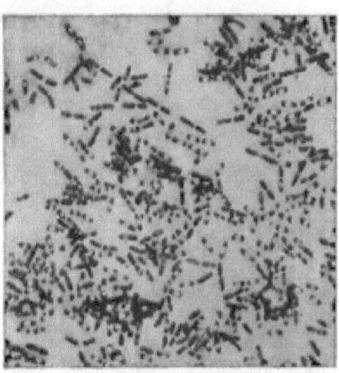

Fig. 139.—Bacilli of anthrax (700 diameters). (Koch.)

Fig. 140.—Formation of spores in anthrax bacilli (700 diameters). (Koch.)

The blood passing through the adjacent arterioles and capillaries seems also to become altered; the red blood-corpuscles become packed and finally stagnate in the capillaries and smaller arteries. The walls of these vessels, including the veins, lose their impermeability, and a number of white and often red blood-corpuscles emigrate into the surrounding tissues, densely infiltrating their interstices, thus producing the characteristic *swelling*, "*turgor*."

As a consequence of the increased blood-supply, possibly also of the active chemical process, a marked increase of the local temperature is observed—"*calor*." And, if we add that *pain* of the parts thus affected is never absent, we have completed the classical cycle of the four cardinal symptoms of inflammation—"*rubor, calor, turgor, dolor.*"

NOTE.—The causes of local pain may be several. The initial pain is very likely due to a direct influence of the ptomaïnes upon the sensory filaments. Direct pressure caused by the dense infiltration may also have some influence; but the most acute pain is undoubtedly effected by the actual destruction of the nerve-tissue during the advanced stages of suppuration.

Stagnation and dense infiltration finally produce a very high degree of tension, leading to compression of larger afferent vessels. The infiltrated portions, devitalized by suppression of the normal circulation, readily succumb to the inroads of the millions of micro-organisms, and actual necrosis rapidly follows. The last stage of textural destruction is the final liquefaction of the tissues and infiltrating leucocytes, aided by the exudation of large quantities of lymph-serum from the adjacent unobstructed blood-vessels, and thus the *formation of an abscess* or a cavity filled with lymph-serum, myriads of dead white blood-corpuscles (pus-cells), and quantities of shreds of necrosed tissues, is accomplished.

The veins also participate in the disturbance. Coagulation of their contents—thrombosis—takes place, and existing stagnation is materially augmented.

The deleterious part played by thrombi in the causation of metastases will be later mentioned.

When a septic inflammation of sufficient extent and intensity has been well advanced, the great tension of the parts will necessarily cause an overflow of the most diffusible contents of the focus into the surrounding efferent vessels—the veins and lymphatics. The ptomaïnes, thus entering the general circulation, will at once produce systemic intoxication, manifested by a very marked rise of the body-heat, rigors, sickness, headache, delirium, and general dejection—in short, a deep-going alteration of the nervous system, known as *septic fever.*

## V. SPREAD OF SUPPURATION.

The way of the extension of septic textural destruction is twofold. It takes place, *first,* by a direct infiltration of the tissue-interstices by columns and hosts of the immensely prodigious micrococci—that is, by an immediate growth and extension of the microbial colony; and, *secondly,* on the way of the lymphatics, openly communicating with the focus of suppuration. Into these, bacterial masses, or pus charged with micrococci, are forced by the hydrostatic pressure exerted by the tension within the abscess.

If the parts affected are composed of loose tissues, the spread will be rapid and extensive; if the parts are dense, the inflammation will remain localized as long as the density of the tissues (fasciæ, for instance) will resist the pressure of the secretions. But, as above mentioned, this very pressure, or tension, involves another great danger. The afferent blood-vessels become thereby occluded, and the resulting stagnation generally leads to extensiv necrosis.

As long as new areas of tissue become infected through the lymphatics, constant high fever and increase of the local symptoms is the rule. An incision laid through the parts at an initial stage of the process will expose a honeycombed mass of tissue, containing a number of small foci, some of them confluent, and all filled with pus, the intervening substance being discolored, pale, or more or less broken down and softened, or sloughed.

In direct proportion with the spread of the infection and the multiplication of suppurating foci, is the magnitude of necrosing areas, occasionally involving an entire limb. Organs of scanty vascularity, as, for instance, fasciæ, tendons, and bone, are the first to succumb.

The microbial colony begins to show signs of exhaustion in most cases after a more or less prolonged period of florescence. The parasite becomes less prolific; its direct ingrowth into the tissues is less and less active, and the life of the white blood-corpuscles, densely infiltrated into the marginal parts of the abscess, is not compromised by their invasion with micrococci. They are not converted into pus, but withstand the attack of the parasites and remain a mass of embryonal connective tissue, that forms a dense wall inclosing the suppurating cavity. This embryonal connective tissue uniformly permeates all the adjacent parts, among others the lymphatics and thrombosed veins, *forming a more or less effective barrier to the extension of the septic process* and to the absorption of deleterious soluble substances into the general circulation.

*This self-limitation of the spread of septic destruction* is generally marked by a remission of the intensity of the general and, in a measure, of the local symptoms. At this stage, according to ancient notions, *the abscess has matured.*

NOTE I.—For obvious reasons, the incision of a *matured* abscess is generally followed by a rapid healing of the cavity. The detachment and liquefaction of the contents of the abscess are well completed, the extent of the process is well rounded off, as it were, by the wall of newly organized connective tissue, and repair can commence under favorable circumstances.

Nevertheless, it must be strongly urged that the most dangerous abscesses never ripen—that is, show no tendency to self-limitation—and that the measures ordinarily employed for maturing them, such as vigorous poulticing, only tend to intensify their malignity, and to cause irreparable damage, that an early incision might have averted. A case vividly illustrating the perniciousness of thoughtless poulticing is quoted on page 248.

NOTE II.—Not every bacterial infection leads to suppuration, although the rule suffers very few exceptions indeed. One of the exceptions is illustrated by the following: CASE.—I. N., laborer, aged twenty-four, was admitted to the German Hospital in March, 1885, with a very painful, hard, and massive swelling of the axillary contents, the skin being œdematous and angry-looking. High fever and a good deal of sickness were observed, so that pus was thought to be indubitably present. An incision was declined, whereupon a poultice was ordered, with the expectation that it would hasten the process by stimulating suppuration. For a day or two the intensity of the symptoms increased rather than otherwise, several sharp chills followed with profuse sweating, after which came a marked improvement of all the appearances of the case. The redness and swelling diminished, the fever disappeared, and the patient left the hospital cured, glorying in his triumph of endurance over diagnostic acumen.

To explain such cases, it is necessary to assume that, under the powerful stimulation of the local circulation by the cataplasm, the products of bacterial fermentation, bacteria, or even pus itself, are washed away by the lymph-current into the general circulation, where the pto-

## NATURAL HISTORY OF IDIOPATHIC SUPPURATION. 195

maines provoke constant or explosive symptoms of general intoxication, such as high fever or severe chills; the bacteria themselves, however, perish, the living oxidized blood forming an unfavorable pabulum for their existence and propagation. In accord with this theory is the well-known fact that wounds of very vascular tissues, such as those of the face, for instance, will heal without suppuration even when there is a good deal of inflammation of their edges, with pain and fever, denoting the presence of a certain amount of septic infection. The poorer the blood-supply of a part, the greater the destruction wrought by an infectious process.

If the abscess is not evacuated at the stage of maturity through a fortunate spontaneous or an artificial opening, the relief felt by the patient will be a short-lived one. The marginal wall of embryonic connective tissue—that is, the area of *granulations*—will continue to shed lymph and detached leucocytes into the abscess cavity. The intramural pressure will steadily increase until it rises to such a degree as to overcome, on hydrostatic principles, the resistance of the soft plugs of living leucocytes, which occlude the orifices to the adjacent connective-tissue planes and lymphatics or veins. One or another of these offering the least resistance, will be forced out of the way, and a new invasion of hitherto unaffected regions results, with a repetition of all the initial local and general symptoms, marking an extension of the process.

NOTE.—The notion that the law of gravity alone regulates the spread of abscesses is an erroneous one, as it is well known that many forms of suppuration extend in a diametrically opposite direction to the force of gravity. The local spread is prescribed by the direction of the loose connective-tissue planes separating and connecting the different organs, and is mainly influenced by hydrostatic law. Perforation always takes place where resistance is the least.

The infiltration of the tissues by micrococcal colonies sometimes extends to the close vicinity or into the very walls of larger veins. Thrombosis is the direct result, and, if the microbial invasion includes the thrombus, after the detachment of the slough of the vein and the liquefaction of the thrombus, a direct communication of the general circulation with the abscess cavity may be established. The slightest external pressure may serve to throw enormous masses of pus and micro-organisms into the general circulation at this critical period, causing rapid death by explosive septicæmia. In these cases the microscope will demonstrate the presence of micrococci in the entire blood-mass.

In other cases, either spontaneously or in consequence of active movements or external manipulations, a portion of a septically infected thrombus may be detached. The blood-current will at once carry it into the right auricle and ventricle, whence it will find its way into one or another branch of the pulmonary artery, to be there arrested in the shape of an embolus. Around this a hæmorrhagic infarction of the adjacent pulmonary tissues will form, within which a new bacterial colony will become established, leading to the formation of a secondary or *metastatic abscess*. Its appearance is always signalized by a severe rigor.

Thrombosis of adjacent pulmonary veins, and detachment of portions of the new thrombus, followed by its transportation into the left side of the heart, and hence into distant smaller-sized arteries of the body, will lead to a repetition of the metastatic process and its febrile accompaniment, until a number of joints, lymph-glands, the liver, in fact, almost all the organs, become the seat of secondary abscesses.

This is the classical type of well-developed *pyæmia*, formerly so common in all surgical hospital wards, but now become a rare phenomenon wherever the leaven of the Listerian spirit has permeated surgical practice.

This form of microbial colonization of the entire human body baffles every plan of treatment, and almost invariably leads to the destruction of the organism. It is as good as incurable, *but it can be prevented;* hence it is the moral duty of every physician to do everything in his power to avert this form of mischief.

NOTE.—*Recovery of a case of well-developed pyæmia* is so rare that recording the following case seems permissible. The notes were kindly furnished by Dr. A. Caillé, with whom the author saw the patient in consultation at his home in Williamsburg:

"Henry Huhn, an elderly man. Enormous carbuncle over left scapula; necrosis of fasciæ and subcutaneous connective tissue from clavicle to seventh rib posteriorly, the result of three weeks' neglect (poulticing).

"Energetic treatment (by Dr. Caillé) with knife and irrigation (carbolic). Well-marked symptoms of pyæmia; general furunculosis of trunk.

"*August 16, 1880.*—Consultation with Dr. Gerster, who advised tonic treatment and daily *full baths in weak bichloride-of-mercury solution*, together with frequent irrigations with camphorated water. Temperatures at this time on an average 102° Fahr. Pulse, 120 to 140. Dyspnœa, chills, and sweats. Improvement noticeable, but slow. In September, suppuration of almost all the lymph-glands took place within one week, without redness or tenderness, so that at one time a tenotomy knife introduced almost anywhere would draw pus. Subsequently extensive and painful periostitis and abscess at upper third of right tibia developed. About this time *examination of urine* revealed a large percentage of sugar. The patient's diet was properly regulated, and his urine was free from sugar five months later. Mr. H. has since been, and is to-day (December 23, 1886), in excellent health."

It will be noticed that a methodical use of a mercuric lotion was advised by the author several years before Kuemmel's and Schede's experiments brought corrosive sublimate so prominently to the notice of the medical world as an excellent disinfectant. The recommendation was based upon the long-known good influence that corrosive sublimate has upon acne pustulosa of the face. Its application in the shape of a full bath suggested itself by the extension of the affection to almost the entire skin, and by the enormous difficulty in cleansing and dressing the innumerable sores of the patient. Since that time the author has employed the *permanent bath* in another similar case, to the great relief of the patient and his attendants. Twice daily the bath was charged with corrosive sublimate (1 : 5,000) for an hour, after which the solution was drawn off, and substituted with a weak salicylic lotion. The remarkable relief brought about by the immersion of the entire body was due to the circumstance that, *first*, the frequent and extremely painful change of dressings could be dispensed with; and, *secondly*, that, according to hydrostatic law, *the buoyancy of the immersed body relieved to a very great extent its pressure upon the couch spread in the bottom of the bath-tub.* The spread of the bed-sores ceased. Before his attack, the patient had been in very weak health. After three or four seizures by collapse, relieved by increase of the temperature of the bath to 110° Fahr., he succumbed to heart failure.

The contents of the preceding pages have in a rough way illustrated the essence of cellular phlegmon, or the *suppuration of connective tissue,* inelegantly denoted in text-books as "*cellulitis.*"

For obvious reasons *lymphatic glands* very often become the seat of microbial proliferation. Their direct communication with a numerous set of lymphatics and their filter-like structure naturally lead to ready absorption and detention of noxious substances. In this characteristic is to be sought a by no means insignificant protective quality of the lymphatic glands against general invasion of the body by microbial masses.

The difference exhibited by lymph-gland abscesses in comparison with the ordinary forms of phlegmon is due to their anatomical structure and situation. Their strong capsule will resist destruction for a comparatively long time, thus preventing for a while invasion of the vicinal tissues. But the internal tension of a glandular abscess soon becomes very great, and will lead to extensive mortification by compression of vessels.

The anatomical situation of many lymph-gland abscesses, their deep seat and close vicinity to large vessels, the pleura, the fauces, and larynx, invest them with additional importance, both as regards the danger peculiar to their locality, and the technical difficulty of their treatment.

*The skeleton* is fortunately a comparatively rare seat of bacterial infection. The fearfully dangerous and destructive character of *acute infectious osteomyelitis*, or "bone phlegmon," is due to the rigidity and unyielding nature of the periosteum and bone tissue, which lead to rapid occlusion of the blood-vessels, and extensive, often widely disseminated necrosis. The deep situation of the bones renders the symptoms of this form of suppuration extremely violent and dangerous, and increases the difficulties of treatment.

NOTE I.—The so-called habituation of butchers, cattlemen, and anatomists to infection seems to be based rather on structural changes of the skin of their hands frequently exposed to contamination, than to a real habituation, such as is, for instance, brought about by vaccination against the small-pox. That the system of these persons does not become hardened or accustomed to the septic virus is proved by the fact, that phlegmonous processes will readily establish themselves, and develop in the ordinary way, *if the infection occur elsewhere than on their hands*. A more plausible explanation of this apparent immunity will be found in the state of the lymphatics of the integument. Having been the seat of frequent more or less intense attacks of inflammation, they become obliterated and distorted, as it were, by cicatricial changes in and around them. That recent or old cicatricial formations do not possess large-sized lymph-vessels is well known, hence absorption through them of corpuscular elements into the deeper lymphatics will be difficult and scanty. In short, the chronically inflamed state of the skin covering the hands of these persons offers in its infiltrated condition an effective protection against the deep-going or massive implantation of micro-organisms through superficial lesions.

Parallel with this state of things seems to be the well-known fact that children subject to frequent attacks of septic tonsillitis or diphtheria rarely succumb to the disease. Penetration by bacterial elements of the dense cicatricial tissue left behind by many preceding attacks is difficult, and absorption of the ptomaïnes through the scanty lymphatics is very limited. Hence the process soon becomes exhausted through lack of pabulum to the microbial growth. A certain quantity of viable spores remain imbedded in a follicle, to again develop their activity as soon as a simple catarrhal inflammation of the pharynx will have prepared the soil for their renewed growth.

Diphtheria in children who never had been subject to the disease is a much more serious matter. Unchanged tissues with open lymphatics are attacked here. The conditions for local microbial proliferation and invasion of the tissues, and for absorption and systemic intoxication, are much more favorable then, and, as is well known, often lead to unavertable death.

The comparative safety of all operations performed within the limits of a preceding but terminated inflammation—that is, within recent or older cicatricial tissue—is very well known to all surgeons. Reamputations, many joint exsections, almost all necrotomies, rarely give any serious trouble, even if the antiseptic measures taken were not very complete. The infection of an amputation wound made through healthy tissues is much more serious, and its avoidance more difficult, as countless lymphatics and large, newly opened, intermuscular, loosely knit connective-tissue planes offer numerous recesses and countless channels for the reception and unimpeded extension of infection.

Therefore the statistics of amputation wounds have been very appropriately selected as a uniform and reliable test of the value of the different forms of wound treatment.

NOTE II.—Infection through *minute injuries to a granulating surface* by inoculation of active micrococci is the frequent cause of suppurations interrupting the course of repair. Rough treatment of a granulating wound by tearing off the adherent dressings will necessarily lacerate the tender granulations matted into the meshes of the fabric, thus causing minimal hæmorrhage. If an unclean probe, or finger-nail, or nitrate-of-silver stick, previously used on a virulent case, and then applied to the granulations, should carry and deposit some active micrococci into one of these minute lesions, an ulcerative process of the granulations will ensue, and, if the ulceration extend into adjacent tissues, phlegmon will develop. *Granulations should always be covered by "protective" before the application of gauze or other dressings.*

## Conclusions.

Suppuration is always undesirable and dangerous, and, if possible, should be avoided by all means. Its essence is textural destruction and death, and systemic intoxication. The phrase "healing *by* suppuration" is an absurdity, is misleading to the student, and should be banished from text-books. As a matter of fact, healing never takes place while active suppuration lasts; it occurs only after the limitation and termination of suppuration, not *by* it, *but in spite of it.*

The expression "laudable pus," as applied to the contents of an abscess during one of its stages of spontaneous limitation or maturing, is also misleading. *Pus is never laudable;* it always is a menace to the health and integrity of the animal organism. Suppuration is a treacherous ally, and its aid should never be invoked by the modern surgeon, or at least should be shunned as long as other ways of curing an ailment remain untried.

### VI. DIAGNOSIS AND TREATMENT OF PHLEGMON.

#### 1. *General Principles.*

The way to the cure of phlegmonous processes is indicated by the manner in which unaided nature occasionally accomplishes it. If the direction in which suppurative destruction progresses should luckily be outward—that is, toward the skin—perforation and spontaneous evacuation of the abscess cavity will occur. If by another lucky accident this perforation should happen at the time of "maturity," or the comparative repose of the destructive process, a complete evacuation of the deleterious contents will take place, followed by a decreasing sero-purulent and bland discharge, and by contraction and final occlusion of the cavity.

But nature unaided is a very poor surgeon. Very often destruction does not tend toward the skin; its natural tendency is to spread in the direction of least resistance, that is, along the cellular tissue, and, by the time that spontaneous openings establish themselves, the damage to deep-seated organs may be very extensive. The coincidence of maturity and perforation is also rare. In its absence the perforation will not lead to complete evacuation, and the septic process will persistently extend in one or another direction, not relieved by such incomplete drainage. Lastly, natural drainage by perforation will often be located in the most unfavorable place, and will not be ample enough for the escape of large masses of pus and of sloughing tissue.

The most direct indications for the cure of phlegmon are offered by a clear understanding of the natural history of its causation and development, as presented in the foregoing pages.

*One or more properly made incisions, followed by effective drainage*, will at once empty the focus of most of its infectious contents, relieving at the same time the dangerous amount of tension.

Infected tissues not yet liquefied, and still adherent to the walls of the abscess, must be disinfected by more or less frequent or permanent irrigation with a germicidal lotion. Finally, all conditions tending to impede free arterial and venous circulation must be eliminated by proper position—that is, elevation of limbs, removal of constricting dressings or clothing.

The necessity of rest—that is, the avoidance of all mechanical injury—is a matter of course.

(*a*) **Superficial Suppuration, or Septic Ulcer.**—Inspissation of the discharges of an infected superficial lesion will, by the formation of a crust, often prevent proper drainage, causing a more or less complete occlusion or retention. The gentlest way of detaching these is by the application of a warm dressing of gauze moistened with a two-per-cent solution of carbolic acid, evaporation of which should be guarded against by an external layer of rubber tissue or oiled silk. After due softening under this warm, moist dressing, the overlapping epidermidal masses, hiding small recesses, should be laid open by cautiously clipping away their undermined edges with curved scissors. *This can be done without causing the least pain.* Thorough disinfection by the lotion contained in the dressings will thus be possible, and the diffusible qualities of carbolic acid will not fail to exert their beneficial disinfecting influence upon the germs scattered through the vicinity of the ulcer. Its yellow coating, consisting of a superficial layer of mortified tissues, will be cast off, the angry look of the neighboring skin will disappear, and the remaining healthy granulations will soon be cicatrized over.

*Streaks of lymphangitis* extending toward the pertinent lymphatic glands should be well salved with mercurial ointment. But if their cause—the septic state of the ulcer—be removed, they will disappear without special treatment.

(*b*) **Cutaneous and Subcutaneous Phlegmon.**—This graver form of suppuration is marked by violent local and general symptoms. High fever, with rigors, the general sense of sickness, headache, and a foul tongue and breath are present. The skin over the focus of infection becomes deeply inflamed, œdematous, and shows dense infiltration, manifested by hardness and pitting. The constant gnawing pain puts sleep out of the question, and the spreading of the affection over new areas of tissue is evident.

## Cataplasm or Incision?

The question whether resolution of the gathering by topical applications, hot or cold, should be attempted, or immediate incision should be resorted to, is of great practical importance, and not always easy to determine.

*The intensity and extent of the process should be herein the main guide.* The consideration that an incision is after all the most effective antiphlogistic measure, affording relief from tension, evacuating a very large proportion of the noxious substances, and permitting the direct application of antiseptics—in short, that it promises prompt success, conserves a large part of the affected tissues, saves much pain and suffering, and averts local and general danger—should stand foremost in the surgeon's mind, whose persuasive authority ought to gain the patient's consent to an early operation. Especially where the rapid spread of the affection and grave general symptoms make prompt relief urgent, dilatory measures and cowardly temporizing are improper. *The cataplasm is resorted to not only to allay the patient's pain and fear, but often serves as a convenient mantle to hide ignorance or indecision.*

*Carbuncle* represents the most pronounced form of cutaneous phlegmon, and its treatment, given hereunder, may, with due modifications, serve as a type of the therapy for the entire class of cutaneous suppurations.

Out of motives of humanity, and because it offers the surgeon time and deliberation, so necessary for thorough work, anæsthesia is always advisable,—in many cases indispensable. After the usual preparations for an antiseptic operation, a free incision should be made through the middle of the inflamed area, penetrating through the skin to the fascia. One or more small foci filled with pus will be thus opened. If their number be great, two or three more parallel incisions should be added. The engorgement or hard infiltration of the adjacent skin will be admirably removed by *Volkmann's multiple puncturing* (Fig. 141). The blade of a narrow, straight bistoury or tenotomy knife is grasped about one third of an inch from its point, and is thrust in quick succession thirty, forty, or, in very extensive cases, a hundred times through different parts of the infiltrated region. The punctures should be evenly distributed. A large quantity of bloody lymph, or occasionally, if a vein be hit, pure blood will escape, and the swelling and hardness will at once be markedly reduced. No attempt should be made to check this escape of blood or serum, as coagulation will soon stop the flow. Thorough irrigation with corrosive-sublimate lotion, packing of the deeper incisions with strips of iodoformed gauze, and an ample *moist*

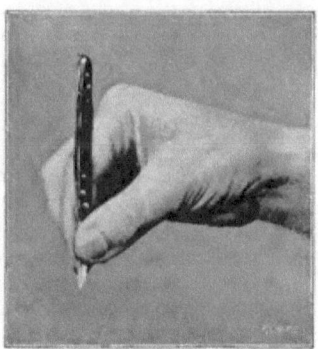

FIG. 141.—Attitude of the hand for multiple puncture.

*dressing*, held in place by loose turns of bandage, will complete the work. An immediate fall of the temperature, with marked local and general relief, will reward both patient and surgeon. Daily, later on, a rarer change of dressings will lead to a rapid cure.

If the patient declines an operation, topical applications are in order. *Cold*, in the shape of iced compresses, or the ice-bag, will be proper where the affection is superficial and accompanied by lymphangitis. On the whole, it may be said that cold is beneficial in the initial stages of most phlegmonous affections, and is often very well borne and efficacious in the milder forms. To many it becomes unbearable from the time that suppuration is well established, and often induces a severe chill, the real cause of which, however, is always to be sought in the presence of pus.

NOTE.—Cold is badly borne by elderly or run-down subjects, or those prone to rheumatism.

*Dry or moist heat* is very soothing to many patients, and is a powerful stimulant to the local circulation. Occasionally it undoubtedly averts threatening suppuration, and may aptly be employed as a tentative or initiatory measure. However, if the local and general symptoms continue to increase, it should not beguile the surgeon into procrastination. Especially if a gathering become so massive as to cause *fluctuation*, incision should not be further delayed.

NOTE.—The main effect of the curious and often incomprehensible combinations of substances entering, at the recommendation of laymen and some physicians, into the composition of *poultices*, seems to be upon the faith and imagination of the patient. *Moist heat* is their active property, and, the simpler and cleaner its employment, the better it will be. The nauseous practice of smearing the skin, or, still worse, a wound, with hot linseed dough, is not yet extinct. Even a well-inclosed poultice is not a proper covering to a wound, unless a clean cloth and clean mush be taken for each application. Certainly a mixture of soured linseed with ichor and pus, inclosed in a foul rag, is the worst of all abominations that a decaying era of surgery has left behind as its legacy. A *clean cloth* dipped in and *wrung out of hot water*, covered over with a piece of oiled silk, is the best, the cheapest, and the least unappetizing of all cataplasms. The cataplasm should never be placed in actual contact with a wound. The interposition of a thin, moist dressing will protect the wound from mechanical insults unavoidably connected with the change of poultice, and the poultice itself will thus remain unsoiled by the secretions of the wound.

For special treatment of carbuncle, see page 224.

*Subcutaneous phlegmon*, left to itself, or treated by too long poulticing, will assume very large proportions. The form of the abscess cavity is rarely globular, but mostly irregular and sinuous. This is partly due to confluence of several smaller abscesses, partly to irregular extension, caused by the varying density of the subcutaneous connective tissues. *Fluctuation soon appears*, and without delay one or more incisions should be placed so as to drain every recess in the most direct manner. Volkmann's punctuation of the peripherical infiltration of the skin, a thorough irrigation of the cavity, and a moist dressing, constitute the treatment of these cases. The first incision is made where fluctuation is most marked; the index-finger of the left hand is then cautiously inserted, and carefully explores the interior of the abscess. This examination is very important, and upon its result depends the locating of the drainage-tubes. Counter-incisions are made over the tip of the left index, which pushes up the skin from within. *All squeezing of the abscess at this stage of the operation should be carefully avoided.* After the placing of the drainage-tubes, and a thorough irrigation, no pus should be contained in the abscess. If, therefore, gentle external pressure causes the escape of new masses of pus, *this is a sign that one or more recesses, communicating by small openings with the main cavity, remain undrained, and need further attention.* They must be located, and separately incised and drained.

If fluctuation persist over one or more places in the vicinity of the central abscess, it will be found that unopened, independent abscesses require additional incisions.

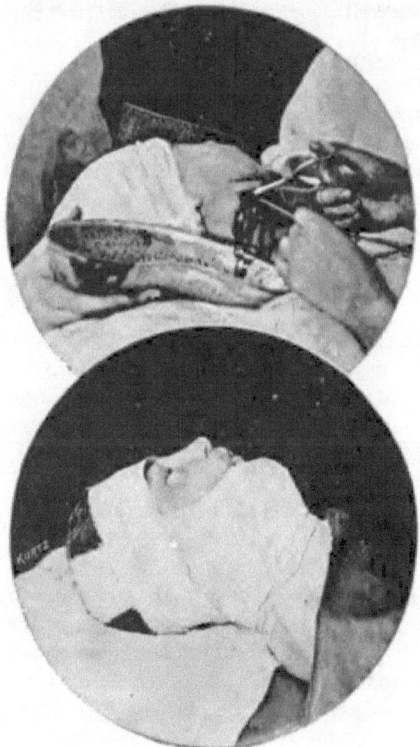

FIG. 142.—Hilton-Roser's method of incising a deep-seated abscess.

FIG. 143.—Completed dressing of cervical abscess.

The rough tearing and breaking down of septa of tissue within the abscess by the surgeon's finger is unsafe, on account of the unnecessary hæmorrhage it provokes, and because it may lead to pulmonary embolism. It is better to make a sufficient number of counter-incisions.

*The squeezing out* of abscesses through an insufficient spontaneous or artificial opening constitutes what may be called *surgical barbarism.* If the opening is too small or improperly placed, the abscess can never be drained by the aid of the law of gravity alone. External pressure must be employed to remove its contents, and this must be often repeated to prevent refilling of the abscess. As "squeezing out" is a very painful process, the patient will naturally shrink from it, and will let matters go. The abscess becoming nearly filled, only the overflow will escape through the insufficient aperture. The result is slow extension of the suppurative process, with continuous fever. Dressings of any kind will only make matters worse, and no relief will follow till another more properly located artificial or spontaneous opening supply the defect of drainage.

The best proof of the adequate treatment of an abscess is the fact that at change of dressings *the cavity is found empty, and all the secretions are contained in the dressings.*

The frequency of the change of dressings should be regulated by the amount of the discharge.

(c) **Deep-seated or Subfascial Phlegmon. Lymph-Gland Abscess.** — Still more serious than subcutaneous suppuration is a phlegmonous inflammation of the superficial or deep-seated lymphatic glands, or the submaxillary or the parotid salivary glands. The danger of these forms of septic tissue-decomposition consists in the great tension which their poisonous contents attain; the difficulty of their spontaneous evacuation on account of the massive barriers interposed between them and the surface of the body, and last, but not least, the likelihood of their perforation into the mediastinum, pleura, or peritonæum, or the erosion of large vessels situated in their immediate vicinity.

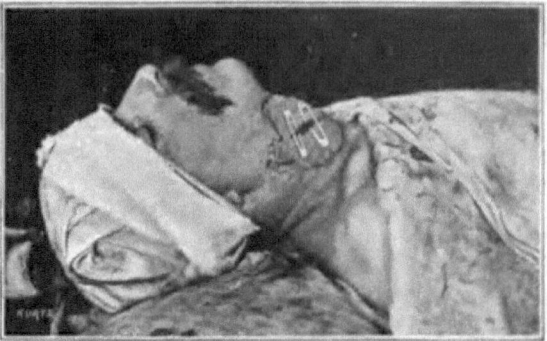

FIG. 144.—Underpadding of safety-pins thrust through drainage-tubes after incision of cervical abscess.

Deep-seated phlegmon is characterized by the extremely hard and deep-going infiltration of the superjacent tissues, *a general and massive œdema* of the soft parts, extending far beyond the limits of the inflammatory process, so that a limb, for instance, attains double its size; marked functional disability of all organs, even distantly related to the focus of disturbance, and very violent symptoms of systemic septic poisoning.

In the beginning the skin covering the affected locality is œdematous but pale; gradually it flushes up and becomes hard and brawny.

Incision and drainage is the sovereign therapy in these cases. No time should be wasted in attempts at an abortive treatment, as every hour of delay may cause irreparable damage. The distant hope of resolution, or the desire to produce "maturing" by poulticing, should not be allowed any weight in the face of the knowledge that extensive necrosis is the unavoidable consequence of the rapidly increasing dense infiltration characteristic of this condition. *Relief from excessive tension is the first and most urgent indication*, and this can be reached only by an incision.

The objection that these abscesses can not be opened safely while they are small, is erroneous, as will be shown directly. But, even if the surgeon should not succeed in opening the small cavity, cutting through the integument and fascia will do material service by averting the greatest danger.

*Hilton-Roser's method* offers a safe and easy manner of evacuating these foci. Anæsthesia is, of course, indispensable. A free incision through the skin over the most prominent part of the swelling should expose the fascia, which should also be divided by easy strokes of the point of the knife to a sufficient extent, say an inch or two. After this the knife is laid aside. If a small aspirator be at hand, search for pus can be made by puncturing and aspirating different parts of the swelling. This, however, is not necessary. A grooved director is inserted into the center of the incision, and is briskly thrust into the swelling, or, if large vessels be near, is gradually insinuated by steady rotating pressure. At a certain point resistance will suddenly cease, and a drop of ichor or pus will be seen exuding from the groove of the instrument. A dressing-forceps should now be placed in the groove of the director, and should be pushed into the focus. The grooved director can now be removed, and the forceps withdrawn while its branches are held as wide open as possible. A gush of bloody pus will follow the instrument. If the opening be too small, dilatation with the dressing-forceps should be repeated once or twice, until it becomes large enough to admit a stout drainage-tube. Irrigation and a moist dressing complete the procedure. (Figs. 142, 143, and 144).

If the incision was delayed too long, the relief of the general symptoms will not be as prompt as after early operations. The presence of adherent necrotic tissues explains this fact. But the spread of the mortification is checked, and the fever will abate as soon as the sloughs become detached and expelled.

Very numerous applications have taught the author the great value and safety of this method, which, therefore, can be warmly recommended.

*Fluctuation* is a very late symptom in all deep-seated abscesses, and should not be waited for. An explorative aspiration of a doubtful swelling will generally disperse uncertainty, and the production of pus will induce the patient to consent to the incision.

The hæmorrhage from large, deep-seated abscesses is sometimes copious. It comes from the walls of the abscess cavity, which are very vulnerable; hence rough exploration, squeezing, or any unnecessary manipulations should be carefully avoided.

NOTE.—It is best in cases of great emaciation to open the abscess according to Hilton-Roser—to insert a large-sized tube, and to desist altogether from exploration and irrigation until a few days later. The cavity will contract, its contents will spontaneously escape toward the point of least resistance—that is, through the drainage-tube—to be absorbed by the dressings, and much blood will be saved in this manner.

*Phlegmonous Erysipelas.*—A combination of extensive phlegmon with true erysipelas is not very common. What is ordinarily known as "phlegmonous erysipelas" is generally nothing but a very extensive subcutaneous phlegmon, mostly with, sometimes without, subfascial complications. The worst cases are directly chargeable to prolonged poulticing, and their treatment is rendered very difficult by the frequent occlusion of the drainage-tubes by large tow-like masses of necrosed connective tissue and fascia.

DIAGNOSIS AND TREATMENT OF PHLEGMON. 205

*Gangrenous phlegmon* (Pirogoff's acute purulent œdema) represents one of the highest degrees of microbial poisoning, where the multiplication of the micro-organisms is so rapid and pervading that the establishment of innumerable foci throughout all of the tissues composing a whole limb leads to extensive general infiltration. Board-like hardness, a dusky hue of the integument, blebs and ecchymoses, and finally, thrombosis of veins and arteries, will end in necrosis of the entire enormously swollen and cold limb. Incisions do not yield pus, but only give vent to scanty quantities of turbid ichorous serum. In these cases the prognosis is very bad, and the most heroic incisions rarely succeed in saving the member. If too long delayed, even a high amputation may fail to save the patient's life. (Figs. 145 and 146.)

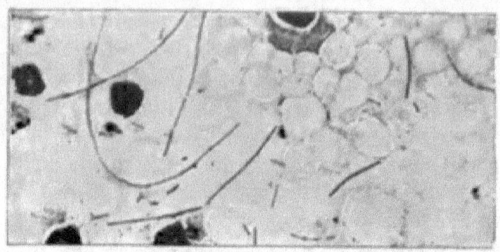

FIG. 145.—Bacilli of malignant œdema or acute progressive phlegmon (700 diameters). (Koch.)

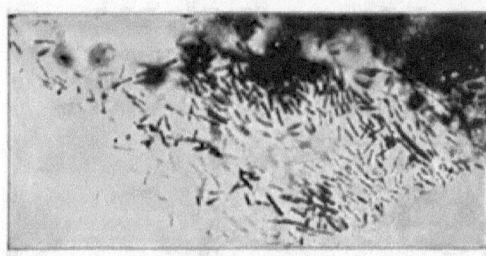

FIG. 146.—Bacilli of malignant œdema in the kidney (700 diameters). (Koch.)

*Emphysematous Gangrene.*—The inoculation of the human organism with a specific bacterium (Fig. 134) is generally followed by the development of a dusky, rapidly spreading infiltration, exhibiting on palpation the peculiar crackling, and on percussion, the tympanitic sound of subcutaneous emphysema. The process is accompanied by profound septic intoxication, with delirium, high temperatures, chills, and dejection, and terminates in gangrene of the affected parts. Resolute measures—that is, timely amputation performed through healthy parts—may succeed in preventing a fatal issue.

(*d*) **Acute Infectious Osteomyelitis.**—Suppuration of the medullary substance of parts of the skeleton represents one of the most dangerous and destructive forms of phlegmon. Its cause is the establishment of cultures of the *gold-colored grape-coccus* in the capillaries or arterioles of the marrow. The manner in which this infection occurs is still matter of controversy. So much, however, is known that it is most common during adolescence, and that a preceding suppuration, followed by exposure to weather, or certain traumatisms, are common provocative causes.

The invasion is marked by a severe chill, followed by a deep alteration of the general well-being. Very high temperatures, with chills, somnolency, a dry tongue, foul breath, intense gastric disturbance, bear witness to the gravity of the disorder. The insidiousness of the local and the gravity of the general symptoms lead to *frequent errors of diagnosis* on the part of practitioners who never have seen this affection, or are careless observers. The favorite locality of the disease is the shaft of the long bones near one or another epiphysis, as, for instance, the lower end of the femur. This, together with the upper part of the shaft of the tibia, is its classical seat. No bone, however, is exempt from the disorder.

The first local manifestation is a deep-seated, unbearable pain, soon followed by a general and deep-going œdema of all the soft parts overlying the focus. The skin is pale. As the soft parts covering the adjacent joint are also swollen, and its movement is painful, *the erroneous diagnosis of acute articular rheumatism* is frequently made.

Often the patient is unconscious or quite listless at the time of the physician's first visit, and the local symptoms escape attention. As a matter of fact, *typhoid fever or meningitis is frequently diagnosticated*, and the affection remains unrecognized until the appearance of a fluctuating swelling or, in extreme cases, spontaneous perforation of an abscess dispel the error.

The essential features of the morbid process are identical with those of cellular phlegmon, modified, however, by the peculiar structure of bone. On account of the rigidity of the osseous lamellæ inclosing the Haversian canals; of the cancellous and cortical substances inclosing the medullary tissue, and of the periosteum, the dense infiltration and massive exudation will rapidly heighten the intraosseous tension to such a degree that, the vessels becoming occluded, more or less extensive necrosis results.

The excessive tension of the noxious exudations penned up within the rigid tissues will cause a copious overflow and absorption of plasm charged with ptomaïnes, which will not fail to cause a profound intoxication, manifested by very grave general symptoms.

*Cortical osteomyelitis,* or what is known in text-books as *suppurative periostitis,* is the mildest form of the affection, and is most amenable to preventive treatment. The necrosis caused by it generally involves the outer part of the bone only, producing a *cortical sequestrum.* When the epiphysis is attacked in the vicinity of a joint, perforation and articular suppuration may occur and very seriously complicate the case.

CASE.—S. C., aged twelve, a somewhat anæmic boy, received, December 19, 1882, a kick from a playmate upon the spine of the tibia, which caused considerable pain for a while, but no discoloration. The next day a severe chill, with intense local pain and an extensive hard swelling of the injured region, set in. The boy became listless and delirious; he rapidly emaciated; the swelling extended in all directions. The author saw the patient December 29, 1882, in consultation with the family attendant, who, two days previous to this meeting, had made a small incision corresponding to one of the many points where perforation of the skin threatened. The boy being anæsthetized, a free incision three inches in length was made by gradual preparation down upon the anterior surface of the tibia, beginning a little below the patella. Every bleeding vessel was carefully tied at once, and thus clear insight and much bloodsaving were effected. A large ulcerative defect of the periosteum was found corresponding to a well-circumscribed greenish-yellow spot of the tibia. This defect extended to the capsule and into the knee-joint, which was found in open communication with the subperiosteal abscess, and was distended with pus. Two incisions were made into the joint for purposes of drainage. The popliteal space, thigh, and calf contained a number of burrowing secondary abscesses, mostly subcutaneous, which were also severally incised and drained. The entire major saphenous vein was found in a state of purulent phlebitis, its course being marked by a chain of small, angry-looking swellings of the skin, which, on being opened, all yielded pus. As it was probable that the entire vein would suppurate, it was slit up, beginning from the ankle, to within a few inches of Poupart's ligament, and the remaining parts of the thrombus were turned out. The hæmorrhage from entering branches was checked by packing with narrow strips of iodoformed gauze. A very tardy improvement followed these extensive measures. *January 10, 1883.*—A third incision into the upper recess of the knee-joint, and two more counter-incisions were made into the popliteal space. Large masses of necrosed connective tissue came away at almost each change of dressings, and, although the febrile disturbance had much abated, the boy seemed to steadily lose ground on account of the enormous suppuration. The cleansing of the wounds was so slow, the pain and suffering at the unavoidably frequent change of dressings so distressing and enervating to the patient, that, January 14th, amputation was thought of as a last resort. The parents, however, firmly declined the step, and fortunately so, as the boy ultimately recovered, with anchylosis of the knee-joint. A few small shells of necrosed bone came away from the epiphysis previous to the definitive closure of the wound.

*Central osteomyelitis* is much more destructive to the osseous tissue than the cortical affection, often causing necrosis of the entire shaft. It frequently extends to the epiphysis, and involves the adjacent joint.

NOTE.—The excruciating pain felt by the patient is principally due to the tension of the periosteum, separated from the bone by more or less pus. Ordinarily, the extension of suppuration by perforation into healthy parts is marked by an increase of the local and general suffering. Not so in osteomyelitis. Perforation of the periosteum, and evacuation into a loose plane of connective tissue, is always marked here by relief of the intense periosteal pain, and often by a temporary decline of the fever, due to the reduction of the enormous tension which first prevailed. With the increase of the tension in the secondary abscess the fever rises again, but the pain never reaches its former intensity.

Similar relations obtain in all forms of suppuration where the seat of the morbid process is confined by dense fascia or the capsule of a joint. Submaxillary and parotid cynanche, septic inflammations within the prepatellar or olecranic bursæ, and all joint-suppurations exhibit the same peculiarity. As long as the suppurative process is confined within the mentioned closed spaces, the tension and its immediate consequences—necrosis and copious overflow of fever-generating poisonous material into the lymphatics, causing intense toxic symptoms—are at their acme. As soon as perforation and partial evacuation of incarcerated pus into the meshes of the vicinal loose connective tissue occurs, a relaxation of the intense pain and a temporary remission of the septic fever are observed.

*Can Necrosis be averted?*—Where the diagnosis is made out early, where the superficial situation of the bone—for instance, the tibia—favors a precise localization of the focus, and where the affection is cortical, a free and early incision may avert, and, as a matter of fact, often does avert, necrosis, or at least will prevent its extension. In the beginning, perhaps, even the ravages of central osteomyelitis could be limited by early trepanning of the medullary space in one or more places. So much is certain and proved by experience, that prompt incision of the periosteum and trepanning of the affected bone admirably relieves the acuity of the local and general symptoms.

CASE.—The author has to quote from memory a very instructive case of recent infectious osteomyelitis of the lower end of the humerus observed in 1880 in the surgical department of the German Dispensary, and operated in the presence of Dr. W. Balser and other colleagues. A young woman, exhibiting an unusual degree of lassitude and a pitiable facial expression of suffering, was led into the place by two of her friends. Her left elbow-joint was semiflexed; it showed a pale, dense, and uniform swelling. Her attendants reported that she had had a severe chill in the morning of the preceding day, and had been very sick ever since then. The thermometer showed 105° Fahr. in the axilla. Extremely acute pain was complained of in the lower end of the humerus, just above the olecranon. Osteomyelitis being diagnosed, the patient was anæsthetized. A good-sized hollow needle being inserted until its point was caught by the bone at the site mentioned, a drop or two of thick pus appeared in the barrel of the hypodermic syringe. An ample incision was carried along the outside of the triceps tendon down to the bone, whereupon about two drachms of pus escaped. The periosteum was found detached, and, being deflected by an elevator, was found turgid and deep red, except at the place of detachment, where it was broken down and greenish-yellow. Profuse oozing took place from the exposed bone and periosteum, excepting an irregular area of bone covering about two square inches just above the posterior supratrochlear fossa. This area was grayish yellow, and did not bleed—in short, was necrosed. The wound was loosely packed with carbolized gauze, and was enveloped in a moist dressing. The patient was taken to her home, whence she was removed the following day to a hospital by her relatives, because she was too sick to be taken care of at home. The author was assured that her incessant moaning due to the excruciating pain had stopped during the night following the operation.

Some years ago the author saw a fatal case of pelvic osteomyelitis in consultation with Dr. H. Kudlich. The patient succumbed to the violence of the initial symptoms —that is, to acute septicæmia. The seat of the disease was the sacrum and os ilium of a very muscular man. Very intense sciatica and high fever composed the initial symptoms. Enormous œdema of the left thigh and inguinal region appeared a short time before death, revealing the nature of the affection, which until then had baffled attempts at diagnosis. The pelvis was found occupied by phlegmon extending below Poupart's ligament. The probable source of the infection was a recrudescent suppurative otitis media of old standing.

The subject is full of difficulty and surrounded by many drawbacks in all its aspects. The impossibility of an early and precise diagnosis as to location, the depth, and often the inaccessibility of the seat of the disease, will render many cases impracticable for preventive treatment.

Secondary abscesses must be incised and drained as early as possible according to rules above given.

(e) **Chronic Suppuration due to Bone Necrosis. Necrotomy.**—The most common seats of acute osteomyelitis and subsequent bone necrosis are the femur and tibia near the knee-joint.

This fact may perhaps be explained by the circumstance that the upper epiphysis of the tibia and the lower epiphysis of the femur ossify much later than the other epiphyses of these bones. The active growth and abundant blood-supply near the knee-joint seem to favor the importation and deposition there of active micrococci circulating with the blood.

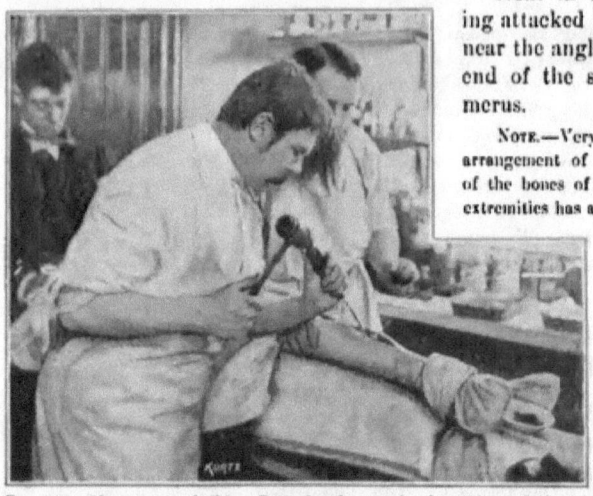

Next in frequency of being attacked is the lower jaw near the angle, and the upper end of the shaft of the humerus.

NOTE.—Very likely the different arrangement of the nutrient vessels of the bones of the upper and lower extremities has a certain influence upon the frequency of the location of osteomyelitis near the knee and shoulder joints. *The nutrient vessels of the femur and tibia diverge from the knee-joint; those of the humerus and the bones of the forearm converge toward the elbow.*

FIG. 147.—Necrotomy of tibia. Leg placed on a hard cushion. Irrigator playing from the right.

The direct and abundant blood-supply of the malleoli and the coxal end of the femur seems to cause an earlier consummation of the osteogenetic process at these localities, and also makes them liable to a form of infection peculiar to the infantile period of life—namely, tuberculosis. Tubercular affections of the ankle- and hip-joints are more common in children than white swelling of the knee. During adolescence, when the physiological fluxion toward the knee-joint preponderates over that toward the ankle and hip, the tendency to osteomyelitis near and tuberculosis near and in the knee-joint becomes more pronounced. Similar relations seem to prevail in reference to the upper extremity. During infancy white swelling of the elbow is more common than that of the shoulder and wrist-joints; in adolescence the upper end of the humerus is the common seat of acute osteomyelitis; in adults the shoulder and wrist are more frequently attacked by tuberculosis and osteomyelitis.

Whenever an attack of osteomyelitis terminates in the formation of an abscess and the establishment of one or more fistulæ, the acute features of the initial stages of the disorder disappear. The abundant discharge of pus is followed for a while by a gradual decrease of secretion, which again increases as the separation of the sequestrum becomes more and more complete. This is explained by the fact that, as the dead bone becomes gradually detached, the pus-generating surface of the cavity containing the sequestrum becomes proportionately larger. In the mean time new osseous substance is thrown out by those portions of the adjacent bone and periosteum which were not destroyed by suppuration, and thus a more or less perfect *involucrum* is formed around the sequestrum. After complete detachment of the sequestrum, suppuration is generally profuse.

* Hyrtl, "Descriptive Anatomie," 1870, p. 209.

If the affection is extensive and no spontaneous or artificial relief is vouchsafed for a long period, a deep deterioration of the general health will follow, characterized by emaciation, anæmia, albuminuria, and in extreme cases by amyloid degeneration of the liver and kidneys.

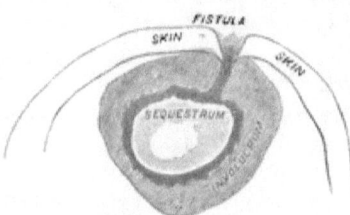

Fig. 148.—Diagram of a transverse section, showing relations of sequestrum, involucrum, fistula, and skin.

The *diagnosis of the presence of a sequestrum* can be made by noting the diffuse thickening of the affected bone, the profuse secretion from one or more fistulæ, and by direct probing. If the direction of the sinuses be straight, the silver probe will strike bare and roughened bone-surface. The latter symptom, however desirable for the establishment of a positive diagnosis, is not absolutely necessary to it. Indeed, the cases are quite common where tortuous channels prevent direct probing.

*Detachment of the sequestrum* is indicated by its mobility under the pressure of the probe-point, or, when probing is impracticable, by the long duration of the trouble and the increasing or profuse discharge.

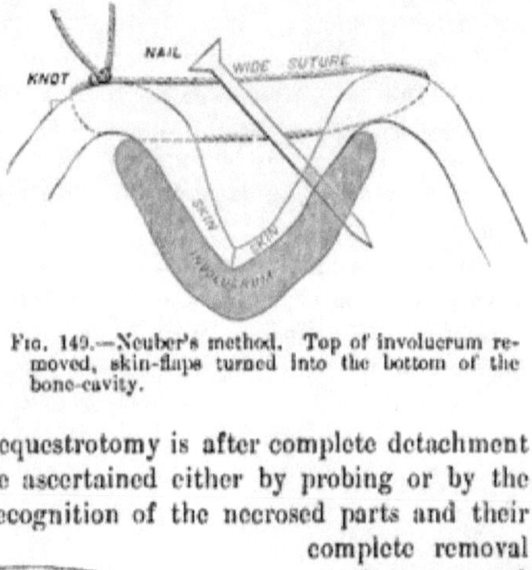

*When to Operate.*—It may be laid down as a general rule that the best time to perform sequestrotomy is after complete detachment of the dead bone, which can be ascertained either by probing or by the general aspects of the case. Recognition of the necrosed parts and their complete removal are then easy, and will be followed by a rapid cure. This rule, however, admits of important exceptions.

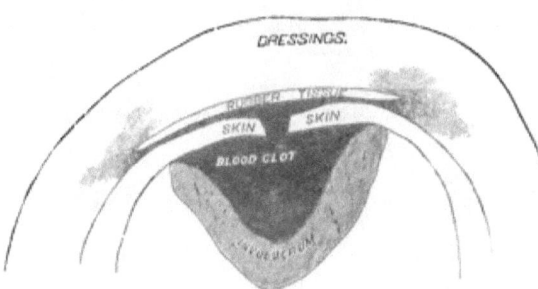

Fig. 149.—Neuber's method. Top of involucrum removed, skin-flaps turned into the bottom of the bone-cavity.

Fig. 150.—Schede's method. Diagram showing relations of organizing blood-clot.

NOTE. — Extensive necroses of the lower jaw are frequently accompanied by a profuse discharge of fetid pus into the oral cavity. This and the inability to masticate food, do frequently render early relief by operation very desirable. The objection that to perform a complete operation will necessitate the sacrifice of healthy bone is not tenable, as it may be urged that even an incomplete operation, if it only accomplish the removal of the greatest portion of the sequestrum, will be followed by a decided improvement of the patient's condition. After a while, a secondary operation can be done under more favorable circumstances. Similar considerations may also indicate an **early** sequestrotomy in other regions.

NECROTOMY.—Artificial anæmia by Esmarch's band and antisepsis have marked important changes in the technique of sequestrotomy. Control of the hæmorrhage, and the possibility of healing even the largest sequestrotomy wounds without suppuration, justify a deliberate search after detached foci containing sequestra by thorough exposure of the interior of the affected bones. *Long incisions and a free use of mallet and chisel* **are** *proper. A compressive antiseptic dressing will insure against secondary hæmorrhage.* The formation and maintenance of a moist blood-clot in the wound will bring about rapid filling up of the cavity by new-formed bone, and will terminate in firm and speedy cicatrization.

The introduction of the use of Esmarch's band **has** deprived extensive necrotomies of **their chief** danger—profuse hæmorrhage. The danger of **septic** disturbances following necrotomy was slight even before **the** adoption **of the** antiseptic method, as the densely infiltrated state of the adjoining tissues made absorption of septic matter from the wound difficult, and their rigidity rendered efficient **drainage** very easy. The chief advantage of the antiseptic method is to **be sought in the** possibility of effecting a cure without the long course of **suppuration** formerly characteristic of the healing of these cases.

**Neuber's implantation of skin-flaps was** the first step **in the** direction of **accelerating the cure** of necrotomy **wounds.** But *Schede's methodical and successful utilization of the protective properties* **of the moist** *blood-clot is* **the simplest** *and most perfect means to the end in* **view.**

The indispensable conditions for a successful employment **of Schede's** method are laid down in the following propositions :

*First.* Thorough exposure of the seat of the disease by incision and by the use of mallet and chisel.

*Secondly.* Complete **removal** of the *whole* sequestrum, *or all the sequestra,* and of the *entire pyogenic membrane* lining **the** cavities and sinuses, by scooping and scraping with the sharp spoon.

*Thirdly.* Thorough disinfection of all the nooks and crevices of the **wound by a** vigorous **use of the irrigator** and corrosive-sublimate lotion, **and by wiping it** out with **a clean sponge.**

NOTE.—The final flushing and mopping out should always be done with the strongest solution of corrosive sublimate used by surgeons (1 : 500). Residua of this strong lotion are then washed away by a **mild solution to** prevent mercurial poisoning.

*Fourthly.* The formation of a blood-clot which **should** fill up the wound to the level **of** the skin, and its preservation from putrefaction and exsiccation by **a** suitable antiseptic dressing (page 10).

NOTE.—Leaving behind the smallest spiculum of undetected dead **bone, or a shred of the** pyogenic membrane, will partially or totally compromise the success **of this procedure, and no** amount of irrigation will avert suppuration. Fulfillment of the second **proposition is not difficult** except in the *disseminated form of necrosis*, where a number of small foci, each containing its sequestrum, and all connected by more or less narrow and tortuous channels, are scattered within a wide area of the affected bone. But even these difficulties can be overcome by the exercise of circumspection and painstaking, favored by artificial anæmia, which renders detection of discolored bone and the entrance to bone sinuses comparatively easy.

*What Chisels to use.*—The chisels generally sold by surgical cutlers have little to commend them for efficient and rapid work. Their shape and size are unsuitable. "Albert Buck's warranted chisels," as sold by most hardware dealers, and generally used by carpenters and joiners, are well tempered and excellent. They should be fastened to an ordinary, smooth, wooden handle, without indentations, to insure the possibility of perfect cleansing. The author has found a set consisting of a one-inch, a half-inch, and a third-inch chisel, and of a one-inch and a half-inch gouge, to answer every purpose. A light wooden mallet, perfectly smooth, its head made of boxwood, can be bought in any house-furnishing establishment, and is much preferable to the small metal mallets of the instrument-makers.

*The Modern Manner of Performing Necrotomy.*—The following description may serve as an elucidation of the technique of a sequestrotomy. The parts being well cleansed with soap and hot water, shaved, and disinfected by mercuric irrigation, after Esmarch's band is applied, an incision is carried down to the bone over or near the fistulæ. The length of the external incision should be proportionate to the extent of bone thickening. The thickened bone should always be attacked where it is most superficial, *the site of the incision being determined rather by the question of accessibility than by the location of the sinuses.* Where the bone is superficial, as, for instance, the tibia, the incision may be at once carried down to it. Where there is a thick mass of overlying soft tissues, the incision should be gradual and preparative, and all cut vessels should be at once ligatured. The periosteum is pried up on both sides of the cut with an elevator, and, where it is found adherent by cicatricial tissue, is cut away, until the entire affected area is well exposed. Integument and periosteum are held back with a pair of Volkmann's retractors, and the roof of the cavity containing the sequestrum is chiseled away. This can be done very rapidly by a workmanlike use of the mallet and chisel, until the sequestrum is *completely exposed*. This being done, the sequestrum is lifted out of its bed with a pair of forceps. The irregular edges of the cavity are next smoothed off, overhanging parts are removed, so as to permit a careful and thorough ocular examination of all its recesses. Care must be taken not to leave behind any dead bone. The sharp spoon should be used in vigorous strokes to clear away all granulations or softened osseous tissue, until the entire wound-surface presents a bleeding, clean, and healthy appearance. *Débris* and shreds of granulations are flushed out with a strong irrigating stream, and, to make sure that no detached particles of tissue are left behind, the cavity should be mopped out with a clean sponge.

Where the operator is not certain of having rendered the cavity perfectly aseptic, it is safest not to apply suture, but to fill it with a loose packing of iodoformed gauze, and to swathe the limb in a moist compressive dressing. The dressing should be ample, and should contain externally a good layer of elastic material, as, for instance, absorbent cotton. The turns of the roller bandage

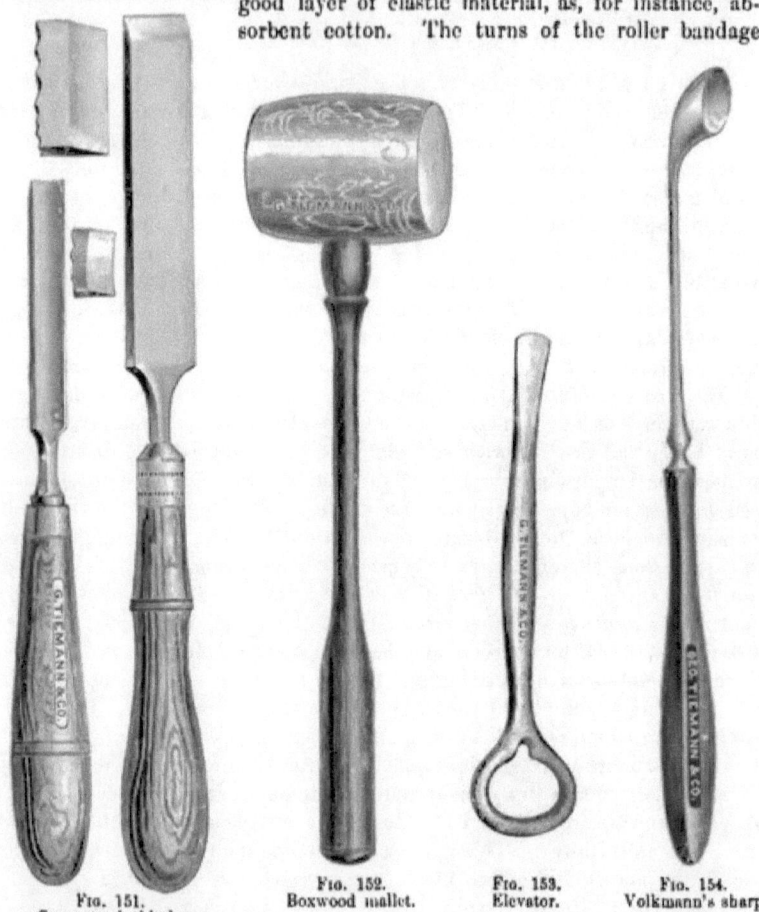

Fig. 151. Carpenters' chisels.
Fig. 152. Boxwood mallet.
Fig. 153. Elevator.
Fig. 154. Volkmann's sharp spoon.

should be tight and close, to insure a sufficient amount of elastic compression as a safeguard against secondary hæmorrhage. Ample padding will prevent strangulation. After the dressing is finished, the limb is held vertically while Esmarch's band is removed.

NOTE.—No alarm need be felt if the finger-tips or toes do not turn pink at once. A momentary lowering of the limb will immediately produce the flush indicative of the hyperæmia due to paresis of the vasomotor nerves.

Vertical elevation by suspension or propping up should be maintained for two or three hours, till a firm clot form in the wound. Should some blood permeate the dressings and appear on their surface a short time after the operation, then sufficient pressure was not employed. Suitable-sized compresses of iodoformed and sublimated gauze should at once be laid upon the blotch, and should be firmly held down by a clean elastic or flannel bandage. This additional pressure by the elastic bandage should not last more than an hour.

CASE.—Herman Albertin, school-boy, aged nine. Central sequestrum of lower end of shaft of humerus and disseminated necrosis of lower epiphysis due to acute osteomyelitis. Necrotomy performed April 12, 1884, at German Hospital, under chloroform. A longitudinal incision five inches long, commencing at the upper third of the posterior aspect of the left humerus, was successively carried through the skin, fascia, and triceps muscle, until the musculo-spiral nerve was exposed and freed from its bed. It was taken up and held aside by a blunt hook. The periosteum was incised, turned aside, and held up by a pair of Volkmann's four-pronged hooks. The posterior face of the thickened shaft of the humerus was chiseled away, exposing an irregular-shaped central sequestrum, three inches long. The overlapping parts of the involucrum were further chiseled off, until the entire sequestrum could be easily lifted out of its place. Two small, round sequestra were removed from the lower epiphysis, and the entire trough-shaped cavity was carefully scraped out with a sharp spoon. A small strip of iodoformed gauze was placed into the most dependent part of the bone defect, and was brought out at the lower angle of the wound. The triceps, fascia, and skin were united by three tiers of continuous catgut suture. A compressive gauze dressing was bandaged around the limb, and the constricting band was removed. The arm was held in vertical suspension for two hours, and after that was placed in the semi-elevated posture on a pillow. The temperature remained normal throughout. The first change of dressings was made April 26th, a fortnight after the operation. The dressings contained only a small quantity of dried blood. The fillet of gauze being removed, a new dressing was applied. The patient was discharged from the hospital April 30th, with a small, superficially granulating wound corresponding to the place of drainage. He returned for another change of dressing May 12th, when the wound was found entirely cicatrized over.

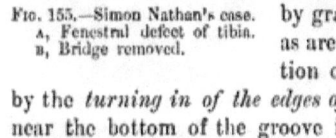

FIG. 155.—Simon Nathan's case. A, Fenestral defect of tibia. B, Bridge removed.

In cases where the surgeon is reasonably sure of having produced an aseptic wound, either Neuber's method of implantation of skin-flaps or, what is better, Schede's treatment can be employed.

*Neuber's Method of Implantation.*—Neuber's idea consists in the endeavor to cover up with skin, if possible, all the raw surfaces left by the operation. Primary union is the object, and a minimum of uncovered raw tissues is left to heal by granulation. *Longitudinal bone defects*, such as are caused by the removal of a necrosed portion of the shaft, are partly or entirely covered by the *turning in of the edges of the cutaneous wound* till they meet at or near the bottom of the groove in the bone (Fig. 149). It is necessary for this purpose to dissect up laterally the skin on both sides of the incision to a goodly extent, so as to render it movable and easily held in the new position. One or more wide sutures of catgut are passed through the skin at the points of reflection (Fig. 149), to retain the flaps in position; and, where this is not sufficient, a well-disinfected nail is driven through the edge of the flap into the bone. The groove thus formed is loosely packed with strips of iodoform gauze, and the limb is incased in an aseptic dressing.

NOTE.—Nails are disinfected either by boiling in water or by being passed through an alcohol-flame till they assume a dull-red heat. After this they are dropped into the vessel holding carbolic lotion and the instruments.

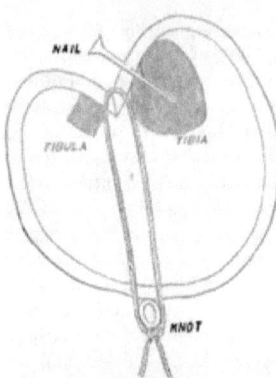

FIG. 156.—Simon Nathan's case. Implantation of cutaneous edges into the defect by transfixing catgut suture.

CASE I.—Simon Nathan, clerk, aged nineteen, admitted to the German Hospital April 18, 1886. Had been operated on three years ago for necrosis of tibia by Prof. Schönborn, of Königsberg. A fistula remained on the anterior aspect of the leg, that closed up and broke open several times every year. The probe detected exposed but smooth bone. April 22d.—The patient was anæsthetized and the tibia was exposed. It was found that the sinus led into an oblong defect (Fig. 155) of the shaft, through which the probe could be passed, so as to be clearly felt beneath the soft tissues of the calf. The length of this defect was a little more than an inch, its width half an inch, and its walls were formed by very hard condensed bone. Apparently the sclerosed condition of this bone and its scanty blood-supply was the cause of the frequent ulceration of the deciduous granulations forming within the track. The bridge of sclerosed bone, together with the adjacent condensed parts of the shaft, were removed by mallet and chisel; the edges of the cutaneous wound were dissected up sufficiently to admit of an easy adjustment within the gap between the tibia and fibula (Fig. 156). Two stout catgut sutures were passed through both edges of the skin-wound, and were brought out by a Peaslee's needle on the under side of the calf, where they were firmly

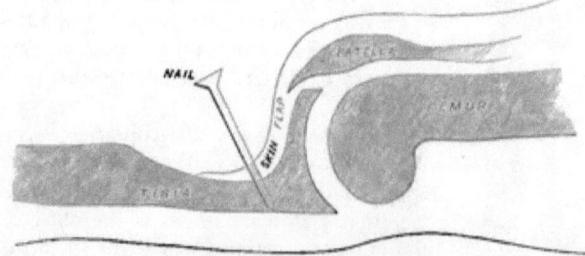

FIG. 157.—Neuber's method. Frank Nagengast's case. Implantation of triangular flap into the defect of the head of tibia.

knotted over a piece of stout drainage-tube. Thus the edges of the skin-flaps were well drawn into the bottom of the defect. To somewhat relieve the pressure by the drainage-tube upon the skin of the calf, a nail was driven through one of the flaps into the tibia, and the leg was dressed antiseptically. Slight elevations of the temperature without general or local discomfort were observed on the two successive days, after which the normal standard remained unchanged. The dressings were removed May

9th, and the skin-flaps were found firmly adherent in their new position. Some cutaneous ulceration of the skin on the calf had taken place. The nail was removed. The patient was discharged cured June 1st.

NOTE.—A sclerosed and ill-nourished state of the involucrum will often lead to a repeated breakdown of the granulations lining an old sinus. Stimulating injections will sometimes effect a cure, but in rebellious cases success can be had only from a thorough removal of the condensed portions of the bone and sinus.

CASE II.—Frank Nagengast, aged eight, a very anæmic boy. Necrotomy of tibia, November 2, 1885, at Mount Sinai Hospital. Extraction of a large central sequestrum

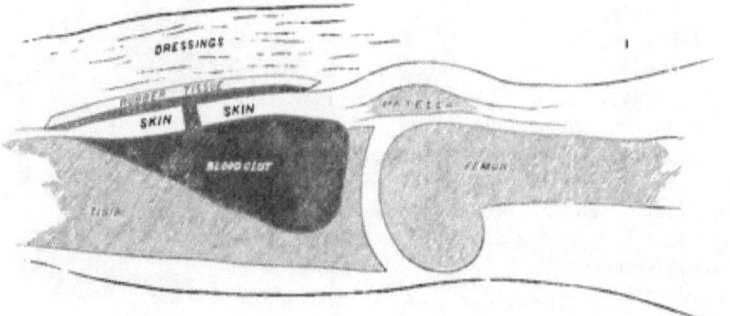

FIG. 158.—Diagram illustrating Schede's method applied to a case like that of Frank Nagengast.

comprising the entire thickness of the upper half of the shaft, a narrow extension reaching down to the lower epiphysis. Three small sequestra, together with a lot of softened granular cancellous tissue, were removed from the head of the tibia. The remaining posterior portion of the involucrum was so slender and brittle that it broke into several fragments during the operation. Lateral implantation of the skin by means of transfixing sutures by Peaslee's needle. Antiseptic dressing and a lateral splint. First change of dressings November 23d. Healing of the wound by adhesion corresponding to the shaft. Sinuses leading into narrow cavity in lower portion of tibia, and a larger cavity in the head of the bone. Fractures united with some sagging of tibia downward. *December 17th.*—Bloody reinfraction of tibia; scraping of upper and lower cavities. *January 10, 1886.*—Lower sinus closed; up-

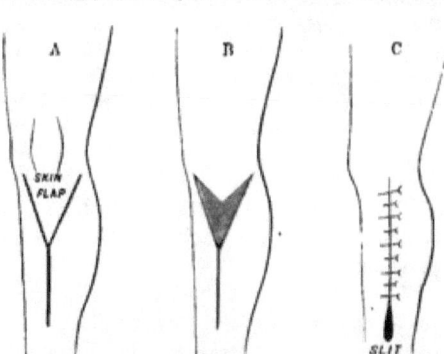

FIG. 159.—Frank Nagengast's case. A, Triangular skin-flap. B, Skin-flap turned into the cavity; the dark space to heal by granulation. C, View of necrotomy wound treated according to Schede's method.

per cavity shows no tendency to heal. *February 22, 1886.*—*Osteoplastic closure of cavity in head of tibia according to Neuber.* A triangular skin-flap, containing the insertion of the quadriceps tendon and the periosteum, was raised from the anterior aspect of the tibia. The remaining roof of the cavity was removed by mallet and

## DIAGNOSIS AND TREATMENT OF PHLEGMON.

chisel. Previous to this the capsule of the knee-joint was carefully exposed to avoid entering the joint. The granular lining of the cavity was gouged away, and only a shell, consisting of the articular surface and the posterior portion of the head of the tibia, remained intact. The triangular skin-flap was turned down into the bottom of this cavity, and there attached by a nail (Figs. 157–161). The remaining uncovered Y-shaped portion of the wound was left to granulate. Under an antiseptic dressing firm union of the flap to the underlying bone took place, and the granulating part of the wound was firmly cicatrized over by the middle of April.

*Schede's Method* (Fig. 162).—Schede's plan has the great advantage over Neuber's method that it can be employed successfully under the most varying conditions. *Its simplicity and independence of the presence or absence of a sufficient covering by skin commend it to the attention of the surgeon.* The author found Neuber's plan inadequate where much integument had been lost, and was replaced by an extensive cicatrix.

CASE I.—Frank Hyman, aged twelve, received, in May, 1886, a blow on the left tibia, after which central osteomyelitis developed. *August 9th.—Necrotomy.* Two large sequestra were removed from the upper half of the shaft, requiring three separate parallel incisions for their extraction. The wound was very carefully evacuated of all granulations, and disinfected with a 1:1,000 solution of corrosive sublimate. Simple suture of the cutaneous incisions; a small drainage-tube was placed into the upper angle of the longest incision. All the incisions were covered with strips of disinfected rubber tissue, and the limb was dressed with sublimated gauze. The first dressing remained unchanged for four weeks, when only a shallow fistula remained at the place where the drainage-tube had lain. This was scraped, and it promptly healed.

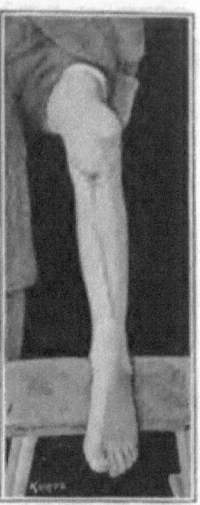

FIG. 160.—Anterior view of Frank Nagengast's leg after completed cure.

The large cavity became filled with a blood-clot, which organized without suppuration.

The treatment of the osteomyelitic processes of the *femur* and their sequelæ, notably of necrosis, presents peculiar difficulties of technique mainly due to the deep site of the bone. Long incisions are usually indispensable, access to the remote portions of the bone is difficult, and the necessary injury to many muscular branches of the femoral artery, and the difficulty of effective compression of the muscular masses, render the question of after-hæmorrhage rather serious. It is, therefore, advisable not to deplete the limb by

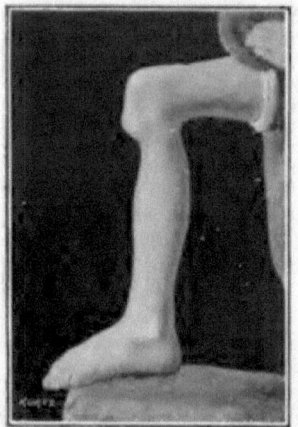

FIG. 161.—Lateral view of Frank Nagengast's leg.

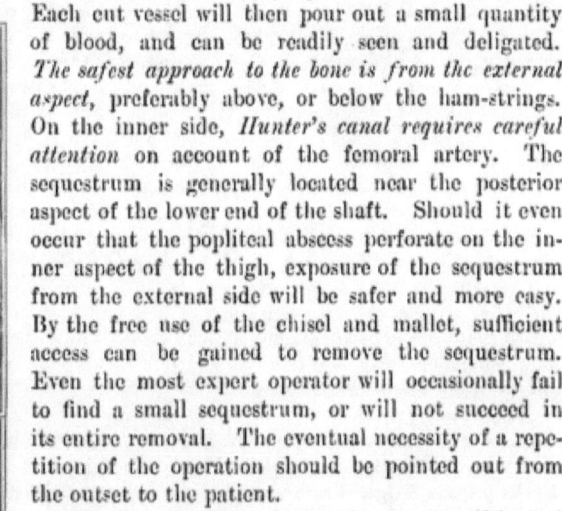

an elastic bandage of all its blood before applying Esmarch's constriction. Each cut vessel will then pour out a small quantity of blood, and can be readily seen and deligated. *The safest approach to the bone is from the external aspect,* preferably above, or below the ham-strings. On the inner side, *Hunter's canal requires careful attention* on account of the femoral artery. The sequestrum is generally located near the posterior aspect of the lower end of the shaft. Should it even occur that the popliteal abscess perforate on the inner aspect of the thigh, exposure of the sequestrum from the external side will be safer and more easy. By the free use of the chisel and mallet, sufficient access can be gained to remove the sequestrum. Even the most expert operator will occasionally fail to find a small sequestrum, or will not succeed in its entire removal. The eventual necessity of a repetition of the operation should be pointed out from the outset to the patient.

*Inferior Maxilla.*—As a rule, osteomyelitic foci of the lower jaw communicate with the oral cavity. This makes the preservation of the aseptic condition of the wound rather difficult, and sometimes, notably in the presence of a neglected and foul set of teeth, an impossibility. Where the process is extensive, an external incision is preferable, as it lessens the danger of the entrance of blood into the respiratory tract, and facilitates complete and clean work.

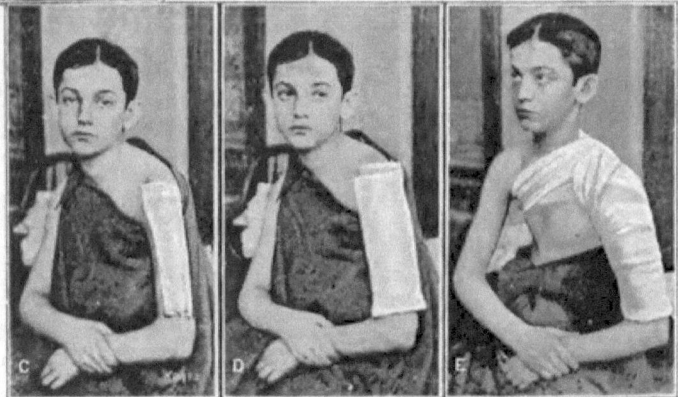

Fig. 162.—Illustrating successive steps of Schede's dressing. A, Necrotomy wound. B, Protective. C. Iodoformed gauze. D, Sublimate gauze. E, Complete dressing. (Case of Samuel Krongold. Photographs taken ten days after operation.)

CASE.—1. Eckert, tailor, aged twenty-three, contracted **traumatic acute osteomyelitis** of the horizontal ramus of the left side of the lower jaw, after the extraction of a carious tooth, done November 2, 1886. The **intense pain** of the beginning was relieved by a spontaneous discharge of pus into the **oral cavity**. The author saw the patient November 23d, when the thickening of **the jaw**, the profuse secretion, and direct probing put the presence **of a** sequestrum beyond doubt. *Sequestrotomy performed November 25th.* The mouth had been prepared **for a day** or two by frequent rinsings with salt water; the face had been shaved. The **back** of the anæsthetized patient's head was rested on a low, hard roll made of a **blanket.** The hair was wrapped **up in a** hood made of a towel dipped in **corrosive sublimate,** the **chest** protected **by another** wet towel. The skin of the jaw was well soaped and rubbed off with mercuric lotion. Then an incision two inches and a half in length was made along the lower edge **of the horizontal ramus.** The facial artery was exposed, separated, secured by two **pairs of artery forceps, cut through between,** and doubly deligated. The periosteum **was** incised to the entire length of the external cut, and was reflected upward with an elevator. Before opening into the oral cavity, a sponge held by a long **sponge-holder** was thrust into the mouth to the vicinity of the fistula, to receive any blood **that might** escape that way. An oblong quadrangle of the external lamella of the alveolar process and body of the ramus was chiseled away, exposing a cavity containing three sequestra and a mass of ulcerating fetid granulations. The cavity was carefully scraped out by the sharp spoon, irrigated with corrosive sublimate, the **soiled** sponge in the mouth having first been substituted by a clean one. The opening freely communicating with the oral cavity was plugged with **a strip** of iodoformed gauze, **that** reached just within the focus; the external wound was closed by a number of catgut stitches, a short **drainage-tube being first** placed in its posterior angle. *December 2d.*—First change of dressings. **No reaction;** no fever. External wound was found closed, the drainage-tube was shortened, and was found **still containing a dark-red blood-clot.** The iodoform plug was left undisturbed, and was **removed by the patient's family attendant** at the end of the second week. Discharge **was scanty throughout. Patient cured December 20th.**

*Bone Abscess.*—Circumscribed acute osteomyelitis of minor intensity, caused very likely by infection with a very limited number of micrococci deposited in the medullary substance from the blood, does not have a pronounced tendency to induce massive necrosis. Breaking down and emulsification of the affected parts are tardy, and thus opportunity is given to the surrounding tissues for throwing up around the focus a protective wall of granulations. The extension of the abscess is slow, and the local as well as general disturbance effected by it is of a chronic character. Nightly exacerbations of fever, with occasional chills and sweats, and localized, deep-seated pain of a throbbing nature, gradual hypertrophy of the bone, with atrophy of the pertinent muscles, trophic changes of the skin, as glossiness and local sweats, and increasing emaciation, are the characteristic symptoms of the affection, which extends over months and even years. The marked thickening of the bone, the spontaneous local pain, augmented by pressure on percussion, and the absence of fistula are mainly to be considered as to diagnosis. Therapy consists in doing what is to be done with all abscesses—*evacuation and eventually drainage.*

The conspicuous thickening of the bone serves as a convenient guide to the purulent focus. After the application of Esmarch's constrictor, a free incision, made according to the rules described in the paragraph on necrotomy, exposes the bone, the surface of which is generally found covered with osteophytic excrescences, that somewhat impede the raising up of the periosteum. All the soft parts being held away by sharp retractors, the thick layer of new-formed bone is pared off with the chisel, layer by layer, until the cavity containing pus is exposed. Sometimes a number of discrete or communicating foci are present, and the surgeon must make sure of not overlooking any of them. It is best, accordingly, to expose the medullary space throughout the entire extent of the thickening. By entirely removing the roof of the cavity, it is converted into a more or less shallow trough, all parts of which are exposed to ocular inspection. The smooth pyogenic membrane lining the abscess is carefully removed to its last shred by vigorous scraping and gouging with the sharp spoon, and by subsequent irrigation. A final flushing of the wound with a strong (1 : 500) solution of corrosive sublimate will make sure of the destruction of all lingering germs. The wound is sutured and dressed according to Schede's plan, and, if the removal of all diseased tissues and infectious secretions was thorough, rapid and uninterrupted healing under the blood-clot will take place.

CASE I.—Richard Boss, metal-worker, aged thirty-eight. Chronic painful thickening of the shaft of the humerus of two years' standing. Glossy skin, atrophy of the muscles of the arm and forearm, formication, and hyperidrosis, together with paretic symptoms affecting principally the musculo-spiral nerve. Nightly exacerbations of local pain and hectic emaciation. *February 2, 1887.*—At the German Hospital, exposure by chisel and mallet of a bone ab-

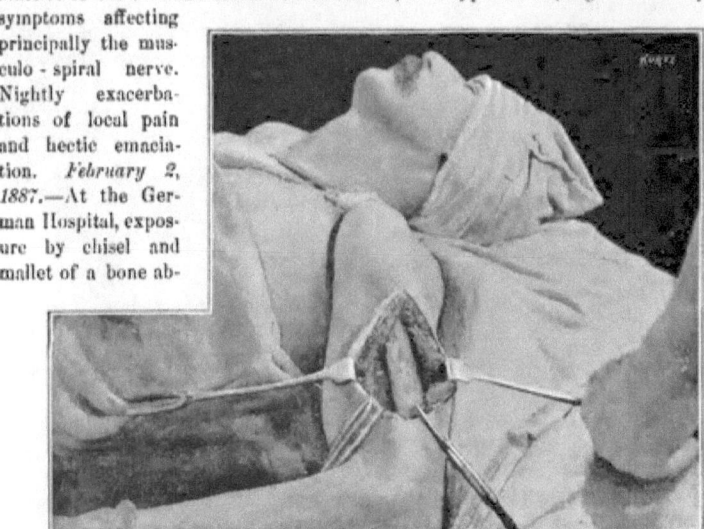

FIG. 163.—Exposure of thickened humerus containing a central bone abscess. Elastic constrictor tied above the acromion, and thence passed around thorax into the opposite armpit, where it is secured by another ligature.

scess occupying the middle and upper part of the medullary cavity of the left humerus. Schede's method of dressing the wound. *February 17th.*—First change of dressings. Wound united by the first intention. Two superficial drainage-tubes were removed. *March 6th.—* Patient discharged perfectly cured with improving function of the extremity. (Figs. 163, 164, and 165.)

CASE II. — Samuel Krongold, school-boy, aged twelve, had had, several years ago, compound dislocation and acute suppuration of the left elbow-joint, complicated with acute osteomyelitis of the lower epiphysis of the humerus, in consequence of which several sequestra had to be removed by the author. Three months ago a painful thickening of the shaft of the humerus appeared, causing marked deterioration of

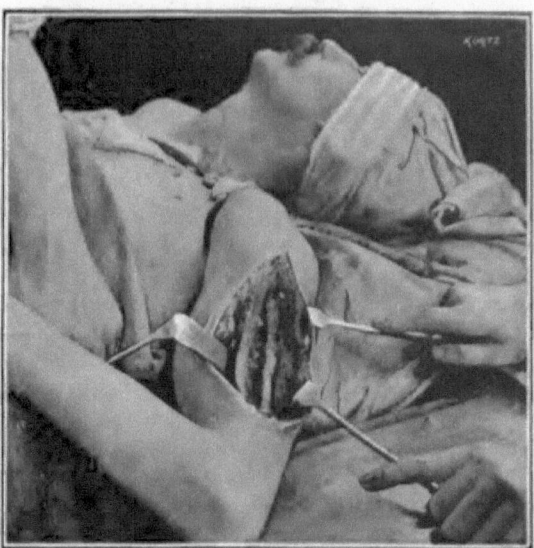

FIG. 164.—Cavity chiseled open. Its contents removed with the sharp spoon. (Richard Boss.)

the boy's health. *February 18, 1887.—* At the German Hospital, a central bone abscess occupying the middle portion of the medullary space of the humerus was exposed and evacuated, and was treated by Schede's method. *February 26th.—* The first change of dressings took place, and the entire wound was found healed with the exception of the slit left open for drainage at the lower angle of the wound, which was occluded by a

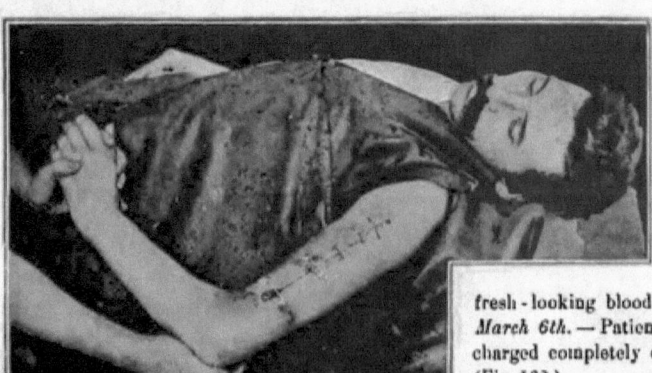

FIG. 165.—Richard Boss's wound treated according to Schede's method. Photograph taken February 17th, fifteen days after operation.

fresh-looking blood-clot. *March 6th.—* Patient discharged completely cured. (Fig. 162.)

The remarkably short and complete cure of both of these cases is undoubtedly to be attributed to the adoption of Schede's plan. Plugging of and introducing drainage-tubes or any foreign substance into the bone cavity are done away with, and organization of the massive blood-clot goes on uninterruptedly to the greatest advantage.

### Conclusions.

*Prevention of infection* contains the spirit and aim of *aseptic surgery;* the *object of antiseptic surgery* is *disinfection* and the *conservation of infected tissues.* The first object is attained by a severe discipline of *cleanliness;* the second by the still more severe discipline of *early incisions* and adequate *drainage* and *disinfection.*

A clear comprehension of the processes determining suppuration must result in the firm conviction that an early and free incision of every focus of septic inflammation is the most conservative form of treatment. It prevents local death and general intoxication, the latter only too often the cause of general death. If this conviction will have entered into the "*succum et sanguinem*" of every physician, public opinion will gradually yield to a better understanding of individual and the public interest.

NOTE.—The change in the surgeon's attitude toward the employment of incisions for septic inflammative processes is characterized by these sentences:
*Formerly,* topical applications were the main reliance, incision only a last and extreme resort. *The surgeon had to show cause why an incision should be made.*
*At present,* relief from tension and escape of the noxious substances through incision and drainage is the clear indication to be fulfilled. *The surgeon must show cause why an incision should not be made* in the presence of septic inflammation.

### 2. *Phlegmonous Affections of some Special Regions.*

*a.* **Face. Floor of the Mouth. Neck. Temporal and Mastoid Regions :**

*Anatomical Arrangement of the Connective-Tissue Planes of the Neck.*—Henke's classical essay is the best guide for the clear comprehension of this subject. He injected the different interspaces of a cadaver with liquid gelatin, and studied the manner of its extension between the several organs by exposing the congealed masses, and examining their relations *in situ.* The chief interspaces of the neck are classified by Henke as follows:

1. *The Capsule of the Submaxillary Salivary Gland.*—It forms a completely closed envelope to the gland, from which continuations extend to the superficial and deep cervical fasciæ.

2. "*Previsceral Interspace.*"—The connective-tissue plane or interspace situated between the prelaryngeal group of longitudinal muscles (hyo-thyroids, sterno-hyoids, and sterno-thyroids) anteriorly, and the larynx, thyroid gland, and trachea posteriorly. It communicates with the anterior mediastinum. Perforation of a suppurating thyroid gland leads to invasion of this space, with subsequent compression of the trachea. (Fig. 166, c.)

CASE.—S. C., aged seventeen. The patient was treated by Dr. C. Lellmann for typhoid fever in the German Hospital. In the third week of the disease severe dyspnœa developed, with a peculiar wheezing sound accompanying respiration. On examination, a diffuse swelling was noted in front of the neck. Incision evacuated an abscess communicating with the interior of the thyroid gland, whence perforation must have taken place. Immediate relief followed.

3. "*Retrovisceral Interspace.*"—The interspace between the pharynx and œsophagus in front, and the vertebral column behind. It communicates with the posterior mediastinum. (Fig. 166, A.)

## DIAGNOSIS AND TREATMENT OF PHLEGMON.

4. "*Perivascular Interspace.*"—The interspace containing the carotid artery and jugular vein. It communicates with the anterior mediastinum along the course of the large vessels, and is important on account of the frequent suppuration of the group of lymphatic glands situated in front of, and externally to the jugular vein. Abscesses of this interspace displace the sterno-mastoid muscle outward; they extend along the vessels downward, and, left to themselves, either perforate through the deep and the superficial fasciæ and the skin near the clavicle, between the lower end of the sterno-mastoid muscle and the trachea, or make their way along the vessels into the anterior mediastinum. (Fig. 167.)

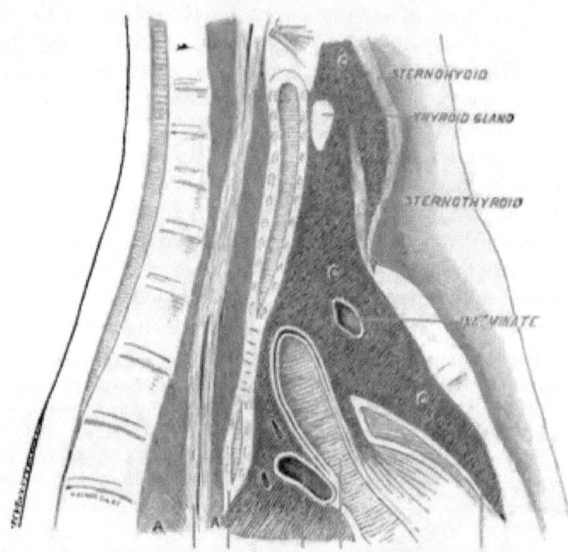

FIG. 166.—c, Previsceral space. A, Retrovisceral interspace. Antero-posterior section. (From Henke.)

5. "*Intermuscular Space.*"—An interspace situated at their crossing, between the lower third of the sterno-mastoid and the omo-hyoid muscles. This space owes its origin to the sliding of these contiguous muscles upon each other, and is limited posteriorly by the scaleni. It contains a group of lymphatic glands, seated near the posterior edge of the lower third of the sterno-mastoid muscle (supraclavicular glands), and communicates inward and upward with the retrovisceral space, and along the subclavian vessels with the axillary cavity. Supraclavicular abscesses usually extend into the arm-pit. (Fig. 168.)

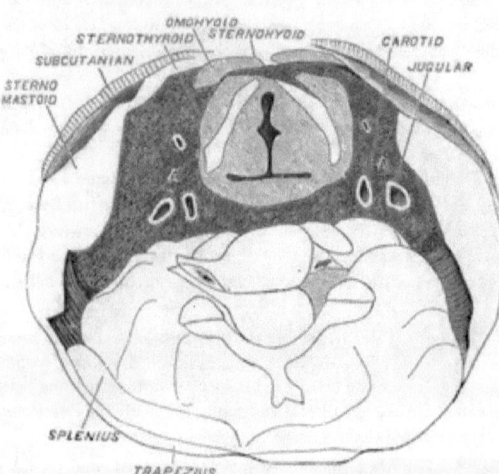

FIG. 167.—Perivascular interspace. Transverse section. (From Henke.)

(*a*) FACE. — The most serious form of cutaneous and subcutaneous phlegmon observed on the face is *the carbuncle*. It is characterized by a dense, hard swelling of conical shape, extending far into the subcutaneous connective tissue. It has a dusky red color, and its apex is marked by one or more yellowish discolored spots, which are surrounded by a bluish halo. Septic thrombosis extending through the jugular veins into the cranium is to be feared in this affection. The systemic intoxication is generally very intense, high fever being the rule. In some of the worst cases the intoxication is so deep as to cause symptoms of collapse, with low, sometimes even subnormal, temperatures.

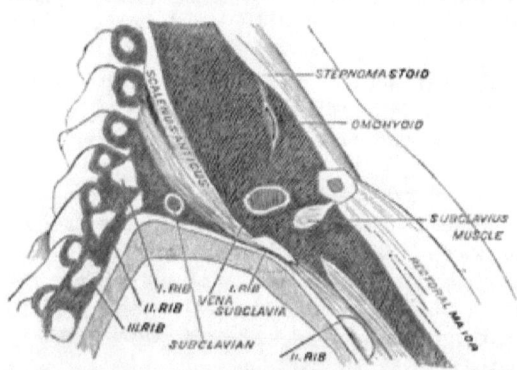

Fig. 168.—Intermuscular space. Lateral antero-posterior section. (From Henke.)

In this condition an early and most energetic treatment is urgently indicated, and is almost always followed by elimination of the infectious process.

A crucial incision, or, in extensive cases, a number of parallel incisions, carried in length and depth beyond the indurated area, will relieve tension and permit the escape of the contents of many smaller or larger incarcerated foci. The incisions should be packed lightly with strips of iodoformed gauze. *In cases of anæmia*, where loss of blood would materially increase the danger, *the actual cautery* should be so applied as to convert the entire infected area into a dry eschar. This or the incisions should be enveloped in a moist dressing, which has to be renewed according to the amount of secretions.

NOTE.—The following bloodless treatment applied by Slesarewskij in forty-four cases of carbuncle seems to deserve trial, as it yielded very good results in his hands: Inspissated crusts are first removed, then the diseased surface is sprinkled with from thirty to sixty grains of corrosive sublimate powder. The dusky halo surrounding the center of the sore is thickly covered with blue ointment, and the whole is enveloped in a compress soaked in carbolized oil (1 : 10), fastened with a roller bandage. In case of severe pain, an ice-bag is placed over the dressing. The following day, corresponding to the application of the mercuric salt, a gray, very dense eschar will be visible, which will separate ten days later, and will be followed by rapid healing. Slesarewskij never observed mercuric intoxication during or after the application of this method of treatment. ("Centralblatt für Chirurgie," 1886, p. 805.)

CASE.—The author lost, of a considerable number of cases treated by incision, only one by septic phlebitis of the right lateral sinus. The patient, a middle-aged cigar-maker, was seen in consultation with Dr. L. Weiss, and an enormous carbuncle occupying the right side of the upper lip and cheek was found, with extensive œdema of the eyelids and the right side of face and neck, which was due to general thrombosis of the pertinent veins. The patient was semi-comatose, somewhat cyanosed, and had a poor pulse. He had obstinately opposed any incisive treatment for six days, and the case seemed clearly beyond the reach of surgical skill. The incisions caused very little hæmorrhage, as most of the divided tissues were necrosed. He died of collapse on the seventh day of his illness.

The author has never tried any of the "maturing" forms of treatment in this affection, and would unhesitatingly declare measures which are apt to stimulate suppuration, such as poulticing, to be always risky, and sometimes positively dangerous.

(*b*) NECK.—(a) *Fauces and Pharynx.*—The tonsils and the connective tissue in which they lie imbedded are the most favorite site of superficial and deep-seated septic processes. *Diphtheria* is very likely a microbial affection due to the colonization of micrococci upon the surface and in the follicles of tonsils, that are in a state of catarrhal or scarlatinal inflammation. It is characterized by superficial or deep-going putrid necrosis of the affected tissues, often extending to the pharynx, larynx, velum, pillars, and the nasal mucous membrane, and is generally accompanied by a serious general intoxication. The systemic intoxication is most prominent when parts having an abundant supply of lymphatics, as the pillars of the fauces, the velum, pharynx, and nasal mucous membrane, are involved. The scantier development of the tonsillar and laryngeal lymph-vessels seems to be the cause of the minor intensity of the systemic symptoms observed in affections localized in these parts. Characteristic intumescence of the deep cervical lymph-glands is a regular consequence of the affection of the first group of localities; it is more rarely observed in purely tonsillar or laryngeal diphtheria. An invasion is apt to leave behind a certain disposition to renewed attacks, which is perhaps due to the fact that quiescent spores of bacteria remain imbedded in the recesses of the follicles, to develop their activity whenever a new catarrhal inflammation and exudative process prepares the ground for their multiplication.

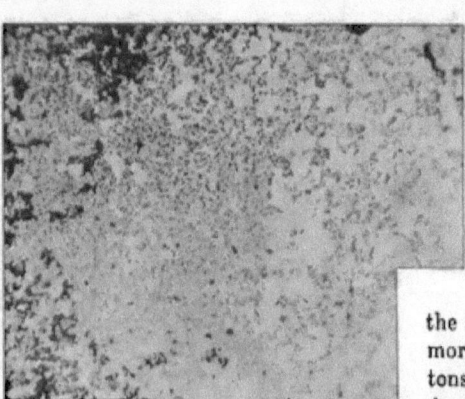

FIG. 169.—Bacteria from case of vesical diphtheria with putrescence (700 diameters). (Koch.)

But, on the other hand, frequent attacks, and the accompanying formation of cicatricial tissue within the textures of the tonsils, seem to lead to a certain immunity from the graver forms of the disease. As a rule, persons who never had diphtheria suffer more severely than those who have gone through many attacks; and diphtheria of children formerly free from the disease is a much more serious condition than the so-called habitual "follicular tonsillitis." While a first attack is usually, habitual follicular tonsillitis is rarely, complicated with glandular enlargement.

The condition of things here is comparable to that which was mentioned as the "habituation of the hands of anatomists to septic infection" (see page 197, Note I). The disease is highly contagious, hence isolation of the patient is imperative wherever possible.

Aided by a sustaining and stimulating general treatment, the disinfection of the local septic state should be most energetically pursued. According to the age and disposition of the patient, this will have to be done differently. In small children of a good disposition, pencilings of the affected parts with milder or stronger solutions of corrosive sublimate repeated every hour, and, in case of nasal diphtheria, hourly syringing of the interior of the nose, should be practiced. A mixture of corrosive sublimate 0·03, alcohol 25·00 (or one-half grain to the ounce), can be safely used for penciling the tonsils and pharynx. A tepid watery solution of 1:5,000 for syringing the nasal cavity will be well borne. Care must be taken to keep the nostrils well anointed with vaseline to prevent eczema, and never to use a sharp, long-beaked syringe. During the struggles of the resisting child the mucous membrane is easily lacerated, and the hæmorrhage and certain infection of the part thus injured are not indifferent in an affection where the least complication may suffice to fatally determine the case. The safest manner of douching the nose is by attaching to the nozzle of the syringe a piece (six inches in length) of soft rubber tubing, such as is used on infants' feeding-bottles, its distal end being first provided with a few lateral holes cut into it with scissors. The syringe is filled with the warm lotion, the well-greased flexible tube is introduced into the nostril and pushed back until it is felt to touch the posterior pharyngeal wall, the child's head is inclined forward, and then the contents of the syringe are briskly thrown into the nasal cavity. The immediate reflex closure of the larynx and isthmus faucium will prevent the entrance of considerable quantities of the lotion into these organs, and the energetic stream will aid the detachment and expulsion of crusts, membrane, and liquid secretions. On account of the swollen condition of the mucous membrane, the entrance of acrid secretions into the Eustachian tubes need not be feared.

The throats of larger children or grown persons can be cleansed by frequent gargling with a tepid solution of (1:5,000) corrosive sublimate, containing one teaspoonful of cooking salt. The principal weight should be laid upon a *frequent* application of the gargle and a stimulating, nourishing, general *régime*.

Whenever the aspect of the malady is very threatening, the application of the galvano-cautery to the affected parts may be advisable. It is, with cocaine anæsthesia, a safe and rational process. That only a portion of the patches are accessible, some of them being beyond the surgeon's reach in the nasal cavity, is no valid reason why those that are amenable to this very effective mode of disinfection should not thus be treated.

The best way of cauterizing the tonsils and pharynx is the following one :

The head of the anæsthetized patient is drawn over the underpadded edge of the table until it assumes the dependent, or Rose's, position (Fig. 170). The surgeon introduces a bent tongue-depressor, or the bent handle of a tablespoon, well back into the fauces, and instructs the anæsthetizer to keep the tongue out of the way by it. This will expose the pharynx in an admirable fashion to permit of the exact and thorough application of the thermo- or galvano-cautery to the patches thus exposed. If the disease be limited to visible parts of the oral cavity, and all the patches can be thus treated, a rapid improvement of the general state of intoxication will, as a rule, at once follow the procedure. Where only a part of the patches is thus treated, the improvement will not be as complete.

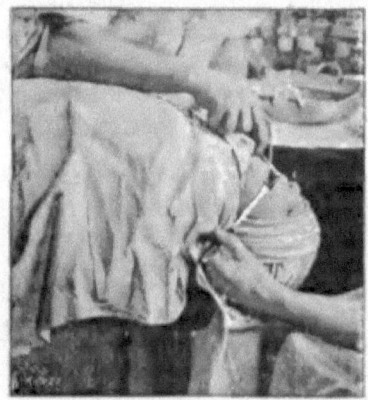

Fig. 170.—Rose's position. Head dependent from the edge of the operating table.

The glandular enlargement also requires attention, and should be treated as was explained elsewhere.

If the process descend to the larynx, very alarming dyspnœa will gradually develop. It should be combated with external hot applications to the throat, and the inhalation of moist, warm air generated in the sick-room. The patient's strength should be carefully husbanded by frequent doses of liquid nourishment, and the avoidance of unnecessary excitement, exposure, and, most of all, *strong emetics*, the abuse of which has cost many a child's life. In most cases the membrane will get detached piecemeal, or will come away in one or more large masses, and relief will follow, perhaps only to be succeeded by another or several suffocative attacks. As long as there is no lung complication, the pulse fairly good, *intubation* offers fair chances of success. Where the patient's strength has been consumed by a very long, ceaseless struggle for air, or the depressing use of emetics, the chances are by far more slender. Yet even the most desperate cases sometimes yield unexpectedly good results. When intubation is not feasible, tracheotomy has to be performed.

*Preventive Treatment of Tonsillitis.*—The tonsils are the points where the first patches become visible in most cases, and whence the local infection extends to other contiguous parts. After frequent attacks of tonsillitis, the surface of the tonsils becomes irregularly indented by cicatricial retraction ; the tonsil itself is enlarged, and often yields on pressure one or more yellowish plugs of a very fetid cheesy matter which were contained within the follicles.

NOTE.—Drs. E. Gruening and S. Cohn called my attention to this fact, which I have repeatedly verified.

These yellowish masses are, as shown by Gruening, swarming with leptothrix and other micro-organisms, and the presence of these is undoubtedly at the bottom of the so-called "disposition" to catch the disease. The reservoir of infecting material is ever there; the patient carries it constantly with him, and a catarrhal hyperæmia, followed by some infiltration and epithelial erosion, is all that is needed to develop a new attack of "follicular tonsillitis," which may not threaten its possessor with great danger, but is just as contagious to others as any case of diphtheria. One observation like the following will carry much conviction.

Two children of the same family had attacks of sore throat one after the other. The first, a boy four years old, who has had tonsillitis a number of times, exhibited the usual symptoms of his affection; the second one, a boy about a year old, and hitherto free from the disease, was carried into the sick-room of the first child by an obstinate nurse, and came down the next day with very alarming systemic symptoms, high fever, and somnolence, exhibiting a small patch on his left tonsil. The first boy recovered in about four days, the usual length of his attack; by the time that he was well, the baby had died under symptoms of most acute septicæmia. A petechial rash, commencing on the nates and feet, extended upward, and gradually flecked the entire skin. The patch on the tonsil had grown and others had developed, the somnolence turned into coma, and was followed by death.

The wet-nurse of this child and the cook of the family, who had kissed the corpse, became seriously ill with diphtheria; especially the latter, whose condition was critical for three or four days. At the same time, a male servant and two more members of the family contracted sore throats of various degrees of intensity, and the house had to be abandoned. A friend and his wife called in the evening shortly after the child's death to pay a visit of condolence. The next morning one of their children was down with malignant diphtheria, and died in a day or two of septicæmia.

Destroying the entire surface of the tonsil, together with the contents of the follicles by the application of the galvano-cautery, would seem to be rational, and has been found a safe and effective measure for lessening the disposition to renewed attacks of diphtheria. It is infinitely safer than a bloody ablation of the tonsils, as the dangers of hæmorrhage and diphtheria of the wound-surface are thereby avoided. The smooth, dense cicatrix thus produced offers a very good protection against new infection.

In adults, or even in half-grown children amenable to control, the reduction of the tonsil can be gradually accomplished without general anæsthesia, the procedure extending over a number of sittings. The throat is pencilled with a cocaine solution until local anæsthesia is produced; then a cold galvano-caustic burner is introduced. It is placed against the part to be treated, the current is turned on, and one fourth or one third of the tonsillar surface is thoroughly seared. For an hour or so, small pieces of ice should be swallowed by the patient to allay the slight pain. The sittings can be repeated about twice a week or oftener.

## DIAGNOSIS AND TREATMENT OF PHLEGMON.

*Quincy sore throat* (peritonsillitis) is a phlegmonous process established in the tonsil itself, or in the loose connective tissue in which it is imbedded. The tonsil is found enlarged, projecting into the pharynx, and displacing forward the anterior pillar and velum. Dysphagia and more or less salivation with high fever are regularly present, and do not terminate until thorough evacuation has taken place. In most cases confluence of a number of small abscesses and simultaneous evacuation is observed. In others, especially when the tonsil itself is the seat of the affection, a number of abscesses develop and open one after another, and retard recovery for a week or two. No local treatment short of incision can effect a substantial improvement, and the different gargling mixtures are only useful in clearing the throat and mouth of the foul, sticky slime aggravating the patient's sufferings by exciting very painful reflex movements at deglutition. Hot salt water (one teaspoonful to a quart, about 6 : 1,000) is the best, as it is the most solvent gargle, and can be easily procured. As the exact location of the abscess can not be ascertained easily beforehand, it is wise to wait with the incision until the swelling is well developed. A digital examination of the swollen region is always advisable, as it is not rare that the tip of the finger detects a pitting spot at which incision will release pus. If pitting can not be detected, an examination with the tip of a silver probe will possibly help to ascertain the most painful spot corresponding to the focus to be incised. The relative distribution of the swelling may also serve as a guide in determining the seat of pus. Acute enlargement of the tonsil itself with diffuse œdema of the pillars and palate indicates suppuration *within* the tonsil. Displacement of the relatively normal tonsil inward is a sign of *retro-tonsillar suppuration*. A combination of both will show the worst association of distressing symptoms.

*Incising Tonsillar Abscess.*—A lancet-shaped pointed bistoury is protected with strips of adhesive plaster to within an inch of its point (Fig. 171), the tongue is depressed with the left index-finger, while the right hand thrusts the knife into the base of the swelling through the anterior pillar at the point previously determined. The antero-posterior direction should be rigidly adhered to

Fig. 171.—Lancet-shaped bistoury wrapped up in adhesive plaster for incision of tonsillar abscess.

on account of the vicinity of the carotid artery. If the first puncture be unsuccessful, a second one should be made in another likely place, and, as soon as pus appears, the blade should be turned *inward*, that is, toward the median line, and should be withdrawn, dilating the incision in that direction. A number of fibers belonging to the levator palati will be thus divided, and their retraction will create a patent orifice, favorable to good drainage.

*Retro-pharyngeal phlegmon* is a comparatively rare suppuration of the retro-pharyngeal connective tissue, due to septic infection of the glands normally imbedded in it. It is mostly observed in small children. The symptoms are those of retro-pharyngeal abscess from tuberculous caries of the cervical vertebræ, but its appearance is much more rapid, accompanied by high septic fever and more acute local distress, causing difficulty of deglutition, regurgitation of food through the nostrils, and alarming dyspnœa. The most characteristic symptom is the peculiarly rigid attitude of the head, which is erect, and thrown back to a certain extent at the same time. The voice is thick and guttural, as though a voluminous foreign body were held in the throat.

In some cases the suppuration extends to the "intermuscular space," and causes the appearance of a lateral external swelling behind the sterno-mastoid muscle. The transverse diameter of the neck then appears widened. Inspection of the pharynx shows that the posterior pharyngeal wall is displaced forward, is densely infiltrated, and sometimes fluctuating.

Incision should be done through the oral cavity if the inflammation is confined to the retro-pharyngeal region, but will be more advantageous if done from without and behind the sterno-mastoid muscle in cases where external swelling of the cervical region is noticeable.

In the first case, the children should be held as for penciling of the throat, and the person having charge of the head should be instructed to throw it forward at a given signal, so as to favor the escape of pus and blood outward from the oral cavity, and prevent its entering the larynx.

If lateral swellings appear, proper incision from without will afford efficient drainage, and at the same time will help to avoid the dangers accruing from the entrance of pus into the larynx.

The manner of incision is best illustrated by the subjoined cases.

Of a large number of cases treated at the German Dispensary, and a few seen at consultations in private practice, only two have terminated fatally, and in both serious hæmorrhage occurred a few hours after the incision.

Case I.—S. P., aged eighteen months, seen May 17, 1883, with Dr. L. Weiss. Retro-pharyngeal and submaxillary abscess developed during the florid stage of a violent scarlatina with diphtheria. Dysphagia and dyspnœa. Small lateral incision through the skin and fascia parallel to, and behind the posterior margin of the left sterno-mastoid muscle. Successful search for pus with a stout hypodermic needle, carried inward and a little backward toward the retro-pharyngeal space. Insinuation of a grooved director along the hollow needle, followed up by the introduction of a small pair of dressing forceps, which were withdrawn half opened. Escape of about one and a half ounce of pus and introduction of a drainage-tube. Two hours after incision copious secondary hæmorrhage set in, and rapidly terminated in death. Giving away of the wall of a sloughing vessel must be assumed to have caused this issue.

Case II.—Henry W., aged four and a half months, a healthy child, developed March 4, 1883, fever and dysphagia, due to the presence of a number of small abscesses situated in the retro-pharyngeal connective tissue. Several of these were incised by Dr. A. Jacobi, with apparent relief of short duration. New foci appearing, the incisions were repeated March 6th and 8th. *March 9th.*—Dysphagia became complete and dyspnœa alarming. Although the incisions through the retro-pharyngeal space continued to bleed, increasing the danger by the addition of hæmorrhage to the other symptoms, the extension of the process to the connective-tissue plane of the large vessels and the alarming dyspnœa left no alternative but death from suffocation or an incision of the abscess from without. *March 9th, at 2 P. M.*—This was done, evacuating about half an ounce of pus. A drainage-tube was introduced into the bottom of the cavity, and, to limit the oozing, a compressory dressing was applied. *At 4 P. M.*—Scanty but continuous hæmorrhage set in from the drainage-tube. This being removed, the cavity was plugged with strips of iodoformed gauze, and the bleeding edges of the incision were seared with the thermo-cautery. *At 8.30 P. M.*—The child died of acute anæmia.

*March 10th.—Post-mortem examination* by Dr. A. **Seibert** in the presence of Dr. L. Bopp and the author. On the neck, close to the posterior edge of the left sterno-mastoid, a cutaneous incision was found one inch in length, its edges marked **by a** dark-red, bloody infiltration. A probe entered the retro-pharyngeal space, where it could be felt with the finger placed in the oral cavity. A skin-flap being raised and turned upward, a couple of intumescent, dark-red lymph-glands, situated near the anterior edge of the sterno-mastoid muscle, were exposed. The sterno-mastoid muscle was cut away at its lower insertion and was turned upward. The vascular sheath **was** opened, and the **deep** jugular vein and carotid artery were carefully examined **and** found intact. A **wall** of tissue one **third of an** inch in thickness was found **interposed** between **these vessels and the track occupied by the silver probe. The prevertebral** interspace **was found distended by a dark, massive, and soft clot, extending upward to** the base of the cranium, **and downward to the level of the third tracheal** cartilage. Cervical vertebræ normal.

Doubtless it was a case **of** hæmophilism.

(A case of retro-pharyngeal infiltration, simulating the symptoms of **abscess, was** seen by the author in the German Hospital, in which *acute infectious* **osteomyelitis** *of the second cervical vertebra* was the cause of the trouble. Henry Ludwig, **bartender,** aged twenty-one. *February 16, 1885.*—High **fever set in** with a chill and **stertorous** breathing. The face was slightly **cyanosed and the voice** had a thick sound characteristic of retro-pharyngeal swelling. The patient held his neck rigidly, and in moving supported it by his hands. A typhoid condition prevailed. The house surgeon of the German Hospital made a free incision into the swelling occupying the retro-pharyngeal region, but no pus escaped. In spite of weight extension, sudden death occurred, March 20th, from compression of the medulla. Post-mortem examination revealed a far-gone destruction of the second, third, and fourth cervical vertebræ. The odontoid process **was detached,** and had fatally compressed the medulla.)

*Acute infectious osteomyelitis of the lower jaw* occurs either in the **adult** after traumatism, such as for instance fracture of its entire thickness by **violence,** or injury to the alveolar process caused by the extraction of teeth ; or spontaneously in the adolescent. The latter form **is quite** frequent, and results generally in more **or less extensive necrosis and the formation of** abscess. Perforation **usually takes place toward the oral cavity, though occasionally invasion of the submaxillary capsule or the vascular** interspace is observed. **Early incision will allay pain, relieve the fever, and will prevent** the extension **of** suppuration.

The treatment of **necroses of the mandible** was **disposed** of elsewhere.

(β) *Submaxillary and Parotid Cynanche.*—Both **the** submaxillary and parotid salivary glands are inclosed in complete and very **dense** fascial envelopes. On account of this **anatomical peculiarity, and in the case of the submaxillary** gland, the vicinity of the tongue and larynx, purulent inflammations of these organs present some peculiarly grave features worthy of special attention.

Human saliva normally contains a chemical substance akin to the ptomaïnes or to snake poison, that, like the latter, seems to play an important part in the process of digestion. Whether an undue development of this albuminoid substance, or exclusively the direct absorption of septic matter from the oral cavity is at the bottom of the septic inflammations of the salivary glands, is not known—suffice to say, that occasionally one or the other of these glands becomes the seat of suppurative inflammation. Their resistant envelope leads to incarceration of ichor and pus, to the development of enormous tension and its deleterious local and general effects—which are dense infiltration and necrosis of the contiguous soft parts, with dysphagia and suffocative attacks, and a highly septic fever.

*Sublingual or Submaxillary Cynanche* (*Ludwig's Angina*).—A painful, deep-seated, hard swelling of the submaxillary region appears, and is quickly followed by chills and high fever, the swelling rapidly increasing in extent and hardness, and the skin over the submaxillary gland turning dusky red. As long as the patient is up, his head is held rigidly in one position, the eyes moving in wide circles if he wants to see an object out of his range of vision. Or, if he be unsuccessful, the entire body is turned round slowly to bring the desired object within sight. The mouth is held slightly open, the tongue is dry, the floor of the mouth somewhat œdematous. Speech is difficult, as can be seen from the painful twitchings of the patient's face whenever he has to say something. After a while he will seek the bed. The face will appear slightly œdematous and cyanosed, the eye has a dull and stupid expression, the dry tongue is found lolling out of the mouth, and saliva escaping alongside of it. The floor of the mouth is very œdematous, and by this time the entire submaxillary region will have become swollen and as hard as a board. The labored snoring respiration of the patient gives warning of the extension of the œdema to the soft palate, fauces, and the vicinity of the larynx. The temperature indicates very high fever, and the patient is unable to allay his burning thirst, as swallowing will have become impossible. At this stage œdema of the glottis may cause asphyxia in some cases, requiring immediate tracheotomy. In other cases extensive sloughing of the involved parts of the neck will supervene, and fatal hæmorrhage may be caused by erosion of large vessels. The grave septicæmia alone, or the extension of septic thrombosis to the cranium or right auricle, may end in death.

All dilatory measures, such as hot or cold applications, will be useless, or positively injurious, and the patient's salvation depends on a quick appreciation of the true character of the trouble, followed by prompt and energetic action.

CASE I.—It was observed by the author during his military service in Garrison Hospital No. 2 at Vienna, Austria, in November, 1872. During convalescence from a severe form of typhoid fever, symptoms of sublingual cynanche appeared in a young soldier treated in the division for internal diseases. Fomentations being employed, the swelling assumed alarming proportions. Suddenly œdema of the glottis appeared, and the case was transferred to the surgical division. The left side and frontal region of the neck were found densely infiltrated and very hard, and tracheotomy had to be performed under unusual difficulties by regimental surgeon Dr. Fillenbaum. A number of abscesses were encountered, and purulent perichondritis was found to be the immediate cause of the œdema of the glottis. Tracheotomy relieved the dyspnœa, but the patient died soon afterward of septicæmia.

CASE II.—Jacob H., farmer, aged twenty-one, admitted to the German Hospital January 19, 1886, presented a circumscribed rod swelling of the left submaxillary region, that had appeared with high fever two days before admission. Face cyanosed, expression dull, breathing stertorous; the mouth half open, tongue protruding, floor of mouth œdematous. Temperature, 104·5° Fahr. Immediate incision according to Hilton-Roser's method in anæsthesia. About half an ounce of thin ichorous pus escaped. The incision was enlarged with a probe-pointed knife, and drainage and a moist dressing were applied. In the night a short suffocative attack appeared. *January 20th.*— Temperature, 101° Fahr. Cyanosis and œdema of the floor of mouth appreciably diminished. Improvement continued, no necrosis following, and patient was discharged cured February 6th.

CASE III.—William B., clerk, aged twenty-two. Sublingual cynanche, characterized by protrusion of tongue and very high fever. The family attendant had treated the case for ten days by poulticing, and April 3, 1884, had incised the swelling in the submaxillary region. Relief followed, but in the night alarming dyspnœa, due to arterial hæmorrhage, supervened, that rapidly distended all the interspaces of the left side of the neck, and threatened suffocation. *April 5th.*—Early in the morning tracheotomy was hastily performed by the author, who found the left side of the neck enormously swollen, and some bloody serum oozing out of the small external incision and from the oral cavity. The source of the latter bleeding was found in a sloughy perforation of the floor of the mouth. As hæmorrhage had ceased, only a drainage-tube was placed into the external incision, and a moist dressing was applied. The patient was doing well April 7th, when he was seen by the author the last time. Later on, the family attendant informed the author that another external hæmorrhage had occurred during the process of detachment of the numerous sloughs, requiring deligation of a spurting, probably the facial, artery. Patient recovered

CASE IV.—C. S., watchman, aged thirty-two. Sublingual cynanche of thirty-six hours' standing. Extensive hard infiltration of anterior and left side of neck. Dysphagia, dyspnœa, tongue protruding. *May 5, 1886.*—Incision by preparation at German Hospital. The thickened capsule of the submaxillary gland being divided, a small cavity containing about a half drachm of ichorous pus and *débris* was exposed and drained. It just admitted the tip of the index-finger. Immediate improvement of all symptoms. Patient was discharged cured May 20th.

*Parotid Cynanche.*—This may develop independently or complicated with orchitis during and after acute infectious diseases, such as typhoid and scarlet fever, small-pox, or the measles, or may be the direct continuation of an attack of mumps. It is not as alarming in rapidity of development as the sublingual form, but is apt to be much more tedious on account of the gradual breakdown of the lobulated structure of the parotid gland. One lobe after another succumbs to the suppurative process, and an interminable series of abscesses make their appearance. Generally perforation outward is the rule; occasionally, however, perforation into the spheno-maxillary fossa, and extension into the intermuscular planes of the neck, with all its dangers, ensues. Necrosis of the interlobular septa is a common occurrence. On account of the necessity of avoiding the temporal artery and facial nerve, long incisions are impracticable. They must be small, and several should be made to afford sufficient drainage.

CASE.—II. S., merchant, aged fifty, commenced to suffer about Christmas, 1885, from a furuncle of the external meatus. This led to suppuration of the lymphatic gland normally found in front of the meatus, and, under a poulticing treatment, to an involvement of the parotid gland. The patient was seen by the author January 11, 1886, and exhibited a large, non-fluctuating, very dense swelling of the right parotid region, with a temperature of 104° Fahr. His right eye could not be closed entirely (paresis of the facial nerve), and he was unable to separate the jaws to the slightest extent. Besides, repeated chills, sleeplessness, and the intense pain radiating to the diverse branches of the trigeminal nerve, had demoralized the man completely. A vertical incision placed just in front of the external meatus by careful preparation released a large mass of pus. The relief was very great, and the patient left the house five days later to be treated at the author's office, where he repaired daily for many weeks longer, as the involvement and breaking down of new lobules of the parotid gland made frequent irrigation and constant drainage a necessity. He was discharged cured March 28th. By October the paresis of the orbicularis palpebrarum had disappeared.

(γ) *Acute Glandular Abscesses of the Anterior and Lateral Cervical Regions.*—They are caused by absorption of active micro-organisms dependent on inflammatory processes of the oral and nasal cavities, the pharynx, larynx, the lower jaw, and the mastoid region. They have to be well distinguished from cold or chronic abscesses of the same region. Their onset is sudden; pain and fever rapidly develop, with deep-seated dense infiltration, and gradually the corresponding side of the neck becomes œdematous. Inflammations in the oral cavity, the tongue, the larynx, and the lower jaw produce an involvement of the glands in the *perivascular* space. They can be felt somewhat in front of the sterno-mastoid muscle, extending upward toward the angle of the jaw, and are commonly known as "submaxillary" glands. Affections of the temporal, auricular, and mastoid regions, and of the pharynx, nasal cavity, and œsophagus, on the other hand, are generally followed by intumescence or suppuration of the glands situated in the *intermuscular* space. They can be felt behind the posterior margin of the sterno-mastoid, and their suppuration is apt to extend in the direction of the supraclavicular space.

The question of when to incise these abscesses should not be made dependent upon the presence of fluctuation, as the worst and most virulent cases will have wrought infinite mischief long before the appearance of fluctuation. In very virulent cases, marked by violent general symptoms and rapid local spread, incision should be made at once after Hilton-Roser's method, as relief from tension is the most urgent requisite to prevent sloughing and possible erosion of vessels. Anæsthesia is indispensable.

Where the symptoms are less violent, the spread less rapid, maturing of the abscess may be awaited in case the patients are very averse to an incision.

## DIAGNOSIS AND TREATMENT OF PHLEGMON.

But the responsibility for the consequences of delay should be declined by the physician.

CASE.—Louis Lebowitsch, aged twenty-seven, presser. *December 15, 1886.*—Painful hard swellings developed in the pretracheal and both submaxillary regions with a severe chill. Previous to this the patient had been suffering from a "sore throat" for a few days. The family physician advised poulticing, which, as usual, was enthusiastically attended to by the patient's female relatives. The swellings continued to grow in size; fever and sleeplessness were unabated. *December 25th.*—Suddenly an enormous increase of the swellings in front and on the left side occurred, with dyspnœa and dysphagia, which induced, December 29th, the patient's transfer to Mount Sinai Hospital. Following a hasty summons the author found the patient sitting up in bed, his head held erect, the neck increased to double its circumference, its skin red, swollen, and shining like a large-sized sausage. Boggy fluctuation everywhere. Most intense thirst with absolute disability to swallow even fluids; wheezing, long-drawn respiration with considerable dyspnœa, which became augmented to an alarming degree by the reclining posture. Examination of the fauces revealed a swelling of the retrofaucial soft tissues, and almost complete contact of the slightly intumescent tonsils. Two incisions, one behind the posterior margin of the sterno-mastoid muscle, the other a little below the thyroid gland, released about a quart of a dark-red gory liquid, streaked with pus. This was followed by an immediate disappearance of the dyspnœa, and the patient was able at once to allay his thirst by copious drafts of water. A digital examination of the cavities opened by the incisions showed them to communicate freely. The pulsating carotid could be distinctly felt, lying exposed behind a large, roundish mass of blood-clot, freely projecting into the lateral cavity, and seemingly attached to the pharyngeal wall.

Two stout drainage-tubes were placed in the incisions, the remaining clots were washed out by gentle irrigation, and a large, moist dressing was applied. The fever fell at once from 103° Fahr. to 100° Fahr., but rose the following day to 103° Fahr., as the incisions were clearly insufficient for the drainage of the enormous cavity. Moreover, there was still considerable oozing present, and therefore it was deemed proper to anæsthetize the patient again, for the sake of a thorough exploration, drainage, and possibly prevention of further hæmorrhage. A fluctuating place just above the clavicle was incised, and was found communicating by a narrow channel with the upper cavity. Both of the lateral incisions were now united by preparation, the external jugular vein being first secured by double ligature and divided, and thus by this long incision the interior of the large abscess was exposed to view. The cavity extended from the clavicle to the base of the cranium. In it lay exposed the carotid artery and the jugular vein, to the upper portion of which anteriorly a large, firm, and irregular clot was found adhering, indicating where the hæmorrhage had come from. The loose clots were all cleared out, but the one adherent to the jugular was left undisturbed. Copious oozing from the abscess walls was observed, and checked by a loose packing of iodoformed gauze, preceded by thorough irrigation. The patient was discharged cured on January 27, 1887.

The preceding case vividly illustrates the dangers of protracted poulticing in deep-seated lymphatic abscesses. Sloughing of the wall of an adjacent large vein caused a most serious complication by secondary hæmorrhage. Arterial hæmorrhage would have undoubtedly produced rapid suffocation.

(δ) *Glandular Abscesses of the Temporal, Mastoid, and Occipital Regions.*—Suppurative processes located in the external ear will occasionally extend to one or more lymphatic glands, subfascially situated in front of the external meatus of the ear, and in close vicinity to the parotid gland. They produce very violent general and local symptoms, and require early attention, as a subsequent involvement of the parotid gland is very apt to occur.

*Suppuration of the mastoid cells* is the most common form of extension of a purulent otitis of the external or middle ear. Its symptoms bear great resemblance to those of acute osteomyelitis, and require prompt attention on account of the possibility of necrosis and the involvement of the meninges, brain, or lateral sinus. Where intense swelling indicates the presence of purulent periostitis of the mastoid process, a free incision of all the soft parts down to the bone will often give great relief. But, where the interior of the cancellous structure of the mastoid process is the seat of the disease, nothing short of a free opening of its interior will avail. Formerly, this operation was done with the aid of the trephine, an instrument the penetration of which is somewhat beyond the supervising control of the surgeon. At present mallet and chisel are used for this purpose with greater advantage. The chisel should be held tangentially to the external surface of the mastoid process, thin layers of bone being pared off in succession, until the suppurating focus is freely exposed. Thus injury to the lateral sinus can be safely avoided. Copious irrigation with a *warm* solution of corrosive sublimate and a moist dressing are advisable. The cases in which early operating has prevented necrosis will heal very promptly. Necrosis will retard the cure considerably, and may require a second or even a third operation for the removal of sequestra.

In neglected cases spontaneous perforation through the periosteum will occur, and an external abscess, located posteriorly to the sterno-mastoid muscle, will appear. The tendency of its extension is toward the "intermuscular space," that is, downward into the supraclavicular fossa.

Occasionally the process extends backward and upward upon the occiput.

CASE I.—Fred. Buths, baker, aged eighteen, admitted to ear department of German Hospital, December 17, 1883, with purulent catarrh of the middle ear and suppuration of mastoid cells. Wilde's incision and extraction of some sequestra from the external meatus were practiced by Dr. J. Simrock. A phlegmon of the left occipital region, starting from a sinus below the mastoid process, having set in, patient was transferred, March 25, 1884, to the surgical department. *March 26th.*—High fever and violent headache with vomiting. Several incisions laid open an irregular cavity situated behind the ear and extending downward toward the neck. On pressure, a large quantity of pus oozed out of a recess between exuberant granulations near the lower anterior angle of the parietal bone. These being scraped away, a sequestrum, about one square inch in circumference, and comprising the whole thickness of the skull, was extracted. Pulsation of the bottom of the cavity thus exposed was clearly discernible. Healing progressed without interruption, the purulent discharge from the middle ear ceased, and patient was discharged cured. April 17, 1884, with a deeply indented scar. In October, 1886, he presented himself, complaining of epileptic seizures that had appeared in July, 1886.

CASE II.—E. N., merchant, aged twenty-five. Had been suffering from purulent otitis media for a long time. Suppuration of the mastoid cells, and formation of an external inframastoidal abscess, led to incision, which was done by Dr. E. Graening, under whose care the patient had been for some time. A phlegmonous inflammation of the neck following, January 22, 1882, a consultation was called, when a number of deep incisions back of the sterno-mastoid muscle were made, and the abscesses were drained. The probe felt bare bone in the mastoid notch. Subsequently a considerable quantity of bony grits passed away with the secretions, and the carbolic lotion injected into the drainage-tubes entered the oral cavity. End of March, the patient was discharged cured, and remained well until September, 1886, when he was seen by the author suffering from dementia.

*b.* **Mammary and Retro-mammary Abscess.**—Excoriations and fissures, so common upon the nipples of nursing women, are the portals through which infection enters the multitudinous lymphatics of the mammary gland. A preparatory treatment of the nipples during the last period of pregnancy is the best preventive of the formation of fissures. It should consist in mollifying, and removal by bathing in warm soap-water, of the thick layers of effete epidermis, usually present around the openings of the lacteal ducts. The tender epidermis thus exposed will be hardened, and will become fit to resist the manifold injuries unavoidable during lactation.

Should rhagades develop, a thorough disinfection with corrosive-sublimate lotion (1 : 1,000), followed by touching of the fissures with a well-sharpened stick of nitrate of silver, will in most cases lead to a cure of the painful disorder. Nursing should be either stopped and the milk removed with the breast-pump, or, if continued, should be only permitted with a nipple-shield, until the fissure is closed.

Disregard of these precautions will frequently lead to suppuration.

A large proportion of the inflammatory processes of the breast are non-suppurative, the intumescence, redness, and occasionally smart fever being set up by a retention of the thickish milk of first lactation. Sometimes fluctuation will be felt, and, if an incision is made, no pus—only milk—will escape. Absence of an infection by micro-organisms must be assumed in these cases, which, as a rule, get well without suppuration by simple topical treatment, consisting of the application of moist heat and methodical compression.

Hence, not all cases of acute mastitis terminate in abscess. Winckel saw, in the Dresden Lying-in Hospital, ninety-one out of a total of one hundred and thirty-six cases of mastitis get well without suppuration. Therefore, topical treatment with the ice-bag or cold-water coil (by both of these the secretion of milk is materially reduced), or, if opposition to these be encountered, tepid or warm applications, aided by support and gentle compression of the breast, should be first tried.

Should, however, fever and the local symptoms persist or increase, and fluctuation become apparent, incision and drainage are the measures to be applied.

Abscesses of the mammary gland proper are either *subcutaneous*, then generally located about the nipple; or are more *deep-seated*, that is, *intra-*

*glandular.* A third form of breast abscess is the suppuration of the loose connective tissue found *behind* the gland : *retro-mammary abscess.*

Its location in the vicinity of the nipple and the early appearance of well-defined fluctuation will readily characterize the subcutaneous abscess.

When the deeper parts of the glandular tissue proper become the seat of an abscess, general swelling of the breast-gland is most prominent. The skin of the mamma becomes red and œdematous, and one or more pitting points can be soon detected. *But the breast is freely movable as a whole upon the pectoralis fascia.*

*In retro-mammary suppuration the breast is immovable,* and firmly attached at its base. The glandular tissue is soft and normal, unless a combination of mammary and retro-mammary suppuration be present. Deep fluctuation can be detected by careful palpation.

*Incision of the more extensive abscesses of the breast should always be done under anæsthesia,* as the unavoidable pain associated with thorough work is too great to be endured ; and the measures must be thorough to give a prompt result, as nothing is more unsatisfactory than an insufficient or improperly placed incision. Suppuration is not limited thereby, new points of fluctuation develop, and the interminable process, with fever, sleeplessness, and the drain upon the system, lead to serious emaciation and lamentable demoralization of both patient and physician. *Antiseptic precautions,* consisting of a thorough scrubbing of the surgeon's hands and of the patient's breast with soap and brush, and subsequent rubbing off with corrosive-sublimate lotion (1 : 1,000), should never be neglected. There are microbial cultures of various intensity of virulence, and the touch of an unclean finger may intensify an otherwise comparatively bland form of suppuration, *or may add the poison of erysipelas to that of simple suppuration.*

All incisions penetrating the glandular tissue should be placed *radially,* so as to avoid injury to the lacteal ducts as much as possible.

A place of fluctuation being marked, the knife is rapidly thrust into the abscess, if the thickness of tissues to be cut through is not too great. In the latter case, Hilton-Roser's method is safer and preferable, on account of the possibility of hæmorrhage from a deep-seated vessel.

NOTE.—Billroth recounts a case in which he caused uncontrollable and very serious hæmorrhage by cutting a large branch of the external mammary artery. The loss of blood was alarming, and so beyond control that, after having unsuccessfully tried a number of the usual measures, he finally injected the abscess cavity with a quantity of turpentine oil, that happened to be within reach. The bleeding was stopped, but a formidable gangrenous phlegmon brought the patient very near the grave. She recovered, however.

As soon as the well-dilated dressing forceps is withdrawn, the index of the left hand is slipped into the cavity, and a gentle exploration of its interior is carefully made. Wherever a recess extends toward the skin, the tissues are raised upon the tip of the left index-finger, the skin and fascia are incised, and the dressing forceps is introduced along the grooved director in the well-known manner. In this way a number of short counter-incisions can be made with very little hæmorrhage. Stout drainage-tubes, reaching just within the cavity, are next introduced, and the abscess is well washed **out with the mercuric lotion.** Oozing from **the abscess** walls, which **is sometimes considerable,** will also be checked **thereby.** After this the **breast should** be grasped and gently **compressed between the** extended hands **as a test, whether all recesses *had been duly emptied or* *not*.** The appearance of **additional masses of pus will be a** proof that something was overlooked, **and renewed search** must be instituted to find and drain the overlooked **recess.**

NOTE AND CASE.—The observance of this simple rule led to the recognition of a very interesting and rare form of suppurative mastitis. Mrs. C. F., primipara, admitted to Mount Sinai Hospital two weeks after her confinement, with abscess of the breast. Had very little fever. She was anæsthetized December 20, 1886, and, four fluctuating spots situated just above and near the nipple being incised, the finger was slipped into one of the incisions, and found the irregular and tortuous cavities communicating with each other. A large number of smaller cavities occupying **the** upper half of the mammary gland were entered, and the intervening bridges of tissue were broken down with the finger. Hæmorrhage **was** very scanty. **The cavity was washed out, and, gentle pressure being applied, an additional** large mass of thick **pus escaped.** A long incision uniting the two most distant primary incisions, and passing through the entire width of the gland, **was now made.** It exposed the cavity, which was found lined with necrosed shreds of glandular **tissue.** The abscess walls exuded on firm pressure from hundreds of invisible openings **separate** drops of creamy pus. A portion of the indurated wall of the cavity was pared off, until seemingly healthy tissue was encountered. Firm pressure being repeated, the same exudation of pus from innumerable pores of the cut surface was observed. The section had a deep yellow tinge, and presented the density of fibromatous tissue. The lower half of the breast-gland was normal and **secreted** milk. An iodoform **dressing was applied,** and remained undisturbed until December 27th, when the **patient complained of pain** and exhibited some fever. The dressings being removed, a new abscess was **found and incised near** the upper margin of the **long** incision. The old abscess cavity was granulating, **but its walls** still exhibited the peculiar **appearance of a** large number **of** distinct pus-drops on **pressure.** The wretched general **condition of the** patient, and the presumably interminable suppuration **to be** expected **under the** circumstances suggested exsection of the affected parts of the breast as the most rational measure. This **step, however, was** strenuously opposed by the patient, and she left the hospital **uncured.**

Apparently **we had in this case a form** of purulent mastitis where the suppurative process was primarily located in the lacteal ducts, the interstitial connective tissue assuming the character of shrinking fibroid **or cicatricial tissue,** as in non-suppurating interstitial mastitis. **The** contraction **of** the interstitial tissue led **to closure of the** lacteal **ducts and** to retention ; this to perforation of the **lacteal ducts and** extension of the suppuration into the interstitial tissue ; this, **finally,** to the formation **of a large number of** disseminated abscesses and **necrosis.** Throughout, **the case exhibited** unusual characteristics well-circumscribed localization, **low** fever with appalling destruction of tissues, and **their curious** permeation with canals, **that** could be nothing **but** lacteal ducts, filled with creamy pus. As drainage and disinfection of the infected lacteal ducts were impossible, ablation of the **diseased part of the** gland was clearly the proper **way to** terminate the **process.**

*Retro-mammary abscesses* **usually** point near the lower margin **of the** breast-gland. They should be treated like other deep-seated abscesses, **by** incision and drainage, care being taken to establish the latter in the most dependent position.

When the operation is completed, safety-pins are thrust through the projecting ends of the drainage-tubes near the surface of the skin, and they are trimmed off short. A small ring of iodoformed gauze is placed underneath the safety-pin around the drainage-tube, to prevent its being overlapped by the edges of the wound, and a *moist antiseptic dressing is applied.* In the absence of fever and pain, and if the dressings remain unpermeated by secretions, they need not be changed before three or four days, when the drainage-tubes can be either wholly removed, or one, having previously been somewhat shortened, can be left in the most dependent incision till the following change of dressings.

Where shreds of necrosed tissue are still adherent to the walls of the abscess, secretion will be somewhat more copious, and permeation of the dressings will require daily changes until the necrosed parts come away. During this time, however, if drainage be adequate, *all the pus secreted should be contained in the dressings, and none in the wound.*

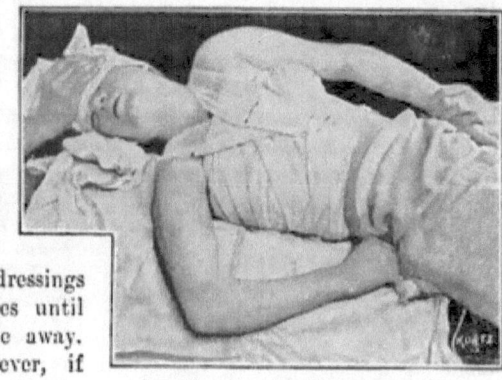

FIG. 172.—Dressing for mammary abscess, or empyema.

After detachment of the necrosed parts, secretion will become scanty and watery in character, and removal of the tubes will be followed by rapid closure of the wound.

In cases where drainage is inadequate, fever and pain will persist, and secretion will remain profuse. The dressings will need frequent renewal, they will be rapidly soaked with pus, and the wound itself will contain more or less of it. This can be easily ascertained by gentle pressure, which will cause a copious flow of pus. Frequent irrigation is a very imperfect substitute of proper drainage; therefore, the making of a well-placed incision should remedy the shortcoming.

*c.* **Empyema.**—Infection of the pleura by pyogenic organisms, either through metastatic processes or by direct extension from the bronchi and lungs; from without by injury, or from purulent affections of the vicinal regions, as, for instance, perinephritic or liver abscess, leads to the formation of empyema—that is, an accumulation of pus within the pleural cavity. The diagnosis of the affection is based upon the fever, dyspnœa, the absence of respiratory murmur, the dull percussion sound, rigidity of the affected side of the thorax, flatness of the intercostal depressions, and more or less marked œdema of the integument over the site of the accumulation.

Probatory puncture with a hypodermic needle will usually yield pus.

The proper treatment consists of timely incision, disinfection, and drainage under antiseptic cautelæ.

*Management of Recent Cases of Empyema.*—The thorax of the anæsthetized patient is cleansed and disinfected, and an incision is made, from two to three inches in length, in the eighth intercostal space, parallel with the ribs, and a little back of the axillary line. The skin and muscles are gradually divided down to the pleura, which is then incised. The sudden gush of pus is checked and moderated by the pressure of the tip of the finger, as too sudden evacuation of the tense accumulation may lead to rupture of vessels, or, in the case of empyema of the left pleural cavity, to fatal embolism of the pulmonary artery. In these cases the heart is displaced to the right side, and any clots that may have formed within the right auricle could be easily detached by a sudden change of the heart's position. This accident has occurred once to the author. However, it did not take place on the operating-table, but happened several days after the operation.

CASE.—Helen Muller, aged eleven. Empyema, with two fistulæ, of six years' standing. Great emaciation; retention of fetid pus; the heart displaced to the right side. *February 27, 1883.*—Exsection of two ribs, multiple incisions, and drainage of the fetid abscess. Daily irrigation produced a marked remission of the fever, and everything seemed to progress favorably, when, March 6th, while playing in bed, the child suddenly became cyanosed, and fell back dead. No post-mortem examination could be had. Death was doubtless caused by embolism of the pulmonary artery.

The pleural incision should be ample, as otherwise voluminous fibrinous pseudo-membranes may clog the exit of pus. A large-calibered drainage-tube, *reaching just within the pleural sac*, is inserted, *and is at once secured with a stout safety-pin*, to prevent its being lost in the abscess. This occurred in one case treated at the German Hospital, and a good deal of trouble was experienced in finding the lost tube.

CASE.—Fridolin Jachle, laborer, aged forty-three, *saccated empyema* of eight weeks' standing. *February 9, 1884.*—Posterior incision in the eighth intercostal space; evacuation of a large quantity of pus. A drainage-tube was inserted, but slipped out of the fingers, and was lost in the cavity. The incision was sufficiently enlarged to admit two fingers, and then a sort of a diaphragm could be felt separating two intercommunicating cavities. A counter incision was made in the mammary line, and the lost drainage-tube was extracted therefrom. Drainage-tubes properly fastened with safety-pins were inserted, and the cavity was irrigated with carbolic lotion. Moist dressings were applied. *April 18th.*—Patient was discharged cured.

Washing of the pleural cavity with warm mercuric solution (1 : 5,000) thrown from an irrigator should be done, until the fluid returns in a limpid state. Then a final flushing with corrosive-sublimate lotion of the strength of 1 : 1,000 should follow, and good care should be taken to drain off the last vestige of the solution by turning the patient so as to bring the incision nethermost. A very ample moist dressing should envelop the patient's thorax.

As long as the temperature remains normal or slightly elevated, and the dressing clean, no change is necessary. Usually, however, the dressings will be soiled within twenty-four hours, and then they must be changed. But irrigation should not be employed so long as the patient's temperature is normal. Only, if renewed fever appear, or the secretion assume a fetid odor, will repetition of the irrigation be necessary. In fresh empyemata, especially of children, *one irrigation thoroughly done at the time of the operation will be found sufficient.* But in some favorable cases of adults the same smooth course of healing may be observed. The discharges will gradually diminish, they will lose their purulent character, and will become watery and scanty. As soon as this is observed, the drainage-tube should be removed, and within four or six weeks from the operation the cavity will be healed by renewed adhesion of the costal and pulmonal pleura. The lung will dilate to its normal extent, and the universal adhesion of the pleural surfaces will gradually give way to constant attrition, until the mobility of the lung and the normal state of things are re-established.

Case.—Henry Fennell, furniture-dealer, aged thirty. Empyema on left side of four weeks' duration. *February 1, 1880.*—Communication with a larger bronchus spontaneously established, giving rise to uncontrollable fits of coughing, which have exhausted the patient to a dangerous degree. *February 6th.*—Incision, drainage, and irrigation with a five-per-cent solution of carbolic acid. The cough stopped at once; the fever fell off. *February 17th.*—Discharge very scanty and watery; drainage-tubes were removed. *February 19th.*—Sudden rise of temperature, with chill. *February 20th.*—*Pleuritic serous effusion on right side. March 1st.*—Effusion on right side begins to be absorbed. Left lung dilated to nearly its normal compass. *March 6th.*—Exudation in right pleura has disappeared. *March 12th.*—Patient was discharged cured.

*Lateral curvature of the spine* is a prominent symptom of long-continued empyema, and is very hard to cure. The moderate amount of lateral curvature that goes along with recent empyema disappears with the restoration of the function of the compressed lung.

*Old Empyema.*—Cases of inveterate empyema with or without sinus throw much greater difficulties in the way of the surgeon's efforts to close the cavity and fistula than recent cases. The retraction and consolidation of the lung, and its envelopment in more or less thick coats of pseudo-membrane, frustrate all attempts at closure of the thoracic cavity. The unyielding lung can not expand, while the contraction of the partially yielding walls of the thorax, accomplished by lateral curvature, by a close crowding together of the ribs, and a corresponding flattening of the affected side of the chest, has its limits. Thus a secreting hollow space is maintained within the chest that can not be obliterated by the unaided efforts of nature, and ultimately the patient's strength and life will be sapped. The injection of irritating fluids, or the packing of the cavity with strips of lint or gauze, are of no avail, and *the only means of effecting a cure is multiple exsection of the ribs according to the plan of Estlander.*

The *rationale* of this plan is to do away with the rigidity of the thoracic wall by removing suitably long sections of as many ribs as are found to be corresponding to the cavity. Thus the limbered thoracic wall may be depressed, and can be brought into actual contact, or nearly so, with the opposite or pulmonal surface of the cavity, where it will be fastened down and retained by cicatricial adhesions that will form before the reconstruction of the exsected ribs.

In due course of time the attached lung may even regain a large proportion of its former functional capacity by distention and aëration, and the more or less complete reestablishment of lung capacity is manifested by the disappearance of lateral curvature.

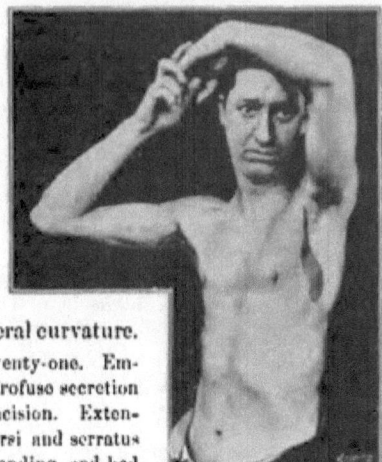

Fig. 173.—Cicatrix in a case of Estlander's operation for inveterate thoracic fistula. (John Springer's case.)

CASE I.—John Springer, clerk, aged twenty-one. Empyema of left side with thoracic fistula. Profuse secretion of pus, escaping through an insufficient incision. Extensive burrowing of pus under latissimus dorsi and serratus muscles. The process was of one year's standing, and had caused lateral curvature and far-gone emaciation. *August 25, 1879.*—Incision and drainage of the external abscesses and of the left pleural cavity at the German Hospital. Exsection of the eighth rib became necessary, as the intercostal space was too narrow to permit of a safe adjustment of the drainage-tube. The operation brought on alarming collapse, which was overcome by energetic stimulation. The external abscesses healed, and, though the secretion from the pleural cavity became much diminished, no tendency to a diminution of the capacity of the sac could be noticed. By New Year, 1880, the patient's general condition had become excellent, and, no improvement being visible regarding the healing of the thoracic fistula, January 3, 1880, Estlander's operation was performed. By an ample vertical incision, commencing in front of the axillary space in the pectoral fold, the third, fourth, fifth, sixth, and seventh ribs were exposed. Their periosteum was slit up longitudinally, and sections of from two to four inches of the ribs were removed, the removed pieces being proportional to the entire length of the several ribs. As soon as the ribs were removed, the thoracic wall could be well depressed into the hollow of the cavity. In order to retard the new formation of bone, the external wound was packed with carbolized gauze, and healed by granulation. The pleural hollow began at once to diminish in size, and April 11, 1880, patient was discharged cured. He has remained well ever since that time, and presented, April 23, 1887, when the accompanying photographs were taken, the following status: A scarcely noticeable trace of lateral curvature; the respira-

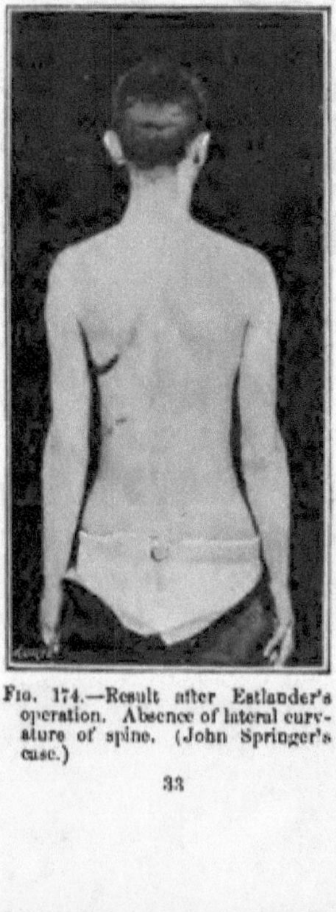

Fig. 174.—Result after Estlander's operation. Absence of lateral curvature of spine. (John Springer's case.)

tory excursions of both sides of the thorax identical. All exsected ribs had re formed and occupied a normal position. Respiratory murmur could be heard all over the left side of the thorax. (Figs. 173 and 174).

CASE II.—Miss Eva C., aged thirteen and a half. Thoracic fistula of two and a half years' duration, leading into a small cavity holding about three ounces of fluid, that had resisted all efforts at cure. *May 12, 1881.*—Exsection of sixth and seventh ribs at Mount Sinai Hospital. *September 20th.*—Patient was discharged cured. In August, 1882, the healed fistula came open, with pain and fever. *September 26, 1882.* —A sequestrum two inches in length, consisting of a portion of the seventh rib, was extracted. The wound healed promptly, and the girl's health remained sound.

The author's rather incomplete record of all forms of empyema of children embraces twenty-two cases. All of these recovered with the exception of two—one died of basilar meningitis; the other of pulmonary embolism.

Of the nine cases of adults, four were cured by simple incision; two by multiple excision of ribs; one, a case of perforation of a tubercular lung cavity into the pleura, died of fatal hæmorrhage into the pleura; and two cases were discharged improved, but not cured.

To conclude, it may be said that the earlier the operation, the safer it is, and the better the results achieved by it.

**d. Phlegmon of the Palmar Aspect of the Hand, of the Arm, and Axilla.** —The hand, on account of its exposed situation, is the most frequent place of small or more serious injury. The necessity of the continued use of a slightly injured hand, and its contact with septic matter, lead to phlegmonous affections of different degrees of intensity.

More serious traumatisms, like incised or lacerated wounds of the hand, become in numerous cases the seat of septic inflammation, in consequence of the improper and uncleanly primary treatment they receive from laymen and some physicians. Neglect of thorough cleansing and disinfection of a small wound often leads to direful consequences, that perhaps the most skillful and incisive therapy can not remedy.

Of the manifold curious practices commonly employed for stanching hæmorrhage and dressing injuries to the hand, only two may be mentioned. *First comes the use of styptic solutions.* They are unnecessary, because digital compression of short duration is capable of stanching even profuse arterial hæmorrhage.

The *second* practice is the favorite closure of soiled wounds about the hand with strips of adhesive plaster or a suture, *without preceding disinfection.*

Some of the worst forms of palmar phlegmon observed by the author were due to similar ministrations by lay or medical advisers.

CASE I.—John McG., liquor dealer, aged thirty-nine. *April 30, 1886.*—Chopped off the tip of his index-finger with a hatchet, and was attended to immediately by a medical quack, who strapped the injured part with a structure of neatly-arranged strips of adhesive plaster without previous cleansing. The wound was a smooth and clean-cut one, and offered the most advantageous conditions for the avoidance of infection. Severe pain, swelling, and fever supervened on the following day, but, at the advice of the medical attendant, the dressing was left on undisturbed for four days.

*May 5, 1886.*—The patient came under the care of the author, who found the wound and its neighborhood tightly compressed by the adhesive strapping, and a phlegmon of the sheath of the flexor and extensor tendons of the index extending into the intermuscular planes of the ball of the thumb. A number of incisions exposed the necrosed tendons, and resulted in a tardy cure after their expulsion. He was discharged cured July 10th.

CASE II.—S. A., laborer, aged thirty-five. Presented himself in January, 1881, at the German Dispensary with an incised wound of the palmar aspect of the thumb, and an extensive subaponeurotic phlegmon of the palm and forearm. The hæmorrhage had been unsuccessfully combated by the patient himself with applications of cobwebs and varnish. Finally, the aid of a druggist was sought, who soaked a piece of lint in perchloride-of-iron solution, and hermetically sealed the wound therewith. Phlegmon set in promptly, and rapidly extended to the palmar bursa. The styptic dressing remained undisturbed, but the palmar swelling was treated with diligent poulticing. At the German Dispensary various incisions were done in anæsthesia, followed by a tedious after-treatment consisting of repeated counter-incisions until cure was effected. The removal of the styptic lint, intimately matted together with living and necrosed tissues, was exceedingly troublesome. The function of the thumb was partially restored.

*Dorsum.*—On account of the loose arrangement of the subcutaneous connective tissue of the dorsal region of the hand, its phlegmonous affections present characteristics similar to those of any other subcutaneous phlegmon. The presence of a large number of hair-follicles favors the localization of septic processes in the cutis, which lead to the formation of typical furuncles or rarely a carbuncle.

*Palmar Aspect.*—The peculiar features of the phlegmonous processes of the palmar aspect of the fingers and hand depend upon the anatomical peculiarities of that region. On the fingers we find, instead of the longitudinal and loose arrangement of the subcutaneous tissue of the dorsum, a dense net-work of short, thick fibers, inclosing a number of small acini of fat. The main direction of the course of these fibers is from the cutis down to the periosteum, or to the sheath of the tendons, to which they are closely attached. The direction of the lymphatics coincides with that of the connective tissue. Upon this centripetal course of the lymphatics depends the pronounced tendency of digital inflammations to penetrate to the bone or the tendons. The well-known tendency to necrosis and the formation of cutaneous, tendinous, or osseous sequestra is, on the other hand, caused by great tension due to the rigid and dense

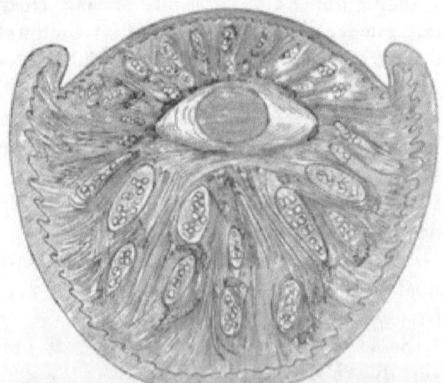

FIG. 175.—Transverse section of terminal phalanx, showing arrangement and direction of connective-tissue fibers. (From Vogt.)

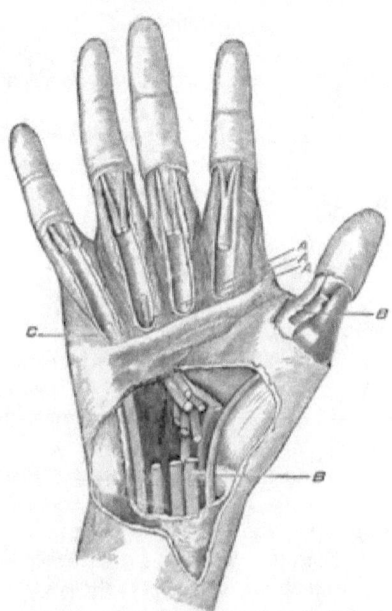

Fig. 176.—A, Blind endings of sheaths of the index, middle, and ring fingers. B, C, Sheaths of thumb and little finger openly communicating with palmar bursa. (From Vogt.)

arrangement of the subcutaneous connective tissue. (Fig. 175.)

The manner of the extension of phlegmonous inflammation within the tendinous sheaths of the palmar aspect of the hand is also prescribed by their special arrangement. Fig. 176 shows the sheaths of the flexors of the *thumb* and *little finger* in open communication with the common palmar bursa, through which pass all the flexor tendons of the fingers to and under the ligamentum capsi transversum, and hence to the forearm. The sheaths of the flexors of the *index, middle, and ring fingers represent separate and closed receptacles,* which terminate on the level of the metacarpo-phalangeal joints. For a short distance beyond these sacs the tendons possess no sheath proper, but are immediately inclosed by loose connective tissue. We see corresponding to these three closed sacs three pointed extensions of the common palmar bursa, into which the tendons enter after passing through the sheathless part of their course. (Figs. 176 and 177.)

*Thumb and Little Finger.*—Upon this arrangement is based the great import of the suppurations of the thumb and little finger, mentioned by the oldest medical writers, and well known to the common people. While gatherings of the index, the middle, and ring fingers often perforate spontaneously near or on the level of the finger-balls (where the blind end of the closed tendinous sheath coincides with the thinnest portion of the palmar aponeurosis), suppurations of the thumb and little finger are very apt to, and as a matter of fact often do, extend at once into the palmar bursa. The knowledge of this peculiarity is of the greatest practical importance.

Fig. 177.—Common palmar bursa injected, and showing extensions toward thumb and little finger. (From Vogt.)

## DIAGNOSIS AND TREATMENT OF PHLEGMON. 247

Aside from the acuteness of the symptoms, phlegmonous affections located on the palmar aspect of the hand and fingers present some peculiarities, the diagnostic significance of which must be mentioned. *Redness of the skin is generally absent*, to appear only when the process has worked its way up to the skin. *Œdema is moderate*, and is often overlooked by inexperienced observers, who are misled by the œdema and redness of the dorsal soft parts to look there, and not on the palmar side, for the focus of the disturbance.

The subjective symptoms are very distressing, high fever and intense pain being the rule.

*Treatment.*—Prevention of phlegmon by guarding against the infection of large or small injuries of the integument is very profitable. Small excoriations and shallow cuts should be cleansed and touched with acetic-acid. Punctures should be well sucked and bled and sealed with an acetic acid eschar; or, if there be the least suspicion of infection by an unclean sharp-pointed object, *dilatation of the small hole*, thorough wiping out of the track with sublimate lotion, and drainage by means of a few short pieces of catgut laid into the bottom of the puncture are to be employed. *In this latter class of cases a moist dressing is appropriate.*

In the presence of an inflammation that is evidently gathering momentum, all attempts at an abortive treatment are risky, as the deceptive relief afforded by hot applications is very apt to induce patient and physician to be tardy with the application of the best and surest antiphlogistic : *the knife*. By the time that the unbearable suffering finally compels energetic treatment, suppuration requires a long incision, and necrosis of a phalanx or tendon may be established. *At first it might have been prevented by a much smaller incision—in fact, by a mere puncture.* The cases where a timely deep puncture with a tenotomy knife released one or a few drops of pus to the most intense relief of the patient were very numerous in the author's dispensary experience, and he can not recommend this truly conservative procedure in warm enough terms. Instead of a terribly painful and tedious illness ending in more or less of destruction, rapid healing of the small wound under the moist dressing will be the rule. And, if we consider that local anæsthesia by cocaine or the ether spray (both more effective if combined with artificial anæmia) has deprived incision of all its terrors, hesitation and poulticing become a culpable offense against the dictates of common sense.

*The diagnosis of the exact locality* of beginning suppuration is easily made by the aid of the unmistakable sensations of the patient. Gentle pressure by a probe upon different points of the affected region, made to cover successively and in a methodical way the entire area in the shape of a spiral, will soon detect the most painful spot. If one or two repetitions of this process confirm the result of the first search, no hesitation need be felt. The point thus found is marked by a shallow scratch or otherwise, the finger or hand is anæsthetized, and the tenotomy knife is boldly thrust down to the periosteum. If a few drops of pus escape only, this will suffice; if more, the puncture should be at once proportionately enlarged, thoroughly irrigated, and covered with a moist dressing. As the affection generally extends to the periosteum or tendon, the incision should always be carried down to one or the other, and should be longitudinal to avoid injury of vessels or tendons.

*Subfascial phlegmons of the palm* should be also promptly and sufficiently incised. The adjoining diagram (Fig. 178) will be found very useful in pointing out the small area which should be avoided on account of the superficial palmar arch. It is situated between the first and last strokes of the capital M that marks the palm. After the aponeurosis has been cut through, any point of the palm can be reached from the lines marked out on Fig. 178, by Hilton-Roser's method.

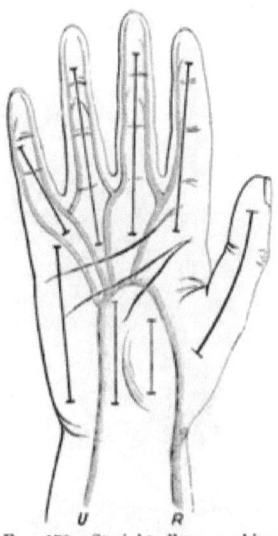

Fig. 178.—Straight lines marking the places where incisions can be safely made. The space between the first and last strokes of the capital M, marking the palm, should be avoided. (From Vogt.)

Incision is advisable even at the risk of cutting the palmar arch, as the hæmorrhage thus caused can be easily stopped by ligaturing the vessel in an ample incision, and Esmarch's band will effectively prevent undue loss of blood during the operation.

There is no region of the human body where senseless poulticing of phlegmons has done more harm, and timely incision can do more good, than in the palm.

CASE.—M. M., saddler, aged sixty-five, had in the latter part of August, 1885, a boil of the face, which he was in the habit of dressing himself. At the same time he infected a small scratch of his right forefinger, from which developed a felon. The family attendant ordered poulticing, *which was kept up uninterruptedly for more than three weeks. Not one incision had been made,* and when the author saw the patient, September 28, 1885, about twenty-four hours before his death from septicæmia, the hand and entire arm presented a terrible condition of phlegmonous destruction. Not one tendon, no joint, was free from suppuration, and a number of phalanges were necrosed; the skin was extensively detached and represented a boggy bag, from which pus flowed copiously through a number of smaller and larger defects due to sloughing. Diphtheria of the throat, tongue, and mouth had also developed the day before the consultation, and the wretched general condition of the patient put any operative measure out of question. The inquiry, how such a state of things could come about, drew the reply that "there were plenty of openings, they seemed to discharge *freely and nicely,* and therefore surgical interference was refrained from."

Neglected cases, where the suppurative process has attained wide proportions, should be treated on general principles laid down regarding the management of complicated abscesses. All recesses should be found out, separately incised, and drained. Where in the course of a long-continued process the soft tissues have been more or less permeated by the septic poison, and multiple small abscesses with a sanious discharge have established themselves, the enormous swelling will render efficient drainage very difficult or even impossible.

*Vertical suspension on Volkmann's arm-splint with continuous irrigation* will often do here very effective service. Its detail is as follows:

After the proper incisions are made and the requisite number of drainage-tubes have been inserted, the arm is enveloped in gauze, is loosely attached to the splint (Fig. 179) by a roller bandage, and is suspended from the ceiling or a suitable frame. One or more irrigators filled with a very weak sublimated or salicylated lotion being also suspended, their nozzles are connected with one or more of the uppermost drainage-tubes. A rubber blanket is so arranged beneath the suspended limb as to catch all the drippings and to conduct them into a bucket placed alongside the bed. The flow of the irrigating fluid is regulated by pushing a match-stick or a straw into the nozzle of the irrigator. In this manner, according to necessity, a free current or the escape of the fluid in drops can be effected.

If the entire limb require irrigation, the use of many irrigators can be obviated by a simple contrivance recommended by Starcke. A tin tube, open at one end, and provided with a number of nipples, is connected with a large irrigator. On the nipples rubber tubes are slipped, and are conducted to the several drainage-tubes, with which connection is established through short pieces of glass tubing. (Fig. 180.)

*Continuous immersion* in a weak antiseptic lotion is a very simple and effective substitute for permanent irrigation, although it precludes the advantages of vertical suspension. The lotion should be changed from three to four times daily, and its temperature is to be regulated by the patient's sensations. Some will have it warm, others will prefer a cool bath. By placing one or two alcohol lamps underneath the tin vessel containing the bath, an even temperature can be maintained.

Fig. 179.—Volkmann's arm-splint for vertical suspension.

CASE I.—Hugo B., laborer, aged twenty-eight, admitted, March 11, 1886, to the German Hospital with extensive phlegmon of the palm, consequent upon an injury to the middle finger. The corresponding metacarpo-phalangeal joint was destroyed. The house-surgeon exarticulated the third finger, and made a number of incisions in the palm, liberating a good deal of pus. By March 12th the temperature had been somewhat lowered, but an ominous swelling of the forearm appeared. *March 18th.*—A number of incisions were made on the flexor side of the arm into the suppurating tendinous sheaths. Moist dressings and elevated posture. Continuous high fever. *March 25th.*—Renewed incisions on dorsum of forearm, exposing the extensor tendons. Swelling of the arm and axillary glands. High fever. The affection proving uncontrollable, on account of the uniform purulent infiltration of the soft tissues, *continuous immersion* of the limb in a 1 : 5,000 solution of corrosive sublimate was resorted to, and was constantly employed during the months of April and May. No mercurial toxic symptoms whatever could be observed during this period of time. The swelling of the axillary glands disappeared a few days after the commencement of this treatment, and a tardy disappearance of the febrile symptoms followed *pari passu* with the detachment of a number of gangrenous muscles and tendons. Toward the end of May all the sloughs were detached, and the little finger was removed on account of necrosis of the phalanges. During June and July a number of small abscesses developed on the hand and along the arm, and were successively incised. End of July all incisions were healed. Active and passive motions and massage restored a part of the motion of the wrist, the thumb, and index. The patient, of whose limb and life we had despaired, was discharged cured and in a florid condition August 26th.

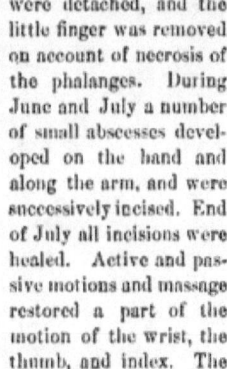

FIG. 180.—Continuous irrigation by means of Starcke's tube, in vertical suspension.

CASE II.—A. W., laborer, aged thirty-two, admitted, August 17, 1886, to German Hospital. *August 7th.*—Sustained an injury of the left forearm. The profuse hæmorrhage was stopped with a tourniquet. The physician left this instrument *in situ*, and ordered to tighten the screw in case of renewed loss of blood. The patient, following the advice of his physician, tightened the tourniquet as directed. *August 9th.*—The forearm swelled up considerably, and assumed a bluish cast; at the same time several chills and high fever set in. Increasing swelling. A homœopathic practitioner of Newark made a few superficial incisions, and, seeing no improvement therefrom, proposed amputation. On admission the patient presented a pitiable condition of septicæmia. Temperature, 105·8° Fahr. The pulse was hardly noticeable, respiration very frequent, the patient cyanosed and somnolent, his body covered with cold perspiration. The entire left arm was enormously swollen, the skin of the forearm extensively discolored, and fluctuation was noted in many places. On account of the collapsed condition of the patient, only a few incisions were made to relieve the pus and to reduce tension. Aside from the large abscesses, a *uniform purulent infiltration of the tissues* was found. *August 18th.*—Numerous incisions were made in anæsthesia, the entire forearm exhibiting a state of ichorous infiltration. Necrosed portions of the skin and of various muscles were ablated, and a number of drainage-tubes were inserted. The arm was kept continuously immersed in a tepid bath for four days without an appreciable improvement of the local or general disturbance. *August 20th.*—The arm was vertically suspended, and continuous irrigation by a weak mercurial lotion was established and kept up until September 18th. This change was followed by slow but unmistakable improvement, interrupted by occasional rises of temperature due to retention. The entire integument of the volar side of the arm was lost by necrosis, and the defect had to be covered by a number of skin-grafts. The patient was discharged cured November 29th, with slight mobility of the wrist and the metacarpophalangeal joints.

By these means many a limb can be saved. The detachment of sloughing tissues should be facilitated by the use of scissors and forceps, and the rule should be upheld *not to sacrifice any part of the hand that is viable.* Even the most sorry-looking, shapeless, and immovable rudiments of this useful organ will be of great value to the patient afterward.

Should all these means be of no avail in checking the progress of suppuration, amputation will have to be considered as a last life-saving remedy.

Case.—Ernst B., shoemaker, aged sixty-nine. Had been for years attended to at the German Dispensary for a chronic fungous affection of the wrist. In the fall of 1885 a phlegmonous inflammation started from one of the many fistulæ present, gradually involving the entire hand, wrist, and part of the forearm. A large number of incisions had been made, but the trouble crept steadily from one joint to another, and along the tendons, until the hand presented one swollen, shapeless, festering mass. *February 13, 1886.*—Amputation of the forearm was done at its upper third. Primary union followed throughout.

*Joints of the Upper Extremity.*—Injury and infection of the *metacarpophalangeal or first interphalangeal joints* frequently take place during a rough-and-tumble fight, when the fist of a fighter hits the incisors of his antagonist. The author has treated four cases of this kind within the last seven years. In one, syphilis followed a very obstinate suppuration of the first interphalangeal joint of the right index.

But often enough secondary suppuration of the finger-joints is caused by extension of a neglected subcutaneous or tendineal phlegmon.

Note.—A very acute *phlegmon of the elbow-joint* came under the observation of the author at Mount Sinai Hospital. A compound dislocation was freshly admitted, and was reduced and dressed so-called "antiseptically" by a junior member of the house staff. Suppuration followed promptly, the sutures had to be removed, a number of incisions had to be made, and a tardy cure was effected, resulting in bony anchylosis of the elbow at an acute angle. (See case of Samuel Krongold, page 207.)

*Suppuration of the finger-joints* usually terminates in anchylosis. In many cases this untoward result can be prevented by *exsection* and subsequent careful treatment by passive and active movements. However, this operation *should never be undertaken before the phlegmonous process has terminated*, and suppuration has assumed a bland character. The author's results achieved by this little operation are very satisfactory, and the procedure can be warmly recommended. As a rule, a more or less movable joint results, which certainly is preferable to a stiff finger. *In one case double exsection was successfully done after a felon of the thumb, involving the metacarpo-phalangeal and interphalangeal joints.* To this end, however, preservation of the tendons is a necessary condition.

CASE I.—Frank P., liquor dealer, aged thirty-six. Seen January 15, 1885, with Dr. H. Balser, on account of a phlegmon of the right index and palm, caused by open injury to the metacarpo-phalangeal joint. The injury was sustained, January 1, 1885, during a fight by violent contact with the antagonist's teeth. The process had lost its virulent character, and subperiosteal exsection, by two lateral incisions, was done January 16th. The cure was uninterrupted. The flexor profundus tendon had sloughed away, hence only the first phalanx could be actively bent. Patient discharged cured February 22, 1885.

CASE II.—S. L., baker, aged twenty-nine. Seen in December, 1882, in consultation with Dr. H. Kudlich. Recent phlegmon of thumb, suppuration of tendineal sheath of flexors and of both the joints of the thumb. *December 12th.*—Three incisions released the tension. After the cessation of the acute stage of the inflammation, December 29th, exsection of metacarpo-phalangeal and interphalangeal joints was done. Uninterrupted cure; good function preserved.

*Phlegmon of the olecranic bursa* is characterized by very acute local and general disturbance due to the great tension maintained by the dense capsule of the sac. Free incision supplemented by Volkmann's punctuation of the infiltrated skin of the vicinity is promptly followed by relief and a rapid cure.

*Suppuration of the cubital or axillary lymphatic glands* is a very common complication of limited or extensive septic inflammatory processes affecting the hand and arm.

*Two forms of suppuration have to be distinguished: One of an acute character*, terminating in the formation of one more or less extensive abscess, the result of confluence of several foci. A spontaneous or artificial evacuation generally leads to rapid cure.

*Another more chronic and very obstinate form*, in which a group of lymphatic glands is attacked in succession, leading to the formation of a series of deep-seated abscesses and a number of sinuses. This form is generally observed in poorly-nourished subjects. The individuality of the glands is not destroyed rapidly as in the more acute form, but their slow and gradual destruction is accomplished by a tedious ulcerative process. Long before the glandular ulceration is terminated, cicatricial contraction of the sinuses leading through healthy tissues will occur, and cause retention. This is followed by an exacerbation of the local and general symptoms, and results in the formation of a new abscess and sinus. The interminable suppuration often leads to serious deterioration of the general condition, marked by emaciation, night-sweats, and loss of appetite. As these cases represent an aggregation of a large number of septic foci imbedded in dense tissue, one or even more incisions will not be adequate for efficient drainage, and in spite of them the process will continue.

*Extirpation of the entire group of affected lymph-glands* by careful preparation is their best therapy. As rupturing of one or more of the broken-down glands, and soiling of the wound by their contents, can not always be avoided, closure by sutures is best omitted. Thorough irrigation with corrosive-sublimate lotion, a loose packing with moist gauze, and a moist dressing are appropriate.

CASE I.—Emma Epple, servant, aged seventeen. Admitted to German Hospital March 31, 1886. As the consequence of a "run-around" treated by poulticing, suppuration of the lymphatic glands of the left axilla developed. The arm-pit was filled with a densely infiltrated large mass of intumescent and very painful glands. The continuous fever and sleeplessness had produced an alarming degree of anæmia and debility, characterized by night-sweats and loss of appetite. As no fluctuation could be made out, and presumably all the affected glands were in a state of suppuration, *extirpation of* the entire glandular mass was advised, and carried into effect April 3d. Dissection of the tumor from the axillary vessels was rather difficult, and, one of the tenacula lacerating one of the brittle glands, a few drops of pus exuded into the wound. After thorough irrigation with corrosive-sublimate solution, the wound was closed by suture, and an antiseptic moist dressing was applied. Previous to this a separate incision was made at the most dependent portion of the cavity for the reception of a stout drainage-tube. A sharp chill and much pain followed the next day after the operation. Undoubtedly, infection of the cavity by contact with the escaped pus had taken place. The dressings being removed, pus was seen oozing out of the drainage-tube. Daily change of dressings and irrigation of the cavity with mercurial lotion was followed by rapid improvement, and the patient was discharged cured, May 7th.

CASE II.—C. H., butcher, aged sixty-two. Slightly cut the dorsum of his left middle finger, October 15, 1885, with a butcher-knife. A phlegmon developed, and was treated by the patient himself with poulticing till October 27th, when spontaneous evacuation took place. For a few days previous to this date, intumescence of the cubital lymphatic glands was noted. *October 28th.*—The patient came under the author's care with an angry swelling of the region of the cubital glands. Incision was proposed and declined. After a couple of wretched nights the patient consented to incision, which was done under chloroform, October 31st. A small amount of pus came away, and a drainage-tube and moist dressings were applied. The momentary improvement soon gave way to renewed attacks of pain and swelling, apparently due to successive suppuration of several glands. Much difficulty was experienced in keeping the drainage-tube *in situ*, the external wound showing a great tendency to cicatrization, while the slow ulceration of the glandular tissue was still progressing. An extirpation of the glandular mass would have been more serviceable in this case than a simple incision. After a tedious and troublesome course of treatment, the case was finally discharged cured, December 27th.

*e.* Suppurative Affections of the Lower Extremity:

(*a*) INGROWN TOE-NAIL.—The most common cause of this distressing affection is the improper care of the toe-nails. Sweating feet, in combination with lack of cleanliness, improperly trimmed toe-nails, and narrow-toed shoes, offer the best conditions for the development of ulcerative processes near the anterior edge of the nail. Whenever the nail is trimmed off too short, the adjacent skin will overlap its angle (Fig. 181). The epidermis being macerated and soft from the profuse sweating, a small amount of friction between the edge of the nail and the skin will be sufficient to cause an excoriation. The pyogenic germs, so abundantly present in the fetid epidermidal masses of sweating feet, will not only come in contact with the raw surface, but will be rubbed into the open lymphatics by each successive step taken by

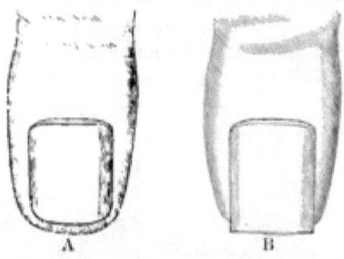

Fig. 181.—A, Wrong way of trimming toe-nail. B, The right way.

the individual. An ulcerative inflammation of the parts will result, which offers poor conditions for natural drainage. Retention of the septic secretions leads to chronic suppuration, and to the extension of the process backward toward the root of and also under the nail, until more or less of it becomes undermined and detached. Exuberant granulations, subject to frequent ulcerative destruction, spring up from the hypertrophied and infiltrated overlapping skin, and, if unchecked, the disorder terminates in the loss of the nail. Occasionally an ingrown toe-nail is the starting-point of phlegmon or erysipelas of the dorsum of the foot. The initial stages of the mischief can often be successfully met with careful local treatment. Disinfecting baths, sprinkling of alum and salicylic powder (alum. usti, ℥ ij; acidi salicyl., ℥ ss; bismuthi subnitr., ℥ ijss) into the stockings, which should be daily changed, and the packing of salicylated or iodoformed cotton or a small piece of heavy tin-foil under the edge of the nail, frequently result in alleviation, if not a cure, of the affection.

More inveterate or extensive cases in persons unable to devote the necessary care and time to the treatment of this trouble will be best cured by operation. After careful scrubbing and disinfection, the toe is rendered anæmic by constriction of its root with a piece of rubber tubing. Local anæsthesia is produced by either an injection of a cocaine solution or the use of Richardson's ether-spray. The point of a bistoury is (Fig. 182) placed against the exuberant tissues adjoining the nail, and is thrust through the margin of the toe. It is carried forward until the integument is separated in the shape of a longitudinal flap. Then the knife is reversed and carried back well beyond the matrix of the nail, where the flap (c) is cut off.

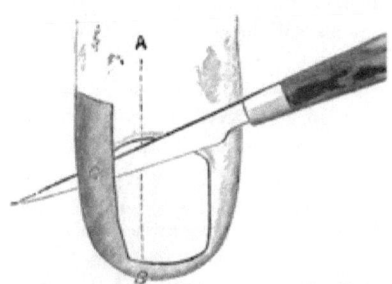

Fig. 182.—Operation for ingrown toe-nail. A, B, Line of section through the nail and matrix.

The pointed blade of a straight pair of scissors is placed under the anterior margin of the nail (Fig. 182, A, B) just beyond the limit of the disease, and, being thrust under it, cuts through the nail in an antero-posterior direction well back of the matrix. One blade of a stout pair of dressing-forceps is next insinuated into the slit in the nail and under the loose segment. This, being firmly grasped, is evulsed with an **outward rotating motion**. Good care must be taken not to leave behind **any shreds of the cut-off matrix**. Any granulations are scraped away **with a sharp spoon, and the wound is well irrigated with mercuric lotion**. **A strip of rubber tissue well soaked in carbolic lotion, and just large enough to cover the wound, is** placed next **to it ; over this comes a strip** of iodoformed gauze and a **small** disinfected **sponge, the latter to** exercise elastic pressure for the prevention of undue **hæmorrhage ; finally comes** a light, compressive *moist dressing*, fastened **by a roller bandage.** While the patient's foot is held elevated, the rubber **band is removed.** The first dressing can be left on for a week or even **two weeks.** Being moist, it **will** peel off easily when removed, and, according to its **size,** the wound will be found either partly or entirely cicatrized over. Care must be taken not to compress the toe too **much, as** necrosis **of the** skin by pressure may develop and retard the healing.

The author has treated several hundred of these cases **in** the manner described **with the best** results, the majority being patients **of the German Dispensary, who** walked **to and from** the institution during **the time of treatment.**

(*b*) CHRONIC ULCERS OF THE LEG.—Neglected excoriations or abrasions **of the skin belonging to the lower third of the leg are the most common starting-point of ulcerous processes. Varices due to stagnation of the venous circulation render the progressive invasion of new areas of tissue by** micrococci, **ever present in the** putrescent discharges, especially easy. Consequently, ulcerative destruction develops. The successful **treatment of this condition must be based upon an** elimination of **the causal factors.** Prevention **or elimination of** decomposition by **antiseptics, and an improvement of the** circulatory conditions by elevation **of the limb or its elastic compression,** form the cardinal points of our therapy.

The affected limb is carefully cleansed with **soap and a soft** flannel rag **until all** the crusts of inspissated secretion and epidermis are removed. This **process will be greatly facilitated by packing of the** parts in strips of **lint** saturated with **vaseline or** unsalted lard the **night** previous to the cleansing bath. Plain **water should never be used on account of** its irritating qualities and its **liability to cause eczema. After the bath** the soap-suds should be simply wiped off with **a soft towel. The ulcer is** well mopped with a 1 : 1,000 solution of corrosive sublimate, **or,** where the stench is very intense, **with** a **4 : 1,000** solution of permanganate of potash. **Iodoform** powder **is dusted over the** ulcer, and a suitable **patch of rubber tissue** is placed next **to it. The eczematous skin in the vicinity** is well anointed with vaseline **or** an astringent salve, and a regular antiseptic dressing **is snugly** bandaged **on to the ulcer, the** roller bandage extending from **the toes** to the knee-joint. **This dressing need not** be removed before **two or three days,** the frequency **of renewal** being dependent upon the **quantity of the discharge.** As soon **as cicatrization** is well advanced, **a simpler dressing, consisting of a strapping of mercurial plaster covered with a pad** of absorbent **cotton, held down by a Martin's elastic bandage, can be** substituted therefor, **and the patient** may be permitted to abandon the recumbent posture and take moderate exercise. When cicatrization is completed, a well-cleansed elastic bandage will suffice to prevent renewed ulceration. It is most convenient to have two elastic bandages, to be worn alternatingly. Under this simple treatment most ulcers of the leg, even those surrounded by callous edges, will develop healthy granulations, and will heal kindly. Due regard should be paid to the general condition of the patient, as on it may depend to a great measure the rapidity of the cure. A marastic state of the system should be improved by suitable nutritious diet; the deterioration of the general health of those addicted to the immoderate use of alcohol should be remedied by a proper regulation of their habits.

In cases of very extensive loss of integument, skin-grafting will give very gratifying results. If this should fail, circumcision of the callous ulcer by a deep cut carried through the fascia, according to Nussbaum, may be tried. The incision should be placed about one third of an inch from the edge of the sore.

(c) ACUTE SUPPURATION OF THE PREPATELLARY BURSA.—Servant-girls and scrub-women, in short, persons frequently subject to house-maid's knee or simple synovitis of the prepatellary bursa, are frequently victims to phlegmonous inflammation of the same organ. The symptoms are those of a subcutaneous phlegmon, heightened by the circumstance that, the phlegmonous focus being encapsulated, great tension is apt to develop. Extensive necrosis and serious septic intoxication must result if no timely relief is afforded.

Dense, hard infiltration and a deep-red flush of the prepatellary region, with œdema, high fever, and marked sickness, are present. The general intumescence may cause errors in diagnosis, as inexperienced observers are apt to look for the source of the trouble within the knee-joint. This mistake can be avoided by noting that in septic bursitis the point of the most intense swelling, redness, and pressure-pain is over the patella, whereas in gonitis pressure over the juncture of the femur and tibia laterally of the patella is most painful, and the patella can be distinctly felt floating on top of the exudation within the knee. A free incision into the bursa, together with Volkmann's multiple puncture of the inflamed skin, is the proper treatment. The cavity should be well irrigated with corrosive-sublimate lotion, loosely packed with strips of iodoformed gauze, and inclosed in a *moist dressing*, which should be daily changed.

(d) ACUTE SUPPURATION OF THE KNEE-JOINT is one of the most formidable types of phlegmon. On its prompt recognition and energetic treatment may depend the safety of limb and life. It should be well distinguished from the more bland, so called, "catarrhal" (Volkmann) inflammations of the synovial membrane, due to tuberculosis or to rheumatic and gonorrhœal influences; and also from metastatic suppuration complicating pyæmia.

It is generally caused by infection of the joint from without through accidental or surgical wounds, or by its invasion of a suppurative process established in the vicinity, as, for instance, acute osteomyelitis **or a subcutaneous** or bursal phlegmon. Idiopathic acute **suppuration of the knee-joint is very rare** indeed.

The invasion is marked by one or more sharp chills, very **high fever,** and a sudden painful intumescence of **the joint.** The limb **is rotated outward,** lying on its outer aspect, is flexed **at an obtuse** angle, and its position is carefully maintained by the patient, **as the constant** pain is terribly intensified by the least **change of** posture. **General** œdema and reddening of the integument soon follow, the septic intoxication frequently producing delirium and a typhoid condition.

The intra-articular tension increasing, perforation of the capsule, generally upward through the bursal extension of the joint beneath the quadriceps tendon, occurs, and is marked by a temporary remission of the intensity of the local and sometimes of the general symptoms. **One or more** subfascial or subcutaneous abscesses, located on one **or both sides of the** quadriceps, appear, and rapidly extend upward and **outward until perforation of the skin permits** the escape of the enormous mass of pent-up pus. Occasionally the matter perforates backward **into the** popliteal space, this way being marked out by the bursæ situated **beneath the popliteus** muscle, which are frequently in open communication with the knee-joint. In this case the abscess will extend **downward along and beneath** the muscles of the calf.

Spontaneous perforation will not **bring about complete and lasting relief, as the** drainage is and must **be inadequate. Profuse suppuration and a con**suming fever, with frequent chills **and** colliquative sweats, will in a short time so depress the patient's condition, that amputation will have to be thought of as the last resort for saving life.

The treatment should be that of deep-seated phlegmon, modified **by the** requirements of the anatomical peculiarities of the knee-joint. The cavity of the knee-joint naturally consists of *three distinct recesses:* one below, the **other above** the patella; the third is an extension of the suprapatellar space, **and** is known by the name of the *bursa of the quadriceps.* In flexion, **where the** knee-pan is firmly **held** down to the condyles, the infra- and supra-patellar spaces become practically non-communicating. Andrews **of** Chicago, to whom we owe **a most excellent** treatise on the subject of injuries to the joints, mentions **a case\* of** traumatic suppuration of the infrapatellar recess of the knee-joint, where, by means of continued flexion and thorough disinfection and drainage of the same space, general infection of the joint was effectually prevented.

To effect adequate drainage of a phlegmonous knee-joint, **each of these** recesses must be separately incised and drained.

A double incision of each **of** these **spaces will be much** more effective **than** a single one, as it will permit more **thorough irrigation.** In very infectious cases two additional incisions will drain **away pus retained** in the reflection of the capsule from the vicinity of the crucial ligaments.

\* Ashhurst's " Encyclopedia of Surgery," vol. iii, p. 723.

The first incision should be made in the suprapatellar space on the inner side, where the capsule is the most ample. Hæmorrhage is generally profuse, hence it is best to penetrate the tissues gradually, and to secure each bleeding vessel as soon as it is cut. As soon as the joint is entered, a dressing forceps is thrust through it to the corresponding point of the other side of the joint, where the second incision is to be made through the tissues raised by the pressure of the forceps. The point of the forceps emerging from this incision, a stout drainage-tube is grasped with it, and drawn into the joint just far enough to clear the synovial membrane. A similar piece of drainage-tubing is inserted into the first incision, and the protruding ends of the tubes, being transfixed with safety-pins, are cut off on a level with the skin. The infrapatellar and submuscular spaces are treated similarly, and, if necessary, the lateral pouches of the joint are also incised and drained. The cavities are thoroughly flushed out with corrosive-sublimate lotion, a large moist dressing is fastened on, and the limb is secured to a posterior splint to insure rest and painlessness during unavoidable changes of posture of the patient. Wherever perforation of the capsule and formation of a circumarticular abscess has occurred, this must be separately incised and drained.

In the great majority of cases, resolute and comprehensive measures of this kind will be rewarded by prompt improvement. Daily change of dressings and irrigation should be practiced until the disappearance of all the inflammatory and febrile symptoms. As soon as the discharges become scanty and serous, the drainage-tubes can be withdrawn one by one. Where the affection is due to osteomyelitis, anchylosis will result as a rule, especially in grown individuals. In children, prompt and adequate drainage frequently results in preservation of mobility.

CASE I.—Charles Hundertmark, aged four. Acute suppuration of knee-joint caused by a blow upon head of tibia. *May 31, 1875.*—Three incisions—one on each side into the suprapatellar space, a third one into the quadriceps bursa. Daily change of moist carbolized dressings and irrigation. Rapid improvement. *June 15th.*—Drainage abandoned. *July 4th.*—Perfect recovery noted, with free active use of the joint.

CASE II.—John S., grocer, aged nineteen. Acute suppuration of knee-joint, with terrible pain and typhoid symptoms. The patient was brought to the German Hospital January 10, 1880, by Dr. Schwedler, who administered chloroform during the transfer, to allay the patient's suffering from the jolts of the carriage. Immediate typical multiple incisions and drainage. The index-finger detected a roughened place on the articular surface of the inner condyle of the femur. Undoubtedly on account of the osteomyelitic process, the febrile symptoms receded very slowly. Permanent irrigation of the joint rendered the frequent, terribly painful change of the dressings unnecessary. A few small sequestra belonging to the cancellous tissue of the femoral epiphysis came away on the twenty-third day. Patient was discharged cured, March 20th, with firm anchylosis.

In exceptionally neglected cases, where the process has assumed the character of a general purulent infiltration, incisions and drainage, supplemented with continuous irrigation, will not be followed by as prompt improvement as is desirable. The continued high fever, the formation of new abscesses, will certainly bring about a fatal termination, unless the limb is amputated clearly beyond the limits of the disease. So-called conservative measures—as, for instance, exsection of the joint—are entirely inadmissible and dangerous under these circumstances. They will fail to remove from the affected parts the elements of contamination, as the most rigid antiseptic measures of the ordinary kind are here utterly inadequate. The phlegmonous process will attack the newly-made wound-surfaces, and the patient's life will be placed in the greatest jeopardy by secondary hæmorrhage. The following case forcibly illustrates the weight of these remarks:

CASE.—Max Loffmann, butcher, aged twenty. Admitted, October 25, 1885, to Mount Sinai Hospital. *October 12th.*—The submuscular recess of the knee-joint was accidentally incised with a filthy butcher's knife. Some synovia escaped from the small puncture; after the accident the patient walked home. Suppuration of the knee-joint set in the following day, with rigors and general dejection. The wound was dressed by a Jersey City practitioner with an adhesive-plaster dressing placed over the incision. The patient was admitted to the hospital in a highly septic condition, large quantities of thin, ichorous pus escaping from the joint on slight pressure. Immediately the patient was anæsthetized, and typical incision and drainage were done. The synovial lining of the joint was coated with a greenish-gray adherent and putrid membrane, in looks identical with the membranous coating in pharyngeal diphtheria. A number of small, purulent foci were opened by the incisions made for drainage of the joint. A moist dressing and dorsal splint were applied. In spite of frequent irrigation, no remission of the high fever or local pain following, amputation of the thigh was proposed, in view of the visible failing of the patient's strength. This, however, was resolutely declined by the patient and his widowed mother, who begged for an attempt to save the limb. The author, against his better judgment, performed exsection of the knee-joint, November 6th. Esmarch's band was applied to the upper third of the thigh without the previous use of the elastic roller bandage, and a continuous stream of corrosive-sublimate lotion (1 : 1,000) was kept playing upon the wound during the entire operation, which was rapidly but carefully performed. Care was taken to operate in healthy parts, and all the involved tissues were removed. The wound was drained and closed in the usual manner, and the dressed limb was fixed upon a dorsal splint. Suppuration of the wound followed, requiring frequent changes of dressing and irrigation, the secretions retaining all the while their peculiar thin, ichorous character noted from the outset. On the afternoon of November 18th, profuse arterial hæmorrhage occurred from the wound, which was temporarily checked by the house-surgeon with the application of Esmarch's band. Being hastily summoned to the hospital, the author found the patient blanched and collapsed. About twenty ounces of a 6 : 1,000 watery solution of cooking salt were transfused into his median vein, and resulted in a notable improvement of the pulse. Amputation of the thigh was quickly done as a last resort. The patient, however, expired before the removal of Esmarch's band.

Post-mortem examination revealed a sieve-like perforation of the popliteal vein and a large oblong defect of the popliteal artery, both of which were found exposed and surrounded by a massive blood-clot. The walls of the cavity containing the clot consisted of broken-down and necrosed tissues.

There is little doubt that an early amputation might have saved the patient's life.

(e) SUPPURATION OF THE INGUINAL GLANDS.—Two groups of lymphatic glands have to be distinguished in the inguinal region—one situated below Poupart's ligament, the other above it. The subinguinal group is frequently the seat of phlegmonous inflammation, due to absorption of septic material from sores caused by the pressure of ill-fitting shoes, ulcerated bunions, ingrowing toe-nail, and excoriations of the lower extremity from scratching in eczema. Their treatment by *incision* does not require special elucidation.

Should, however, their *excision* become necessary, the rules laid down for the removal of tumors from Scarpa's triangle (pages 52 and 55) should be heeded.

Acute suppuration of the *suprainguinal glands* is caused most generally by ulcerative or suppurating processes of the generative organs. Their treatment is subject to the principles accepted for glandular abscesses of other regions, and may be dismissed with the remark that the *best way to incise them is not parallel, but at a right angle with the direction of the fibers of Poupart's ligament*. The edges of the incision will gap asunder, and afford very good drainage even without the use of a tube, and, later on, the edges of the cut will not exhibit the tendency to become inverted, which is the source of much trouble in the after-treatment.

*Interminable chronic suppuration* of the suprainguinal glands frequently indicates their bodily extirpation. The safest way of accomplishing their removal is as follows: Two semi-elliptic incisions should include all the fistulous openings leading into the glandular swelling. They should be gradually deepened until a comparatively healthy part of the swelling is exposed. Here the capsule is incised, and the mass is carefully dissected out with the tip of a pointed scalpel. Blunt dissection should be resorted to only where it is evidently easy, as in using much blunt force the glands may be ruptured, and their contents soil the wound.

This injunction is important, as *intentional or unintentional injury to the peritonæum may become unavoidable*. Should the epigastric vessels be in the way, they must be cut and deligated. Attention ought to be paid also to the seminal cord, which occasionally enters into very close relations with inguinal glandular swellings.

*f.* Perityphlitic **Abscess**:

Up to within a recent period of time it was the prevalent belief that perityphlitic suppuration was located retroperitoneally, and most generally in the iliac fossa, whence it found its way to the surface by pushing aside the peritoneal reflection corresponding to Poupart's ligament. Willard Parker's method of incising perityphlitic abscess was based upon this view.

It can not be denied that the development of most circumappendicular abscesses seems to confirm this view, and that the rules laid down by Parker for the treatment of this group of suppurative processes have yielded, and continue to yield, very satisfactory results in very many instances. Still, it must be said that the exceptions to Parker's type are considerable in number. Formerly they were classed as cases of general or localized "idiopathic peritonitis." Their treatment was non-surgical, and their issue very uncertain and often fatal.

We owe the better understanding of the elements of this phenomenon to Treves and Weir, but principally to McBurney, who demonstrated that in the vast majority of instances the formation of abscess in the right iliac fossa was due to intraperitoneal inflammatory processes, mostly of the vermiform appendix, and commonly accompanied by ulceration, necrosis, and perforation of this viscus. The frequency of the location of perityphlitic abscess near the parietes of the right iliac fossa is explained by the frequency of the superficial *situs* of the appendix in this region. In these cases the type of development so well described by Parker will prevail. But in a very large proportion of instances the vermiform appendix, either congenitally or in consequence of acquired peculiarity, occupies a deep situation, and in these cases an appendicular perforative process is sure to cause a deep-seated intraperitoneal abscess, more or less distant from the surface, hence infinitely more grave and dangerous both as regards its deleterious possibilities and the difficulty of diagnosis and surgical management. As soon as it became clear that widely different intraperitoneal forms of suppuration might be caused by extension from the appendix, and that their manner of development was wholly unforeseen and unaccountable, a violent oscillation in therapy was initiated by those who proposed, in all cases where the appendix was suspected of causing trouble, a bold exploration by abdominal section, and the extirpation of the appendix, or evacuation at all hazards of the purulent collection, wherever it might be found, and all this without delay.

Though this bold course of therapy has, in spite of its experimental character, yielded very good results in the hands of various surgeons, and although its adoption was absolutely necessary for establishing a clearer understanding of the nature of the morbid process in question, nevertheless it must be remembered that a vast proportion of perityphlitic abscesses do not need operative invasion of the free peritoneal cavity for their successful care, and that a sweeping advice to the general profession to open the peritonæum in every case where appendicular trouble is suspected is, for obvious reasons, fraught with much unwarrantable danger.

Formerly it was considered purely accidental whether an intraperitoneal abscess would appear here or there, and the variability of the surroundings and location of these abscesses was deemed so irregular and erratic that, to the author's knowledge, no attempt was ever made to study the question whether a certain order of development did not prevail even in those forms of perityphlitic abscess which could not be classed with the well-known inguinal type described by Parker. If some light could be thrown upon the detailed nature of these seemingly erratic forms of circumappendicular abscess, instead of the crude general advice to "perform laparotomy," more precise, hence safer, methods of treatment would suggest themselves.

Let us first emphasize the fact that all intraperitoneal abscesses are of visceral origin, and that perityphlitic abscess in particular is due to inflammatory processes located in the vermiform appendix. Though not always, this form of abscess is mostly established within the peritoneal sac.

The proof of this assertion has been so manifold that it is only necessary to refer to the numerous cases of early appendicitis reported by McBurney and other observers, in which, on laparotomy, the free appendix was found to be tightly distended by a copious exudate, and more or less erect by dint of its extreme distention; its walls thickened, hyperæmic, occasionally exhibiting unmistakable signs of circumscribed necrosis with perforation imminent. This distention was uniformly produced by occlusion toward the gut. Occasionally decay had progressed to actual perforation and the formation of incipient abscess, surrounded by a protective barrier of recent adhesions of the vicinal serous surfaces. The appendix was invariably found to be the starting-point of the trouble, and the affection, with rare exceptions, always intraperitoneal. Aside from the numerous instances in which the intraperitoneal and appendicular character of perityphlitis was established by positive observation, the following case may serve to show that the retroperitoneal space back of the iliac fossa is not the seat of abscess in typical cases of perityphlitis. In the spring of 1887 Dr. Lellmann, then on duty in the German Hospital, requested the author to operate on a case of perityphlitis pertaining to his service. The operation was delayed twenty-four hours on account of a misunderstanding, and the next day—a dense, painful tumor being found in the right iliac region—incision according to Parker was done, in spite of the circumstance that the size of the swelling had somewhat diminished since the previous day. The peritoneal lining of the iliac fossa was easily stripped up two inches beyond the external iliac vessels, so that the tumor was freely raised with it from the underlying tissues. No sign of inflammation was found, and, as the case was mending, it was not deemed prudent to incise the peritonæum. The very deep wound was drained and closed, but no pus appeared. Simultaneously with the healing of the incision the tumor disappeared, and the man was discharged cured within a fortnight after the operation.

We need not do more than hint at the causes of appendicular inflammation. Let us first mention the impaction of foreign bodies entering from the gut, acute or chronic forms of catarrhal or ulcerative (typhoid) enteritis, transmitted from the colon and leading to simple hypertrophy or to ulceration, both of these causing irregular constriction mostly in the vicinity of the attachment of the appendix. Another not infrequent cause of stenosis is the doubling upon itself and fixation of the appendix in this position. Stenosis by flexion is thus produced.* With the establishment of hypertrophy and stenosis a loss of contractile power is associated, leading to more or less complete retention and to the inspissation of fecal matter, which finally assumes the shape of one or more globular concrements. As long as the communication with the colon is fairly open, no local symptoms need prevail. As soon as the stenosis becomes considerable, the well-known signs of appendicitis make their appearance. If they are due to a passing state of catarrhal hyperæmia, their acuteness will vary in proportion with the intensity of the stenosis. Thus, with the cessation of causal intumescence and the elimination of the stenosis maintained by it, all trouble may seemingly or really disappear. A case reported by Shrady † aptly illustrates this train of symptoms

---

* F. W. Murray, "New York Medical Journal," May 24, 1890, p. 564.
† George F. Shrady, Meeting of Practitioners' Society of New York. "Medical Record," April 26, 1890, p. 479.

A physician had had four distinct attacks of appendicitis, in all of which the question of operation arose. Dr. Shrady had seen the patient at New York in three of the attacks, all of which were well pronounced, while the fourth occurred in Paris, where the patient was seen by a distinguished surgeon, who made a like diagnosis. There also the question of an operation came up. Each attack was attended with all the usual severe symptoms which would appear to usher in the formation of an abscess; there were dullness, tenderness, more or less rigidity, and some œdema in the neighborhood of the cæcum. In each attack the advisability of operation was freely discussed. The patient was willing to take the risks, but in each instance the symptoms gradually disappeared, and he recovered. He asked Dr. Shrady, should he survive him, to examine his appendix, which was done when death occurred, some time subsequently, of another cause. *The appendix was found perfectly sound.* There was not the slightest appearance of any inflammation around it; it was not even thickened.

Where ulcerative processes have led to the formation of a permanent cicatricial contraction, the appendical trouble is apt to persist even after the cessation of the causal disorder of the intestine. Passing states of local intumescence are then more likely to lead to complete occlusion of the communication between gut and appendix, with serious consequences. But even in these cases temporary improvements are possible with the diminution of the acute swelling of the cicatricial mass.

Before attempting a practical classification of the phases of appendicitis and of the localities in which circumappendicular suppuration is to be observed, this fact has to be pointed out: that, unfortunately, the acuteness or mildness of the local or general symptoms is not an invariable index of the ultimate gravity of a given case. Sometimes fatal cases will set in with a very deceptive mildness of appearances. On the other hand, a very alarming beginning may be followed by resolution or a tractable state of affairs. Hence it must be insisted on that, in reference to this trouble, all therapeutic advice has only a conditional value—to be weighed and accepted or rejected by the surgeon in each separate case.

a. ACUTE APPENDICITIS (WITHOUT TUMOR).—(a) *Simple Appendicitis (no Tumor).*—Anatomy teaches that in the supine body the attachment of the vermiform appendix can be found directly underneath a point located two inches from the anterior superior spine of the ilium, on a line connecting this bony prominence with the navel. Whenever acute and persistent pain appears in this region, accompanied by fever and retching, the pain being markedly increased by palpation of this area, trouble of the appendix can be confidently diagnosticated. In women, bimanual palpation ought to exclude the presence of an inflammatory process of the displaced uterine appendages. Though the local and general symptoms may be very alarming, tumor can rarely, if ever, be detected in the early stages of the affection. Meteorism is also absent.

In view of the impossibility of foretelling whether, in a given case, spontaneous evacuation of the contents of the appendix or perforation is to take place, and in the latter case whether a superficial or a deep-seated abscess is to develop; and, considering the fact that laparotomy followed by excision of the appendix has yielded uniformly good results if done before the access of perforation, it is safe to follow McBurney's advice, which recommends laparotomy and removal of the appendix whenever severe symptoms persist and increase for more than forty-eight hours.

The steps of the operation are these: A longitudinal incision, four or five inches long, parallel with and just outside of the outer margin of the right rectus muscle. Having opened the peritonæum, the appendix is found, which will be rendered easy by first ascertaining the location of the caput coli. The mesentery of the appendix is included in a double ligature of stout catgut and divided. Then the root of the appendix is secured by two ligatures, between which the viscus is cut off. The mucous lining of the stump is either seared with the thermo-cautery, or, after careful disinfection, is touched with a few drops of perchloride-of-iron solution and dried off. Then the stump is dropped back and the external wound is closed.

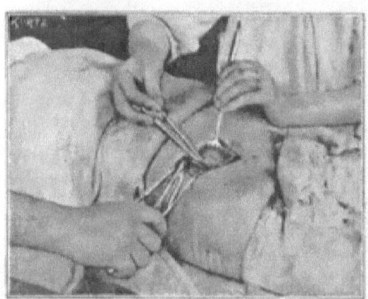

FIG. 183.—Incising perityphilitic abscess.

CASE.—Miss F. L., aged twenty. Has had altogether sixteen or eighteen attacks of appendicitis within two years. Characteristic local pain, irregular fever with temperatures reaching 104° Fahr.; no tumor. Uterine appendages normal. *April 20, 1890.*—Laparotomy. The free appendix is found very much thickened, its distal half distended and bent upon itself, containing a quantity of fetid serum. It was removed. Uninterrupted recovery.

(*b*) *Perforative Appendicitis (no Tumor).*—Sudden increment and extension of the local pain followed by symptoms of collapse, such as profuse cold sweating, a thready pulse, anxious expression, pallor, frequent vomiting, and the appearance of meteorism are indications that perforation and infection of the peritonæum have taken place. This rarely occurs before two or three days after the inception of the trouble. The violence of the symptoms will depend on these factors. If the extent of the perforation is small, and only a small quantity of the infectious contents of the appendix has made its way into the peritonæum, a limiting barrier of protective adhesions may be thrown about the infected area within an hour or so. In this case the alarming features of the case will somewhat subside and a tumor is apt to develop. If, on the other hand, the perforation is large or multiple, a considerable volume of infectious material will suddenly escape. Lively peristaltic action will widely distribute it, and more or less extensive local or, in the worst cases, general septic peritonitis will be established.

*The absence of tumor* in conjunction with very acute local and general symptoms represents an extremely grave combination of things, its meaning being a generalizing peritonitis. In these cases the prognosis is very doubtful, and it will be extremely difficult to save the patient, even by the most resolute measures. If laparotomy is immediately done, the focus laid open, wiped out clean, the appendix removed, and the cavity packed and **drained**, some chances may still be present for the patient's recovery. But where, on account of delay, numerous and widely disseminated abscesses **have** established themselves in the more remote parts of the peritoneal cavity, the patient's death is nearly certain. Prolonged exposure, the impossibility of a sufficient evacuation and drainage of the foci which are found, finally the overlooking of distant foci located in the loins, in front and behind the liver, will sufficiently explain this fact.

CASE I.—William Sachse, aged forty-eight, liquor-dealer. Was treated since September, 1889, in the internal department of the German Hospital for alcoholic neuritis. No habitual constipation. *March 23, 1890.*—Sudden chill. Temperature, 105° Fahr. Slight amygdalitis. No abdominal symptoms. The temperature remained high, although the patient's bowels were well purged with calomel on March 25th. Had a chill in the preceding night, another one in the afternoon, complaining the first time of belly-ache. *27th.*—Pain well marked in ileo-cæcal region. Was **transferred to** surgical service. Temperature, 104·4° Fahr. Meteorism, intense pain in the ileo-cæcal region, but **no** tumor and no dullness. Vomited only once. Laparotomy at 3 P. M. McBurney's incision. Peritonæum filled with turbid serum. Omentum widely adherent to cæcum, in front of which an adherent and very much thickened and elongated vermiform appendix was found. On freeing this, a large, irregular abscess cavity was opened, which did not anywhere approach the parietes, and which was situated below and behind the cæcum, its walls being formed everywhere by intestines. At the root **of the** appendix a large perforation was seen, with three globular fecal concrements lying in front of and outside of it. The appendix contained three more globular concrements of **the size of a small marble.** The appendix was isolated, tied, and cut **off.** Another **large abscess situated** in the median line, and a third one in Douglas's pouch, were opened, irrigated, and drained. Hasty **partial** closure of incision after packing and drainage of the abscesses on account **of** collapse. In the night the temperature rose to 106° Fahr., and the patient expired toward midnight. Post-mortem examination revealed three more abscesses, one **situated** high up behind **the liver.**

CASE II.—David Danziger, **tailor,** aged twenty-two. General peritonitis due **to** perforative appendical trouble of six days' **duration.** Laparotomy January 29, **1889,** at Mount Sinai Hospital. Seven abscesses were opened and drained. Patient seemingly improved, the quality of the pulse improving. Vomiting ceased, but he collapsed suddenly thirty hours after **the operation** and died. Post-mortem examination revealed three perihepatic abscesses.

*b.* ACUTE APPENDICITIS WITH TUMOR; PERITYPHLITIC ABSCESS.— Whenever **perforation of the free** appendix occurs, the invasion of the peritonæum is regularly signalized by the usual symptoms of perforative peritonitis. As before mentioned, a circumvallation by adhesions will form in those cases in which only a small quantity of infectious material has **escaped.** This seems to be the usual course of events. Occasionally, however, **the inflamed** parts of the appendix will first become adherent, and **then be perforated.** In these cases the alarming intermezzo possessing the typical aspect of perforative peritonitis will be missed, **and the** abscess will develop without a tendency to meteorism and collapse, **and** with a gradual but steady growth of the **mainly local** symptoms. The complex of symptoms has little of the character pertaining to peritonitis, and resembles that of an ordinary abscess.

By contiguous extension, which is mostly slow, these abscesses may assume very large proportions. Neglected for a long time, especially if they are limited by intestines only, their secondary rupture, followed by a chill and further extension, or even their generalization, may occur. This, however, is not common in the early stages of the process. The only case of this kind observed by the author occurred nineteen days after the inception of the trouble.

CASE.—H. D., clerk, aged twenty. Subject to alvine sluggishness, contracted, after a more than usually severe spell of constipation, a deep-seated, hard, painful, perityphlitic swelling. Cathartics failed to relieve the bowels, and, high fever with vomiting having set in, the author was consulted. *May 1, 1878.*—Typical swelling of a cylindrical shape was made out in the right groin, and a number of repeated large injections of tepid water into the gut were employed without success. *3d.*—The peritoneal symptoms, notably vomiting, became very distressing, wherefore this therapy was abandoned and opium treatment begun. At the same time an ice-bag was placed over the swelling. The change effected a decided improvement in the subjective symptoms, but the swelling continued to increase and the fever remained unrelieved. *17th.*—Spontaneous evacuation of a large, formed stool occurred. *19th.*—The general condition becoming very poor, incision was urged, but was firmly declined by patient and parents. Suddenly, in the night of the same day, perforative symptoms developed. The patient died, May 20th, of septic peritonitis. Post-mortem examination demonstrated an internal perforation of the abscess, and putrid septic peritonitis. Had the patient consented to the operation, the case might have turned out differently. Perforation took place on the nineteenth day after the invasion.

*The presence of a tumor, which always indicates the existence of protective adhesions, implies a certain amount of temporary security, and, under certain circumstances, may justify a short delay of the operation.*

TYPES OF ACUTE PERITYPHLITIC ABSCESS.—Although the classification of perityphlitic abscess according to location can not be made with geometrical precision, yet it will be found that most cases can be naturally massed in a series of roughly defined groups. The small number of intermediate or transitory forms does not vitiate the practical value of this grouping, upon the right understanding of which must be based some important variations of the operative technique.

It is the author's wish to firmly maintain the importance of the principle that every intraperitoneal abscess should, if possible, be opened and drained without invading the normal peritoneal cavity—that is, through existing planes of adhesion to the parietes. With few exceptions, all perityphlitic abscesses have such an approachable side. To study, to ascertain, and to utilize them is the duty of the conscientious surgeon. It is idle to state that safely incising and draining an abscess through a laparotomy wound—that is, through the free peritoneal cavity—is an easy or indifferent matter. No competent person will believe it.

1. *Ilio-inguinal Type (Willard Parker's Abscess).*—The normal situation of the caput coli and appendix vermiformis near the parietes of the right iliac fossa has the consequence that the great majority of circumappendicular suppurative processes will naturally establish themselves so as to have for one of their limiting walls the parietal peritonæum of that region. This has led to the erroneous belief that perityphlitic abscess is normally located behind the peritoneal lining of the iliac fossa.

This situation involves the great practical advantage that the abscess can be permitted to assume certain proportions so as to render its incision simple and free from the danger of invading the normal peritoneal cavity. Therefore, when an immovable tumor develops in the right iliac fossa soon after the inception of the malady, it is safe to wait a few days until the abscess has assumed a certain size. On the fourth, fifth, or sixth day it may be safely incised. Searching for pus with a hollow needle is superfluous when the abscess is superficial—that is, immediately beneath the parietes; dangerous if it is deep-seated, as the gut might be thus injured or the healthy peritonæum infected.

CASE.—Francisca Bertrand, aged forty-five. Was taken ill with fever early in July, 1882, and developed a deep seated, painful swelling in the left iliac fossa, with high fever and peritonitic symptoms. On the afternoon of August 5th probatory puncture brought out some pus, wherefore, with the aid of the family physician, Dr. Assenheimer, incision was practiced by Hilton's method. A large quantity of pus escaped, and a drainage-tube and antiseptic dressing were applied. In the following night very acute peritonitis set in, to which the patient succumbed August 6th. No doubt the reflection of the peritonæum was injured, and part of the pus must have entered the peritoneal cavity.

The only safe way of opening these abscesses is by methodical and careful dissection, layer by layer being divided by an ample incision placed through the longer axis of the tumor. The vicinity of pus will become manifest by the discoloration and condensation of the tissues. When the abscess is opened and the bulk of its contents has escaped, a gentle exploration by the index-finger is advisable to detect recesses or a foreign body. But all rough treatment of the walls of the cavity by scraping, tearing, or rude squeezing is reprehensible, as it may lead to inward rupture. For the same reason search for and removal of the ulcerated or necrosed appendix from the abscess is to be avoided as unnecessary and dangerous. Two drainage-tubes are slipped into the cavity and fastened in the usual manner. They will facilitate irrigation without causing undue distention. A daily change of dressings will be required for the first week or ten days. As soon as the discharge becomes scanty and serous, the tube should be removed.

The ilio-inguinal type is undoubtedly and fortunately the most common form of perityphlitic abscess, and its time-honored therapy as laid down by Parker will have to be retained as safe and successful.

In sixteen cases of the ilio-inguinal group operated on by the author according to Parker's plan, only one terminated fatally, by erysipelas. The patient was under treatment for hip-joint disease when, unfortunately, the complication with perityphlitic abscess set in.

CASE.—Ernestine S., servant-girl, aged nineteen. Admitted March 2, 1880, to the German Hospital, with the diagnosis of hip-joint disease, the symptoms of which were indubitably present. Emaciating fever, and the characteristic flexion and adduction of the thigh, together with swelling of the gluteal and infrapubic regions, seemed to admit of no doubt. Examination under ether, however, revealed a fluctuating swelling of the right groin, which yielded pus on puncture, and was incised. A large quantity of pus and the stem of an apple or pear were evacuated. Another incision below Poupart's ligament established drainage of an abscess communicating with the perityphlitic gathering. The lower extremity was put into Buck's extension, and the cavities were daily irrigated. Operative measures, directed against the profuse discharge from the lower incision—that is, drainage or exsection of the hip-joint—were contemplated, when the girl contracted erysipelas, and died of it in May, 1880. Post-mortem examination established the fact of hip-joint suppuration, a communication of the perityphlitic abscess with the joint being found, by way of the iliac bursa.

See Case XI, page 129, for necrosed appendix contained in inguinal hernia.

2. *Anterior Parietal Type.*—Next in frequency to the ilio-inguinal form of perityphlitic abscess is the type according to which the bulk of the purulent collection is found immediately behind the anterior abdominal parietes of the right side. Frequently this is associated with a more or less apparent ilio-inguinal tumor, and might be looked upon as its extension. The swelling is generally found behind the right rectus muscle, its shape vertically elongated, its upper limit occasionally extending beyond the level of the navel to the hypochondrium, its proximal margin to or beyond the median line. When an unmistakable continuation of the tumor can be traced into the right iliac fossa, the abscess can be safely opened above Poupart's ligament, as in the preceding group. But occasionally the upper extension will require a separate incision.

CASE I.—Abraham Jacobson, tailor, aged twenty-two. Perityphlitic abscess of six days' duration, the iliac tumor extending inward and upward to the inner margin of the rectus muscle, the space above Poupart's ligament feeling empty. *November 19, 1888.*—Typical incision at Mount Sinai Hospital, a little below and to the inward of the anterior superior spine; drainage. Retention of pus in the upper pocket, hence, November 26th, second direct incision. Rapid improvement. *January 17th.*—Discharged cured.

CASE II.—David Frank, butcher, aged forty-two. Perityphlitic abscess of eight days' duration; tumor extended upward along the line of the rectus muscle to within a hand's breadth of the costal margin. *December 8, 1889.*—Incision two inches and a half to the inward of the anterior superior spine. Evacuation of about a quart of pus; depth of abscess, twelve inches; though the wound was doing well, surgical delirium set in, and the patient was transferred to his home December 24th, where, as his family attendant reported, he soon recovered entirely.

When it is found that the iliac fossa is normal and entirely void of resistance, and a circumscribed tumor can clearly be felt some distance from the ilium and Poupart's ligament, it is necessary to ascertain where to make a safe incision. If the extent of the tumor is great, a direct incision might be confidently made. But if the superficial extremity of the tumor is small, it will be safer to first open the peritoneal cavity in the median line by a small incision, and digitally explore the exact relations and extent of the adhesion. Having thus located the abscess, the exploratory cut is closed, and **the abscess** is incised by a direct route.

CASE I.—Miss Evelyne H., school-teacher, aged twenty-three. Perityphlitic abscess of two weeks' duration. Small tumor to the right of median line, underneath right rectus muscle. Iliac fossa empty. *Per vaginam*, tumor was felt adherent to anterior abdominal wall, and with it bimanually movable backward and forward. *March 7, 1890.*—Exploratory laparotomy in median line below the navel. Just to the right of incision, partly solid, partly fluctuating mass could be felt, its walls being evidently formed of intestine, among which the empty appendix was seen firmly attached. By passing the finger around the attachment of the tumor to the anterior abdominal wall, it was found that the iliac fossa contained healthy intestine, and that the tumor was in no wise connected with it. **Fixation** of tumor by fingers in abdomen; **puncture** through abdominal wall; **fetid pus.** Closure of laparotomy wound by suture. It was sealed with **a strip** of rubber **tissue** moistened with a little chloroform. Incision of abscess along the line of puncture; evacuation of five ounces of pus. Uninterrupted recovery. Discharged cured, April **10, 1890.**

CASE II.—Mark Beermann, **hat-maker, aged nineteen.** Perityphlitic abscess of seven days' standing. Somewhat movable tumor underneath right rectus muscle on a level with umbilicus. Iliac fossa normal. *November 30, 1889.*—At Mount Sinai Hospital, median exploratory laparotomy. Location of adhesion, which was very limited, was established by digital exploration. Closure of laparotomy **wound.** Incision and drainage of abscess. Discharged cured, January 11, 1890.

CASE III.—*Perityphlitic abscess of the anterior type may* **extend to and beyond** *the median line, when it will hold close relations with* **and may perforate into the** *bladder.* Henry Marks, aged seventeen, suffered from habitual **constipation and frequent** attacks of colic. In June, July, and August, 1878, severe **attacks of colic were** noted and overcome by the use of **purgatives.** *August 25th.*—Dr. L. Weiss, **the family** attendant, made out typhlitis and ordered a laxative, which, however, failed **to relieve** the patient. Thereupon opium was methodically exhibited until September 6th, when the **patient** had a spontaneous and copious, formed evacuation. *September 7th.*—The temperature rose to 104° Fahr.; the external swelling in the right groin became very marked. *10th.*—The author saw the patient in consultation with Dr. Weiss. A uniform puffy swelling was found occupying the right groin, and was extending beyond the median line of the abdomen. Frequent urination distressed the patient a good deal, who exhibited the usual hectic symptoms of long-continued suppuration. Deep fluctuation was made out, and evacuation of the abscess was determined **upon. The** transversalis fascia being gradually exposed, it was found infiltrated and **firmly attached to** the underlying tissues. A probatory puncture made in the bottom of the wound, close to the os ilium, gave pus, whereupon the abscess was freely incised, and a large quantity of matter was voided. No foreign body could be found. **Digital** exploration demonstrated a long sinuosity extending toward the median line to a pocket occupying the prevesical space. A drainage-tube was placed in **the main** abscess, another one was carried into the prevesical space, and the wound was dressed **with** carbolized gauze. The patient's wretched condition at once commenced to improve; **appetite and** sleep returned, **and the** profuse night-sweats disappeared. *20th.*—The drainage-tubes became disarranged, and were found slipped out of the wound. Difficulty was experienced in replacing them, and symptoms of retention, with renewed pain and fever, set in again. *23d.*—The author again saw the patient, and replaced the tubes. A considerable quantity of pus was found in the prevesical pocket. From this date on uninterrupted improvement was noted, and the patient got up October 10th. October 20th, the tubes were withdrawn, and October 30th the fistula was closed.

In this case imminent perforation of the bladder wall was prevented by timely incision.

3. *Posterior Parietal Type.*—Whenever perforative processes occur in an appendix located near the posterior parietes of the peritoneal cavity—for instance, near the right sacro-iliac synchondrosis of the lumbar region—the resulting abscess will naturally have a deep situation. Cases will occur in which incision of such an abscess can not be made unless it be done through a laparotomy wound. But there can be no doubt that in a certain proportion of these cases a safe incision may be made from behind.

CASE I.—James Solomon, school-boy, aged thirteen. *April 18, 1889.*—Perityphlitis of five days' standing. In consultation with Dr. W. Morse, an indistinct, very deep-seated, and painful tumor was felt in the region of the sacro-iliac juncture of the right side. By April 22d the tumor had considerably enlarged, and seemed to lie just beneath the right rectus muscle. At Mount Sinai Hospital laparotomy was done the same day over the site of the swelling, which was found to hold no connection whatever with the anterior abdominal wall, but was firmly adherent to the posterior wall of the pelvis. The ascending colon formed the outer wall of the tumor. The appendix could nowhere be found, and was undoubtedly imbedded in the mass of the tumor. The anterior wound was closed, and a long, hollow needle was thrust into the region of the tumor from behind, entering the pelvis a little to the inward of the line of the posterior superior spine, its direction being downward and forward. Pus was gained at great depth, and the abscess was incised and drained from there by a rather long and deep incision. All the febrile symptoms disappeared, and the boy was discharged cured, June 3, 1889.

CASE II.—Samuel Gross, tailor, thirty-three years old. Was laparotomized at Mount Sinai Hospital, January 27, 1889, for internal obstruction of six days' standing. Fecal vomiting was present, with enormous tympanites due to intestinal paralysis. The cause of the obstruction was found in a very long and much distended appendix vermiformis, the apex of which was firmly attached to the under surface of the right half of the transverse mesocolon. Through the loop thus formed about three feet of the ileum had slipped and had become strangulated. Corresponding to the attachment of the apex of the appendix a massive swelling was felt, occupying the space behind the colon, and, when the adhesion was severed, pus welled up from a small aperture corresponding to the site of the attachment. This led into an abscess cavity which was carefully evacuated. The appendix being removed, the intestines were replaced with considerable difficulty. The patient died an hour and a half after the operation.*

CASE III.—Mr. M. G., aged sixty-two. Had been suffering from habitual and very obstinate constipation for years. In May, 1880, profuse diarrhœa set in, and could not be controlled by any of the usual dietetic and therapeutic measures. A grave deterioration of the general condition developed, and the patient lost very much flesh in spite of forced feeding. *August 31st.*—Fever set in, and the presence of a painful swelling in the iliac fossa was made out. *September 3d.*—The author saw the case in consulta-

---

* For complete history, see " New York Medical Journal," May 4, 1889, p. 478.

tion with Dr. W. Balser and Dr. L. Conrad. A large fluctuating **swelling occupied the right half of the pelvis, and tympanitic percussion sound was noted in the lumbar region.** Two incisions were made—one above Poupart's ligament, **another in the lumbar region**—and an enormous amount of gas, pus, and fecal matter **was evacuated.** Profuse secretion and diarrhœa continued, and the patient died **September 22d.** *Postmortem examination* revealed a tight cancerous stricture of the ileo-cæcal valve, and an enormous dilatation of the lower portion of the ileum, which resembled thick gut. Large masses of impacted fecal matter were found in this pouch, which was adherent to the posterior parietal peritonæum, and was freely communicating through a number of ulcerous defects **with the abscess cavity.**

4. *Rectal Type.*—It is a good rule never to neglect to examine the rectum of a patient suffering from perforative appendicitis. A long appendix may become fixed and perforated in the small pelvis, and an abscess is then apt to develop in close vicinity to the rectum, whence it can be safely opened and evacuated. The objection that fæces might enter the abscess has thus far not been **verified by experience.**

**Case.**—August Petry, **clerk, aged eighteen. Was** admitted, November 10, 1887, to **the German** Hospital with symptoms **of** perforative peritonitis. **General** tympanites prevailed, and a tumor could not be felt anywhere, but intense pain was complained of on pressure in the right iliac **fossa.** The poor state of the patient forbade operative interference, and opiates and stimulants were exhibited. By November 13th the patient had fairly rallied. An examination **of the** rectum disclosed the presence of a fluctuating swelling corresponding to its anterior wall. An incision evacuated a large mass of pus, and a drainage-tube was placed into the cavity and brought out through the **anus.** The tube was not borne well. It **excited** tenesmus, and was repeatedly expelled. **As** the patient was doing very much **better, and** the **tumor** had disappeared, it was left off without ill consequences. The **patient was discharged** cured November 27, 1887.

5. *Mesocœliac Type.*—To characterize that most serious form of circumappendicular abscess, the walls of which are composed entirely of agglutinated intestines, and which hold no immediate relation whatever with the parietes **of** the abdominal cavity, the term "mesocœliac" was chosen (from αἱ κοιλίαι, the intestines, and ἐν μέσῳ, between). The abscess is found occupying, as it were, **the** middle of the peritoneal **sac.** Hence, to reach and evacuate this form **of abscess, the** free peritoneal cavity must be opened, and the collection of pus **must be reached** by separating the adherent coils of gut **which** inclose it.

We owe the development of the *technique* of the evacuation of these abscesses mainly **to** McBurney, **whose** procedure is as follows A longitudinal incision, **as** for simple appendicitis, is made parallel to and along the outer **border of the** right rectus muscle. The abnormal cohesion and resistance of **the** implicated intestines will point out the **site of** the abscess. The protruding normal coils of gut should be packed away under a protective bulwark of sponges held *in situ* by the assistants' hands, so that, if the abscess is opened unawares, no pus should soil the healthy peritonæum. Two of the nearest coils are now gently and cautiously separated by gradual **traction, exercised** by the operator's fingers, until a small quantity of pus is seen exuding. It is desirable to let the pus escape slowly, so as to have ample time to sponge it away as it pours out; otherwise the whole field might be overwhelmed and contaminated by a sudden flood of matter.

NOTE.—It seems that exhausting the abscess through a small aperture by means of a syringe would be an improvement upon the mopping up by sponges.

As soon as the bulk of pus has been removed, the cavity is wiped out clean with sponges dipped in an antiseptic solution, and now the adherent intestines are still more separated to permit the surgeon to inspect its interior. If the appendix is loose and easily to be got at, it can be removed, but, if it is found closely adherent and very brittle, it is better to remove only so much of it as will come away easily. A good-sized drainage-tube is placed into the bottom of the cavity, which is, in addition, loosely filled with strips of iodoform gauze. These and the rubber tube are brought out near the lower angle of the wound, and the abdominal incision is closed in the usual manner. If the case is progressing well, the packing can be withdrawn on the third day, as by that time protective adhesions will have formed between the adjoining coils of gut. The drainage-tube is to be removed as soon as the secretions become serous and scanty.

*c.* CHRONIC OR RELAPSING APPENDICITIS AND PERITYPHLITIC ABSCESS.—It was shown how simple catarrhal conditions of the mucous lining of the appendix may lead to more or less complete occlusion of the exit of this viscus. The retention of the secretions will then cause distention and the train of symptoms characteristic of appendicitis. With the diminution of the catarrhal swelling of the mucous membrane, a restitution *ad integrum* will take place. Usually the symptoms produced by this form are mild and tractable. Bland laxatives and opiates, rest in bed, with some form of local applications, generally bring about a lasting recovery.

Where ulceratvie processes, prolonged inflammation, or the doubling of the appendix upon itself, have caused the formation of cicatricial matter—hence permanent **stenosis** of greater or less intensity—the recurrence of severe obstructive symptoms will be more frequent, the intervals between the attacks shorter and shorter, and the tendency to the **formation** of adhesions more pronounced. Thus the very chronicity of the process will yield, in its tendency to the formation of adhesions, a certain protective character. Should perforation occur, these adhesions fulfill a most important function in preventing general septic peritonitis. The number of relapses of appendicitis may be very great; in one of the author's cases sixteen were counted. With the increase of the cicatricial stenosis, the formation of concretions, and the loss of contractile power of the appendix, the tendency to ulcerative or gangrenous lesions becomes more and more pronounced, and finally culminates in perforation.

As we have no means of ascertaining the exact condition of the appendix, frequent recurrence and increasing severity of the disorder clearly justify an attempt at its removal. The term "attempt" is used here purposely to signify that such endeavors may occasionally be baffled by intricate and close adhesions, which a prudent surgeon may prefer **not** to disturb **for fear of lacerating the gut.** It may be said, however, **that, should the first attempt fail, a second one may be crowned with success.** *

All surgeons admit the occurrence of the spontaneous evacuation of perityphlitic abscesses into an adjoining part of the gut. Occasionally perforations into the bladder, rectum, or even the pleura, **have** been observed and described. If such an evacuation into the gut is followed by a perfect obliteration of the cavity and fistula, no relapse will occur. Should evacuation be imperfect, inspissation of the retained pus and a temporary dormancy of the acute signs of the process will result, until some local irritation again provokes rapid intumescence, followed by evacuation of the surplus contents of the abscess. This process may be repeated a number of times, **as a result of which a thick mass of cicatricial matter will be deposited around the focus.** Cases of this order demand surgical interference.

CASE.—Miss Caroline D., aged fourteen. Had had within two years three attacks of perityphlitis with well-marked ilio-inguinal tumor, which never disappeared completely. On April 24, 1888, Dr. L. Arcularius presented her to the author, who advised an operation. A small immovable tumor could be felt occupying the iliac fossa. On May 1, 1888, an incision was made, and a small cavity of the size of a chestnut was laid open. Its walls consisted of a **massive** deposit of cicatricial matter, its contents of a putty-like mass of inspissated **pus, surrounded by a coating of** deciduous granulations. When all the soft matter was scooped **out, a narrow sinus was** traced to a depth of an inch and a half beyond the bottom of the cavity. The wound was packed, and was kept open with considerable difficulty during the entire summer, small quantities of feculent matter escaping from time to time. In the course **of the** following winter the tumor gradually shrank away, the discharge dried up, and, the tube being removed, permanent **healing took place.**

Had the outer opening been permitted **to heal, recurrence of the abscess** would have probably followed, as closure of the communication **with the** gut came about with a great deal of hesitancy. The same state **of** affairs may and does often prevail in abscesses that are evacuated by the surgeon, and in which the outer opening shows **a** more pronounced tendency to closure than the sinus leading from the abscess cavity to the gut. Thus the presence of a however minute fecal fistula that has not healed soundly may bring about a **number of recurrences** in the tract of **the** old abscess. It stands to **reason to say that inadequacy, both** as regards the quality and duration of drainage **of the abscess cavity, has a** most important influence upon the retardation **of the closure of the fecal** sinus. Hence the tendency to relapses will be **very pronounced in cases where evacuation of** the primary abscess took **place spontaneously.**

CASE.—Frank **Kennedy, printer,** aged twenty-five. Had suffered since childhood from a number of attacks of smart pain in the right groin accompanied by fever. In **the** early part **of** 1885 he acquired an oblique inguinal hernia of the right side, and was

---

* I take the liberty of referring to a verbal communication of Dr. F. Lange, who informed me that he once had to abstain from removing the appendix through an anterior incision. Later on the organ was successfully removed through a posterior wound.

ordered to wear a truss, the pressure of which, if the pad became displaced outward, caused intense suffering, so that he had to abandon its use from time to time. In June, 1885, during a severe attack of fever, an abscess broke open two inches and a half below the anterior superior spine. Since then healing and reopening of the sinus had occurred four times. On March 3, 1886, a dense deep-seated tumor could be felt in the right groin, independent of the hernia, which could be easily replaced. Following the existing sinus, the center of the indurated mass was laid open by a large incision running parallel with Poupart's ligament. At the depth of two inches a globular smooth-walled cavity was exposed, within which, imbedded in frail granulations, a stratified coprolithon of the size of an unshelled almond was found. A channel of the diameter of a goose-quill was seen leading from this cavity inward and downward, into which could be slipped twelve inches of a slender drainage-tube. When water was thrown in through this tube, diluted fecal matter regurgitated. Under the microscope this matter was seen containing granules of amylum and fat with fat crystals arranged in the shape of sheaves. The wound was kept packed with gauze till March 25th, and was healed, seemingly from the bottom, by April 14th. On November 15, 1886, the fistula reopened, and the proposition was made to the patient to expose the site of the fecal sinus from within by laparotomy, and to deal with it by extirpation of the appendix or enterorrhaphy. He declined to take the risk, and preferred to wear a tube permanently. Sparse quantities of a feculent, orange-colored serum continued to escape from time to time until the end of 1888, when the tube could not be replaced once, and was abandoned. As it seems, permanent healing then took place.

The proposition made to this patient, to close his fecal fistula by laparotomy and an appropriate dealing with the involved gut, contains the essence of a plan the adoption of which might be necessary in order to bring about the speedy cure of an apparently interminable, most disagreeable, and loathsome ailment. But the necessity for the adoption of such extreme measures must be very rare indeed.

On the whole, it may be said that the recurrence of an evacuated perityphlitic abscess is comparatively rare, and that, if it is due to the presence of a fecal fistula, its lasting cure can in most instances be effected by prolonged and efficient drainage of the outer wound.

Another cause of prolonged suppuration within and around an incised perityphlitic abscess is the formation of one or more extraperitoneal burrows and cavities, located between the several layers of the abdominal wall, which are the direct consequence of inadequate measures at drainage. The primary cause of the abscess may be eliminated, the perforative aperture of the appendix or gut may long since have permanently closed, and yet frequent relapses of suppuration will keep the patient confined to the bed. How to deal with a case of this kind may be seen from the following history:

Mrs. E. T., aged thirty-two. Was operated for perityphlitic abscess by a prominent gynæcologist of this city in the latter part of the summer of 1887. Four weeks after the operation the drainage-tube was withdrawn, and the wound healed promptly, but a reaccumulation and evacuation of pus soon followed, and symptoms of recurrent retention were observed on an average every four or six weeks until January 13, 1889, when, by the same practitioner, bloody dilatation was done with the confident expectation of lasting success. These hopes, however, remained unfulfilled. Up to March 1,

1889, three more recrudescences occurred which were closely observed by the **author**. Each time symptoms of retention were present, though a large and long drainage-tube was constantly *in situ*, reaching to the bottom of the wound. Circumscribed swellings occurred then once above, another time to the inner side of the sinus, and pus was seen **welling** up on pressure from the drainage-tube. It was decided to find **and** remove the cause of this distressing condition by an operation, which was done March 11, 1890, in the presence of Dr. Lange and Dr. Bull, of this city. The tract within which had lain the drainage-tube was exposed to its bottom by an incision nine inches long, **and** running parallel with Poupart's ligament. Carefully examined, it was found to be soundly and firmly closed at the bottom, no manner of communication existing with the gut, though it was evident that only a thin layer of tissue separated the cavity from **the** peritoneal sac. On the lateral aspect of the smooth lining of the old drainage track, and not far from the bottom, two minute apertures were seen inosculating, into which the probe passed for a distance of two and four inches, respectively, **the longer** track leading toward the navel, the shorter upward toward the crest of **the ilium**. When these narrow tracts were slit up, **each** of them was found terminating in a small pocket containing granulations and pus. These sinuses were located within the abdominal parietes, between the muscular and peritoneal layers. Unavoidably, the peri**toneal cavity** was opened in two places, but, as no tumor could be felt within, these apertures were not enlarged. However, a long probe was passed into Douglas's *cul-de-sac* through one of these apertures, where a finger placed in the vagina could distinctly feel its rounded point. **The very** large wound **was** purposely left open, and the dressing consisted in **an iodoform-gauze** packing, which was renewed every twenty-four hours in the beginning, later **on at** longer intervals. Uninterrupted healing followed, though it took a **long** time on account of **the size of the wound**. *June 3d.*—The patient was discharged **cured, and has remained well ever since then.**

*Conclusions.*—1. Mild, presumably catarrhal, **forms of** appendicitis, require no operative measures, but dietetic **and medicinal** treatment by opiates, laxatives, rest, and **local** applications.

2. The more severe and persistent forms of appendicitis may **render excision** of the appendix advisable, especially if frequent recurrence, with **increase** of the violence of the symptoms, is observed.

3. Most perityphlitic abscesses hold close relations with one or another of the abdominal parietes. The location of the parietal adhesions of the abscess is to be first **ascertained, if** necessary, by exploratory laparotomy, and the abscess is to be then incised and drained through the area of adhesion, **thus** avoiding infection **of the sound** peritonæum.

4. Perityphlitic abscesses that possess no parietal **adhesions and have** a mesocœliac situation between free coils of intestine **must be reached by** laparotomy through the uninvolved peritoneal cavity. **Precautions have to** be taken not **to** infect the normal **peritonæum.**

5. Recurrence of suppuration **in the groin,** following spontaneous **or** artificial evacuation of a perityphlitic **abscess,** may be due either to the persistence of **a** small fecal fistula, or to the presence of secondary intraparietal sinuses caused by inadequate drainage and retention.

In the first case prolonged and efficient drainage **is to be employed for** a long time before resorting to artificial closure of the **fecal fistula** by laparotomy and enterorrhaphy or otherwise.

In the second case all sinuses and pockets have to be found by free and careful dissection, and, when they have been slit up and scraped, the wound is to be treated by the open method to effect a sound cure.

*g.* **Abscess of the Liver.**—The diagnosis of hepatic abscess is based upon the presence of a painful and growing intumescence of the liver, accompanied by more or less intense fever, which gradually assumes a hectic character. In the beginning the swelling ascends and descends at respiration; but later on, when the liver becomes attached to the abdominal wall, this mobility disappears. Probatory puncture with a fine aspirating needle can be safely made, and will generally dispel any doubt. As soon as the diagnosis is secured, incision has to be made.

Where adhesion of the hepatic swelling to the abdominal wall is established, or, even more so, where the suppurative process has involved the integument, a free incision can be safely made. A large-sized drainage-tube should be inserted into the cavity, and frequent irrigation should be employed. The wound is covered with an ample moist dressing.

The incision of hepatic abscesses located in the unattached liver require some special precautions. The abdominal wall opposite the tumor is incised under a strict observance of the rules laid down for laparotomy, so as to expose the liver. The incision is packed with iodoformed gauze, and a dry dressing is applied.

In three days firm adhesions of the liver to the abdominal wall will be established, when, the packing being removed, the liver is punctured, and, pus being found, is freely incised and the cavity evacuated and drained.

*h.* **Lumbar Abscesses.**—The significance of acute lumbar abscesses depends upon their causation and upon the locality from which they take their origin. The majority of lumbar abscesses are caused by purulent affections of the kidney or its pelvis—as, for instance, by renal calculus or pyelitis—but in a comparatively large number of cases no affection of the kidneys or their adnexa can be recognized, and traumatism of one or another kind must be assumed as the causative agent.

Contusion and a sudden and unexpected strain of the back were stated to the author by patients as causative factors. The beginnings of lumbar abscess are always obscure and insidious. A deep-seated unilateral pain in the small of the back is first complained of. One or more chills or a low form of hectic fever set in. The patient's back is bent upon the affected side, and is more or less tender. Loss of vigor and emaciation become more and more evident, until a distinct tumor, marked by dullness on percussion, can be made out in the space between the crest of the ilium and the twelfth rib. The way of extension of the abscess is prescribed by the quadratus lumborum muscle, the outer edge of which serves as a landmark for finding and incising it. The presence of pyelitis or pyonephrosis, ascertained by examination of the urine, is very significant, and possible doubts as regards the nature of the trouble may be dispelled by one or more probatory punctures with a well-disinfected hollow needle and the aspirator. A good-sized caliber should be selected, as grumous or flocculent pus is apt to clog a small-sized needle, and a negative result may be arrived at in the presence of a large collection of matter.

CASE.—Mr. I. A., brewer, aged twenty-two, developed lumbar pain and swelling of the right side without any known cause. *April 17, 1881.*—High fever accompanied the seizure, and, though no fluctuation could be felt, the diagnosis of perinephritic abscess was made. *April 21st.*—In the presence of Dr. Heppenheimer, the family physician, four probatory punctures were made with an aspirator needle without positive result, and, unfortunately, the contemplated incision was deferred until the next day, when perforation into the pleura and rapidly fatal pyothorax developed.

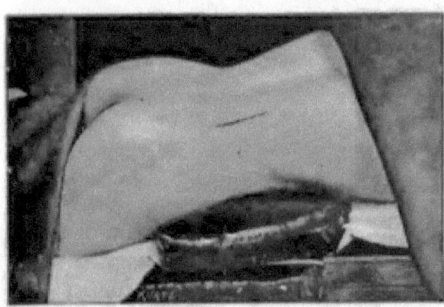

FIG. 184.—Lange's position for renal and perinephric operations.

Had a larger-sized needle been used, pus would have been found, and the fatal termination might have been averted by timely incision.

*Early incision* can never do any harm where perinephritic abscess is suspected, and will be of some use even if pus be not found at the first attempt. On account of the deep situation of the abscess, and the necessity of exploring its interior for sinuosities, which may require separate drainage, an ample incision is advisable. It should be done in anæsthesia under strict antiseptic precautions, and by gradual dissection.

The patient is brought into the position recommended by Dr. F. Lange for nephrotomy. A roll made of a blanket is slipped under the lumbar region, and the body is placed semi-prone upon the affected side, as shown in the accompanying cut (Fig. 184). The vicinity of the swelling is carefully cleansed and disinfected, and the surrounding parts of the body are protected with rubber cloths and towels in the usual manner. A longitudinal incision two or three inches in length is made, commencing about an inch below the last rib, and extending to near the crest of the ilium, and is gradually deepened until the abdominal muscles are all divided. Frequently pus will be reached before the edge of the quadratus lumborum muscle is exposed. Should this not be the case, a grooved director may be inserted underneath the external margin of this muscle,

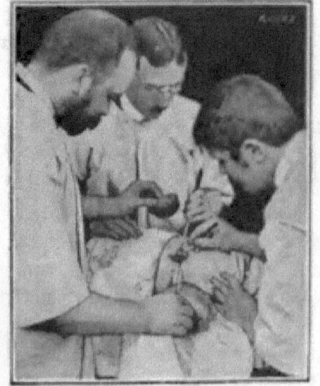

FIG. 185.—Incising perinephritic abscess.

and, being pushed downward and toward the median line, will soon enter the abscess. As soon as pus is seen to appear in the groove of the instrument, a dressing-forceps is insinuated into the cavity, and is withdrawn while held wide open. Blunt dilatation of this kind can be repeatedly practiced until the aperture is large enough to admit the index-finger for exploration.

Should the abscess contain urinous matter or stones, or should the septa of the calices of the renal pelvis be recognized by touch, the causation of the process by perforation outward from a suppurating kidney will suffer no doubt. If found, stones may be then extracted, and the cavity, being well washed with boro-salicylic lotion, is drained by the insertion of one or more stout rubber tubes.

NOTE.—A very efficient mode of draining is the following one: A number of fenestra are cut into the sides of a large-calibered rubber tube, which is placed well within the cavity. Another smaller-sized tube of the same length is provided with a couple of fenestra near its mesial end, and is inserted into the abscess alongside of the larger tube (Fig. 186). A stream of lotion injected into the smaller tube will enter the bottom of the abscess, will wash out its recesses, and will carry away secretions and *débris* through the many fenestra of the larger tube. Safety-pins thrust through the distal ends of the tubes will prevent their being lost in the abscess. An ample antiseptic moist dressing should envelop the entire lumbar region, and the patient should be brought to bed.

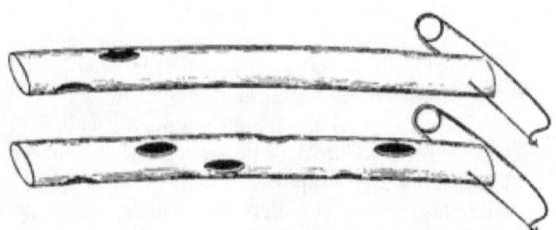

FIG. 186.—Arrangement of drainage-tubes for perinephritic or any other deep-seated and large abscess cavity.

Aside from lumbar abscesses of renal origin collections of pus must be mentioned that depend upon an extension into the circumrenal tissue of purulent processes originally established elsewhere. Perityphlitic abscess, empyema, perimetritic suppuration, and finally cold abscesses due to spinal disease, belong to this order.

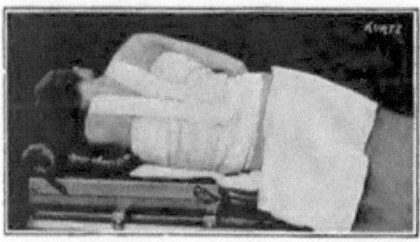

FIG. 187.—Dressing for lumbar or hepatic abscess.

Lumbar abscesses, the relation of which to purulent affections of the kidneys is unlikely or doubtlessly absent, admit of a much better prognosis. They are frequently referred by the patients to traumatisms, and, properly incised, heal very promptly.

CASE.—A. F., pawnbroker, aged twenty-four, sustained, in May, 1885, in jumping and slipping, a severe strain of the left side of the small of the back, which was followed by sharp pain and stiffness for a few days. It subsided spontaneously, but left behind a soreness of varying intensity. *May 20, 1886.*—Fever set in with intense lumbar pain, but swelling came on very slowly. Though looked for, it could not be made out until July 10th, when Dr. E. Schwedler ascertained its presence. The kidneys, gut, and spinal column were found normal. *July 12th.*—Incision by gradual dissection was practiced under ether. The abdominal muscles being divided, the edge of the quadratus lumborum was exposed. Probatory puncture in the bottom of the wound had to be done five times before pus was found high up close to the edge of the twelfth rib, beneath the quadratus muscle. This was drawn aside, and the cavity was opened by Hilton-Roser's method. About an ounce and a half of odorless pus escaped, and digital exploration showed that it had been contained in a small, smooth-walled cavity. Drainage and antiseptic dressings being applied, the wound was irrigated and dressed, daily; later on, at longer intervals. The patient was discharged cured, September 6th.

*i.* Pyonephrosis, Renal Abscess, and Calculous Kidney.—As an exhaustive study of the pathology and diagnostics of the various forms of suppurating kidney would far transcend the limits of this work, it must be sufficient to review the conditions requiring surgical interference.

Whenever cicatricial contraction, of pressure from without, or the impaction of concretions within a ureter impedes or prevents the free exit of the secretions of a normal or diseased kidney, dilatation of the pelvis, or in the later stages of the whole organ, must follow. A tumor will then make its appearance in the lumbar region, the contents of which may vary in character. If a suppurative nephrosis be present, pus will be found intermixed with urinary elements, which will be more or less in proportion with the amount of glandular tissue still performing its physiological function. The longer the retention persists, the more of the secreting elements will perish, and finally the kidney will represent a pus-bag contained within the fibrous capsule of the organ. If the causative factor be the presence of calculi, these will be found floating in the retained fluids.

Impediments to the exit of urine from a normal kidney will be characterized by accumulations lacking purulent elements. When all the secreting tissue has perished, a simple hydronephrosis will be established.

The presence of calculi in the pelvis and calyces of the kidney will generally produce very distressing symptoms, such as local pain, hæmaturia, and pyuria, with fever and emaciation, though the pertinent ureter may be perfectly pervious.

Finally, discrete pyogenic or tubercular abscesses of the glandular kidney-tissue occur, causing all the signs of a deep-seated abscess, which may require operative interference.

The diagnosis must be based on the subjective symptom of pain and objective signs characteristic of the various forms of kidney trouble, as fever, pyuria, hæmaturia, the presence of a painful tumor, and of serum or pus withdrawn by the aid of the aspirating needle.

(*a*) *Nephrotomy.*—The incision of the kidney for the purpose of the evacuation of retained serum, pus, and calculi, is a safe operation often possessing the dignity of a life-saving procedure. It is performed as follows:

NOTE.—Aspiration of the diseased kidney should always be looked upon as a diagnostic and not a curative expedient. The complete exhaustion of the purulent contents of a kidney preceding nephrotomy may be the source of serious embarrassment, as it is much more difficult to find an empty, hence collapsed cavity, than one well distended by pus or serum.

The anæsthetized patient is brought into *Lange's position*, which can not be too warmly recommended for its eminent advantages. (See Fig. 184, page 277.) Contrary to former usage, the patient is put with the belly on a firm roll in the *semi-prone position*, so as to have the diseased side not uppermost, but occupying the lowest level near the edge of the table. The kidney will be pushed well up into the loin by the pressure of the subjacent roll, and will be rendered more accessible. Finally, it will be held without further external aids within easy reach. The patient's body being well protected in the usual manner, a transversely oblique incision, commencing two inches from the spine, and carried midway between the crest of the ilium and the costal margin, is gradually laid through skin, fascia, and muscles, until the fascia containing the circumrenal fat is exposed. With this incision extended far enough outward, ample space can be made for the removal of the kidney, should this become necessary, and injury to the pleura (in the absence of the twelfth rib) need not be feared. After the fatty capsule is incised, masses of loose fat will be seen bulging into the wound, which must be held aside by sharp and later by large, blunt retractors (see Figs. 17 and 19, page 40). A second fibrous septum, interposed between the superficial and deep portions of the circumrenal fat, will then be encountered. When this is divided, the posterior and distal aspect of the kidney will come in view. The question will arise now whether the pelvis or the parenchymatous portion of the kidney should be incised. As it has been observed that wounds of the pelvis do not heal as promptly as those made through the renal parenchyma, the incision should be made through the latter, unless it be found that a large stone is occupying the pelvis. A thermo-cautery knife completes the incision, which need not be larger than sufficient to admit the index-finger, with which the interior of the cavity is explored after most of the liquid contents have escaped. If no calculi are found, a stout drainage-tube is inserted and brought out through the wound. If stones are present, they are extracted by means of forceps, the scoop, or the hooks used by Lange. Preceding this, further dilatation of the renal incision may be required to gain room for the difficult process of extracting irregularly angular stones. This completed, the drainage-tube is inserted, and the cavity is flushed with Thiersch's solution. The external wound is lightly filled with iodoform gauze and inclosed in an ample dry dressing. The drainage-tube is to be brought out through a central slit in the outer dressings, and is connected with a longer tube carried under the patient's bed, where its end rests in a suitable vessel containing a few ounces of carbolic lotion. Thus the necessity for a frequent change of dressings will be avoided, should much urine escape through the wound. As soon as the quantity of urine thus voided becomes small, the rubber tube attachment can be left off. The dressings are to be changed every second or third day; the tube is to be retained for a very long time. The tendency to the formation of a permanent fistula is strong in these cases, except where calculi were extracted from an otherwise normal organ. But, with a scantily discharging fistula, life may be very tolerable indeed.

Nephrotomy was performed by the author eleven times, with two deaths. In seven cases tuberculous pyonephrosis necessitated the operation, which gave the patients eminent relief, freeing them from the presence of large and distressing accumulations of pus in the pelvis of the kidney. Once the kidney was incised for an enormous hydronephrosis. In June, 1890, four years after the operation, the patient was still wearing a cannula in a scantily discharging sinus. Of the two fatal cases nephrotomy was done in one for calculous kidney, in which perforation into the pleura and hence into a bronchus had taken place. Both the thoracic cavity and the kidney were incised, but the patient died of a septic pneumonia four days after the operation. The other case concerned a man whose left kidney had been extirpated for the cure of a urinary fistula remaining after nephrotomy done for pyonephrosis due to cicatricial obstruction of the corresponding ureter. One month after the healing of the nephrectomy wound, renal suppression took place on the right side. The patient was admitted to Mount Sinai Hospital in a uræmic condition. Though nephrotomy was promptly performed, the kidney did not recover its functional capacity, and the man died within twelve hours after the operation. (See case of Moses Cohn, page 283.)

(b) *Nephrectomy.*—When a kidney has become totally disorganized through suppuration, or has lost its functional capacity in consequence of the atrophy of the secreting tissues, as, for instance, in hydronephrosis ; or, finally, where obliteration of a ureter has brought about an incurable urinary fistula of the kidney, extirpation of this organ may come in question. Before proceeding to remove a seemingly useless or disorganized kidney it is very desirable to ascertain whether another kidney be present or not. All the methods of examination hitherto proposed for the establishment of the presence of two kidneys, and the diagnostication of their condition by the catheterism of the ureters, have been found unsatisfactory and unreliable. Hence, if there is any doubt of the presence of two kidneys which can not be eliminated by the ordinary means of physical examination, nothing remains but an exploration through either an abdominal or a lumbar section. Lumbar nephrectomy is performed as follows : Without regard to a pre-existing sinus, the external incision is made as described in the chapter on nephrotomy. When the surface of the kidney is reached, the organ is separated from the surrounding fatty tissue by blunt dissection, most conveniently done by the tip of the index-finger. Occasionally a more resisting band will have to be severed by a touch of the knife or scissors. As soon as the kidney is well separated, it can be brought out of its niche by traction, unless its size is very large, when subsidiary incisions will have to be added. Even then occasionally manœuvres will have to be made resembling the development of the infantile head from the vulva—that is, the kidney will have to be tilted and brought out with its end on. This being done, the vessels and ureter are separated and tied each by itself with stout catgut, and the pedicle is cut off at a safe distance from the ligatures, which are also cut off short. The wound is well irrigated, and, if any oozing be present, is packed with iodoform gauze. Secondary sutures may be then employed, which can be closed after the removal of the packing on the third day. A drainage-tube will be needed after the suture is completed, to prevent retention. If no considerable oozing prevail, the wound can be at once sutured, after a good-sized drainage-tube was slipped into the bottom of the cavity. In separating the kidney, the peritonæum may be accidentally injured. In this case the rent ought to be at once closed by suture, if possible ; if not, a strip of iodoform gauze ought to be stuffed temporarily into the rent, until the kidney is removed, when the peritonæum can be more conveniently stitched.

The after-treatment by packing and secondary suture will be the safer procedure in all those cases where unavoidable contamination of the deep cavity by pus, escaping either from a pre-existing sinus or in consequence of the rupture of the wall of an intrarenal abscess, had taken place.

Though not commonly, yet it occurs that, in consequence of long-continued inflammation, the fibrous capsule has entered into such an intimate and firm union with the condensed and shrunken circumrenal fat, that the enucleation of the kidney becomes a very difficult and hazardous undertaking. In these cases the proper mode of procedure is this : After having exposed the kidney, the fibrous capsule is split open along the outer edge of the organ, which now can be readily stripped out of its fibrous coat. On developing the gland, the rather stout pedicle is secured in an elastic ligature, and cut off. If need be, the section of the pedicle can be carried through the renal tissue, in order to prevent slipping of the ligature. The wound is drained and packed, and the ligature is brought out near the inner angle of the wound. The sloughing pedicle will come away in about ten days, when the size of the wound can be reduced by secondary suture.

CASE I.—Solomon Posner, aged thirty-seven, an emaciated, anxious-looking tailor, had been suffering from cystitis since January, 1888. Two years previous to this had had an attack of renal colic of the left side. Frequent and very painful urination with blood and pus, no renal elements, but a trace of albumen in filtered urine. Intermittent attacks of high fever. No pain on pressure in the loins. *November 2, 1888.*— Temperature 102° Fahr. Suprapubic cystotomy at Mount Sinai Hospital. The interior of much congested bladder-wall studded with miliary tubercle, and bleeding at the slightest touch. T-tube inserted, outer wound packed. Two hours after operation, profuse capillary hæmorrhage from the bladder was observed. It was checked by tamponade of the viscus with iodoform gauze. The fever continuing, and a painful tumor having developed in the left loin, this was aspirated December 14th, when mucopus was withdrawn. *December 21st.*—By nephrotomy done in chloroform anæsthesia a large quantity of pus was evacuated. A drainage-tube was inserted into the pelvis of the kidney, from which no urine ever escaped. *December 25th.*—Forty ounces of urine were collected from the bladder. The fever subsided somewhat, but there was an exacerbation every evening. As there was good reason to suppose that the other kidney was fairly healthy, and in view of the fact that the patient's strength was being steadily sapped by the nightly fever, nephrectomy was performed January 25, 1889. The very large kidney was exposed by an ample T-shaped incision. Its separation was very difficult, and though the eleventh and twelfth ribs were resected, lack of space led to the injury of the peritonæum. After the development, deligation, and removal of the organ, the peritoneal rent was closed by suture. Wound packed, no external suture. The pelvis of the kidney was lined with closely adherent cheesy masses ; the cortical and pyramidal substance studded with a large number of smaller and larger caseous abscesses. The rather collapsed patient rallied well, and the temperature fell off and did not range after this above 100° Fahr. *January 27th.*—Passed thirty-six ounces of urine in twenty-four hours. Up to February 4th everything went well, so that the patient sat up on the afternoon of that day, and retired after a hearty supper at 8 p. m. The evening observation gave temperature 100° Fahr.; pulse, 90; urine, thirty-six ounces. At 11 p. m. suddenly stertorous breathing set in, the pulse ran up to 120, the patient was comatose, with insensible conjunctiva, and a deeply flushed face. There had been no vomiting or headache. Urine was still seen dripping out of the catheter placed in the patient's bladder. The wound was examined and found in good order. Death ensued at 5.15 a. m. No autopsy could be secured. As the assumption of uræmia was hardly justified, it is probable that a thrombus found in the stump of the renal vein became detached and gave rise to pulmonary embolism.

CASE II.—Rosaly Cronn, housewife, aged fifty-six, began in 1883 to have rigors, paroxysmal pains in the left hypogastric region with painful and frequent voiding of turbid urine. These attacks recurred every few months for four years till 1887, when a tumor made its appearance. *August, 1888.*—A large quantity of pus was evacuated by an incision made in the left loin. General condition was somewhat improved, but a discharging sinus remained behind. *October 8, 1889.*—On her admission to Mount Sinai Hospital a dense resisting and painful tumor could be felt in the left loin. A probe introduced into the existing lumbar sinus led down toward this swelling. The woman was poorly nourished; her urine contained pus, blood, and a little albumen, but no casts. *October 21, 1889.*—*Nephrectomy.* The sinus led into the small, shrunken and lobulated kidney. The swelling felt before the operation was accounted for by a dense cicatricial deposit in which the organ lay imbedded. A number of calculi were struck by a needle thrust through the kidney, which was found converted into a cicatricial bag. The fibrous capsule was divided, and the organ was stripped out of it. The very short pedicle was ligatured in mass and the kidney was cut away. A cylindrical calculus was found caught in the ligature, but was easily withdrawn from the stump. The peritonæum was accidentally rent during the first attempts to separate the organ. The rent was stopped up with a strip of iodoform gauze which was left *in situ* till the dressings were changed on the fourth day. *October 23d.*—The temperature had not risen above 100° Fahr. Patient had passed twenty-four and a half ounces of urine in twenty-four hours. It contained granular hyaline and blood casts. General condition was good. *October 24th.*—Urine forty-two ounces; temperature normal. From October 28th on the casts disappeared from the urine, but slight quantities of pus were still observed. *November 16th.*—The ligatures and stump came away. Secondary suture had to be done twice to hasten the closure of the large wound. Discharged cured, December 15, 1889.

CASE III.—Moses Cohn, tailor, aged forty-two, had had within the last four years a number of severe attacks of renal colic, accompanied by rigors and turbid urine. Three weeks before his admission to Mount Sinai Hospital another attack set in with vomiting, repeated chills, and severe pain in the left loin. The fever continued till his admission, November 12, 1889, when the temperature was 101°, *the* urine absolutely normal, but in the left loin a painful tumor was felt, which could be well separated from the somewhat enlarged spleen. *November 14th.*—By an exploratory puncture sanguino-purulent, urinous smelling serum was withdrawn. The kidney was exposed, and was found considerably distended. From an incision about twelve ounces of matter were evacuated. The kidney was drained so as to catch the discharges in a vessel placed below the bed. The patient's condition was immediately improved in every way, but the same quantity of urine continued to escape from the drainage-tube as from the bladder, averaging about twenty ounces from every side.

Apparently the left ureter was completely blocked, and, as there was no improvement noticed until December 14th, it was decided then to explore the left ureter. A slender elastic bougie was passed into the ureter, and was arrested at a distance of five inches, the channel appearing to be impassable. Thereupon the kidney was removed, though it was apparently healthy. *December 15th.*—Patient did well; passed twenty-five ounces of urine in twenty-four hours; temperature normal. *December 20th.*—Passed forty-six ounces of urine containing traces of albumen and *a little pus. February 4th.*—Secondary suture. *February 16th.*—Patient was discharged cured and in excellent health. *March 13th.*—He was readmitted with obstructive symptoms of the hitherto unaffected right kidney, which, however, yielded to treatment. Discharged at his own request, March 21st. *March 24th.*—He was readmitted with absolute renal suppression, which was not influenced by medication, wherefore nephrotomy was performed, March 28th, on the uræmic patient. The evacuation of much urinous pus was of no avail; the intoxication was too far gone, and led, in spite of diligent attention, to his death, March 31st. A number of small renal calculi were extracted, and proved the mechanical nature of the obstruction of the ureter. The autopsy revealed softened and much swollen parenchyma of unusually light color, the ureter obstructed by calculous detritus.

CASE IV.—Oscar Hettler, barkeeper, aged twenty-seven, has suffered from acute attacks of pain in the right lumbar region since three years, the pain radiating to the glans penis. Ten days before admission to the German Hospital, fever set in with much sweating. *February 5, 1889.*—On admission, marked anæmia, a movable tumor in the right loin and urine containing much pus. *February 11th.*—Dr. W. Meyer, then on duty, evacuated a considerable quantity of pus by nephrotomy. *February 13th.*—The author took charge of the patient. In spite of free drainage he continued to fail. *March 8th.*—High and constant fever set in, the temperature rising to above 103° Fahr., and a careful physical examination did not reveal any complication by involvement of other organs. *March 14th.*—Nephrectomy was done. It was an easy, short, and comparatively bloodless operation, from which the patient rallied well. During the first twenty-four hours thirty ounces of urine were voided. The high temperatures continued unchanged. *March 23d.*—Patient passed twenty ounces of urine. Temperature still 103° Fahr., and remained high until the patient's death, which occurred April 11, 1890. The excised kidney contained a number of smaller and larger tuberculous foci, most of them not communicating with the pelvis. Autopsy revealed almost *general miliary tuberculosis.* The lungs, left kidney, liver, and spleen were studded with innumerable spots of miliary tubercle of recent origin. Apparently their development caused the patient's death. Repeated careful examinations had failed to reveal the slightest physical signs of the presence of this extensive process.

*k.* **Anal Abscess. Fistula in Ano.**—The anus, the final strait through which all excrementitious matter must pass, is subject to a great number of traumatisms from within and without. Foreign bodies, such as pits and kernels, chicken and fish bones, are frequently caught by and imbedded in the mucous lining of the sphincter muscle. The rough introduction of syringe-points for the application of enemata, scratching and manipulation of itching and bleeding piles, the surgeon's digital exploration, sodomy, and the forcible expulsion of massive fæces, lead to superficial injuries of the mucous membrane and outer skin of the anal region. Persons whose hands and faces are habitually unclean do not scruple much about the untidy condition of their breech. And the fæces of even the most cleanly swarm with bacteria. In view of these facts, the frequency of ulcerative and suppurative affections of the anal region must appear very natural.

*Anal abscesses* are generally located in the ischio-rectal fossa. This is the space limited by the **rectum on the median side, the tuberosity of the ischium externally, the levator ani muscle above, the superficial perineal fascia below.** It is very rare to meet with a periproctitic abscess situated above the levator ani. If such is the case, we have to deal with graver affections involving the pelvic organs, or with abscess from ulceration due to stercoral **impaction caused by** cancerous rectal stricture.

CASE.—Mary Steiger, aged fifty-nine. Far-gone cancer of rectum. Stenosis very tight, causing great difficulty at defecation. A profuse purulent discharge from the anus indicated the presence of ulcers or an abscess above **the** stricture. Exploration of the rectum above the cancer was absolutely impossible. High temperatures **were** noted. *August 13, 1885.—Anterior colotomy* in the German Hospital. No diminution **of** fever after the operation. *August 16th.*—Wound healed by the first intention. *August 17th.*—Patient **delirious.** Discharge from anus very profuse. *August 18th.*— Patient died with symptoms of septicæmia. Post mortem revealed firm union of colotomy wound throughout and a normal peritoneal cavity. In the sacral excavation, just above the massive ulcerated cancer, a very large fetid abscess was found.

The presence of anal abscess is the source of intense suffering to the patient, and ascertaining of its precise location by the surgeon is generally not very difficult. By digital examination of the rectum a resistant, hard, or sometimes fluctuating swelling can be felt protruding laterally into the gut. Early incision is very urgently indicated, as upon it may depend the avoidance of the formation of fistula, or of a dissecting or "horse-shoe abscess," which may detach almost the entire lower gut from the adjacent connective tissue. This latter form of abscess is especially to be feared, as its healing is extremely difficult. But, where fluctuation is absent, successful evacuation of a deep-seated periproctitic abscess is no easy matter.

After a purge and enema, the patient should be anæsthetized and brought into Bozeman's or the lithotomy position. (See Fig. 122, page 167.) A sponge tied to a piece of stout silk is pushed well into the rectum, and the lower end of the gut and the anal region are flushed with corrosive-sublimate lotion. Then the index-finger is introduced and placed against the swollen side for fixation. A stout exploring needle is thrust through the skin into the **swelling repeatedly** from without until it **strikes** the suppurating focus. It is left *in situ* for a guide, and an ample incision is gradually extended until the **abscess is** freely opened. The wound should have **the** shape of a funnel, **its apex** being in the abscess. This **will** secure natural drainage. The **wound is** loosely packed with **iodoformed** gauze, and the anus is inclosed in a moist dressing, which should be renewed every day. Daily irrigation, **or in** very irritable **patients a sitz bath,** will have to maintain cleanliness.

In cases where extensive detachment of the rectum or perforation into the gut has taken place, simple incision will be insufficient, and division of the intervening bridge will be necessary.

By spontaneous evacuation outward, *external incomplete fistula* will be established. Some of these cases can still be cured by a free bloody dilatation of their orifice, and a careful antiseptic treatment as above indicated. But most of them are complete fistulæ, the inner openings of which can not be found on account of their minuteness.

Cases of incomplete internal and of complete fistula should be cut.

*In incomplete inner fistula* a Sims' vaginal speculum is used for exposing the entrance to the sinus. A bent probe and alongside of this a bent grooved director is introduced into it, and is pushed well outward toward the skin, which is incised over the point of the instrument. After this the intervening bridge is divided.

*Complete anal fistula,* especially where several sinuses exist, should always be carefully explored before the incision is made, as otherwise pockets and branching sinuses may be overlooked. A silver probe should be introduced into each sinus and left *in situ* until its turn for cutting should come. A grooved director is carried into the gut along one of the probes, is caught up by the tip of the left index-finger, and turned out of the anus. The bridge of tissue taken up by it is then divided. The edges of the cut are well drawn apart by four-pronged sharp hooks, in order to facilitate securing and tying of spurting vessels. The next sinus is taken up after the first, and every nook and recess is carefully examined and split open until natural drainage is secured everywhere. Free irrigation of the wound should be employed during the whole process. When hæmorrhage is properly attended to, all the old granulations should be forcibly scraped away with the sharp spoon, and the wound should be packed with narrow strips of iodoformed gauze. After this the sponge is withdrawn from the rectum, and a moist dressing is applied and held in place by a T-bandage. (Fig. 126, page 170.)

NOTE.—When the internal orifice can not be found, or a burrow extends upward beyond it, the grooved director should be inserted as high up as the cavity or sinus permits, and thence should be thrust through the mucous membrane into the gut.

The length of time required for the cure of fistula in ano will depend on the extent and form of the wound made by the surgeon. In simple cases a fortnight or three weeks will suffice; complicated ones may need months. In favorable cases, that is, where the fistula is straight and single, *cure can be very much hastened by excision and suture of the entire fistulous track.* The restitution of the parts to their normal condition will at the same time insure against incontinence. The callous lining of the sinus is carefully excised with forceps and curved scissors, and the remaining wound is united by several tiers of buried catgut sutures, the ends of which should be clipped off short. The uppermost tier of sutures should not inclose the mucous membrane, but the curved needle should be introduced close to its edge on one side, and brought out in the same manner on the other side. Thus inversion of the mucous lining will be avoided, and the stitches, being buried under the overlapping edges of the mucous membrane, will be protected from infection by intestinal contents. The external, that is, cutaneous, part of the wound can be closed by silver-wire stitches. Free irrigation of the wound during the entire time of the operation is indispensable to preserve asepsis. Iodoform is dusted over and rubbed into the line of union, and the anus is inclosed in a moist dressing.

CASE.—Simon Schulhof, laborer, aged forty-three and a half, received, during the Austro-Prussian war of 1866, a bayonet wound near the anus. Suppuration and the formation of fistula followed, and resisted three operations which had been performed since that time. *February 5, 1887.*—Under ether, the fistula was slit up at the German Hospital. Its external orifice was nearly two inches from the anal margin; the internal one, one inch and a half up the rectum. The direction of the track was straight, and no lateral sinuses were present. The entire cicatricial lining of the fistula was excised with forceps and curved scissors, and the internal defect was united with three tiers of fine catgut sutures. The external

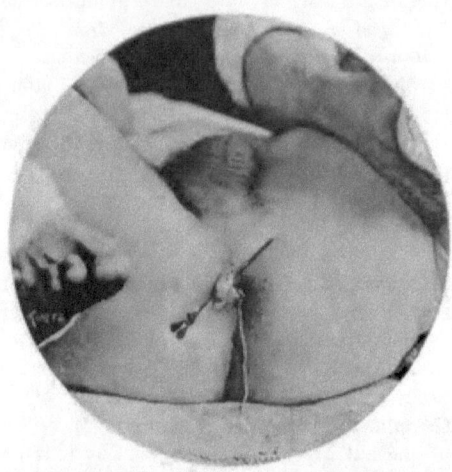

FIG. 188.—Operation of fistula in ano. Grooved director passed through fistula and brought out of the anus, from which is seen depending a thread holding sponge pushed well up the rectum. (Simon Schulhof's case.)

wound was brought together with two silver-wire stitches. Into the outer angle of the skin-wound a short piece of slender rubber drainage-tube was placed. A pledget of iodoformed gauze was placed into the anus, and the wound was dressed with gauze and a T-bandage. No reaction followed. In the afternoon of February 7th, four ounces of sweet-oil were injected into the gut, and the oil-soaked gauze was withdrawn from the anus. An hour after this a large enema of soap-water was administered, and brought away a liquid stool. The next morning a saline laxative was given, and was continued every day, each stool being followed by irrigation of the anus to free it from excrementitious matter. *February 10th.*—The silver stitches and rubber tube were removed. The accompanying cut shows the condition of the wound on the tenth day after the operation. The action of the sphincter was perfect. (Fig. 189.)

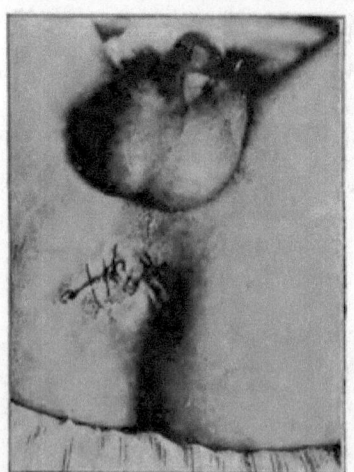

FIG. 189.—Result after excision and suture of fistula in ano. (Simon Schulhof's case.)

Regarding the management of the first and subsequent evacuation of the bowels, the reader is referred to the chapter on hæmorrhoids (page 169).

In very extensive cases of fistula of long standing, where the inner orifice is very high up, say two inches or more above the anal opening, and where avoidance of hæmorrhage is rendered imperative on account of the anæmic condition of the patient, *the elastic ligature can be successfully substituted for the knife.* The grooved director is carried through the sinus into the gut as usual, and, if possible, its point is turned out of the anus. Where this is impossible, a slender, soft, silver probe is armed with a fillet of stout silk, to the end of which a piece of elastic ligature or a small-sized drainage-tube (the size used on infants' feeding-bottles is very good) is firmly tied. The silver probe is next carried along the grooved director into the gut, its point is caught up by the tip of the left index-finger, and being bent upon itself is grasped with a stout pair of dressing-forceps and withdrawn. Thus the silk thread will be placed into the sinus, and with a seesaw motion will clear a way for the elastic ligature, which is drawn through after it. The ends of the elastic ligature, being firmly held each by one hand, are well drawn upon, and become tense and attenuated. Thus stretched, they are crossed over each other in front of the anus, and are secured in this position by a ligature of silk. As soon as the rubber is released, it crowds up against the silk ligature, and is held securely in place. Its ends are trimmed off short.

The elastic ligature is in every way preferable to the silken one, as it cuts through more rapidly, and does not require retightening.

Where the external orifice of the fistula is not close to the anal opening, the intervening skin must be cut through with the knife before the tightening of the ligature, to avoid the intense pain due to strangulation of the cutaneous nerves.

*Incontinence* is occasionally produced by fistula operations requiring single or multiple division of the entire sphincter. In these cases a secondary *proctoplasty* offers fair chances of partial or complete recovery of the function of the muscle.

CASE.—Barto Weil, brewer, aged fifty-six, suffered from distressing incontinence of the rectum, caused by four extensive fistula operations, performed successively for the horseshoe variety of this affection. At the last operation the author applied two elastic ligatures, one of which reached three inches, the other three inches and a half up the rectum. An irregular gaping aperture remained, from which rectal mucous membrane protruded in a number of folds. One granulating oblong surface was still extending nearly two inches into the gut. *May 28, 1886.*—Under ether, the entire irregular cicatrix was excised, and the remaining flaps of mucous membrane, together with the lower end of the uncut rectum, were dissected up and drawn well down. By a large number of catgut stitches the cylindrical shape of the anal opening was re-established, and the new anal ring was sewed to the external skin. A triangular defect remaining on the right of the anus was covered by a skin-flap shaped out of a shrunken integumental caruncle found posteriorly. Two small drainage-tubes were placed well up between rectum and ischio-rectal connective tissue. Primary union followed through the greater extent of the wound, and ultimately continence was fairly re-established. The patient was discharged cured July 24, 1886.

## CHAPTER VII.

### *ERYSIPELAS AND PSEUDO-ERYSIPELAS.*

THE rules of aseptic management described in former chapters are the best safeguard against the infection of operative wounds by the specific coccus of erysipelas. (Fig. 131, page 183; Plate II, Figs. 5 and 6; and Fig. 190.) The author has observed only four cases of wound erysipelas in ten years both of public and private practice. In one of these, in 1879, erysipelatous infection was transmitted from a case of so-called idiopathic erysipelas of the face to the genitals of a woman in childbirth by the author's hands, in spite of ordinary measures of cleanliness. Had disinfection been applied after the usual washing of the hands, the patient might have been living to this day.

The other case of erysipelas was observed after the first visit of a new member of the house-staff of Mount Sinai Hospital, at which the dressing of a nearly healed wound was changed by the young physician in question. The case was cured.

NOTE.—The time of changes in the house-staff of the surgical wards of hospitals is generally signalized by unexpected suppurations. The author has learned to dread the loss of a good and well-trained assistant, who is occasionally replaced by an inefficient, uncleanly, and indolent personage. Disaster can be averted at such times only by increased vigilance and redoubled diligence on the part of the visiting surgeon in personally supervising the details of the service.

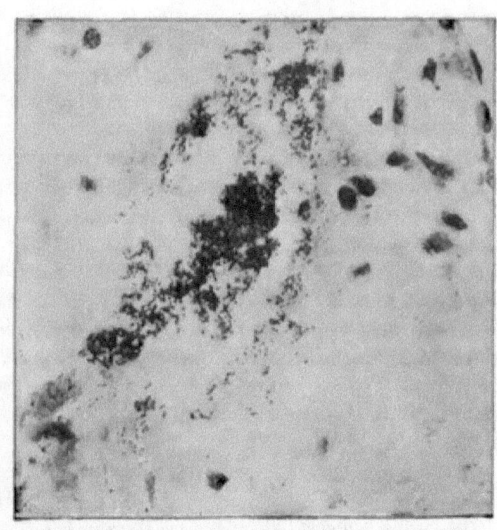

FIG. 190.—Section of erysipelatous skin of head (700 diameters). (Koch.)

The third case was mentioned in the paragraph on perityphlitic abscess.

The last case of erysipelas within the author's experience was that of a young woman suffering from caseous cervical glands. For cosmetic reasons the glandular swellings were punctured with a narrow bistoury, and, a small curette being introduced into the broken-down center of the gland, its caseous contents were scraped out. The small wounds were drained with catgut. Erysipelas, commencing from one of the punctures, set in, but ended in cure. Undoubtedly either the bistoury or, more likely, the sharp spoon was the carrier of the virus.

There is not one among the many topical remedies recommended by the writers for erysipelas that is pre-eminent in limiting or stopping the affection. The author's local treatment consists in moist antiseptic dressings inclosing the affected parts, with a general supporting treatment by proper nourishment and stimulants. The much-praised specific effect of the tincture of iron is, to say the least, very problematic.

NOTE.—Lately Kraske has published a series of cases in which *multiple scarification and puncture of the affected parts*, especially along the line of the spread of the disease, has led to prompt cure. The little operation is followed by the application of a moist antiseptic dressing. As the principle of this mode of therapy is rational, consisting in depletion and disinfection, it would deserve extended trial.

An unmixed infection by the coccus of erysipelas will never cause abscesses. Whenever abscesses form with erysipelas, we have to deal with a mixed infection, namely, by the coccus of erysipelas, and by one or another of the pus-generating cocci.

Phlegmon and erysipelas also represent a mixed form of infection, but this combination is rare. What is generally called *phlegmonous erysipelas* is commonly no erysipelas at all. It is a phlegmon produced by the pyogenic chain-coccus, the spread of which along the lymphatics resembles that of true erysipelas.

*Pseudo-erysipelas* is an erysipelatoid skin affection of the fingers and hand that resembles true erysipelas in most of its morphological features. But it presents this important clinical difference, that it never is accompanied by fever. The affection is very tractable, as the application of a three-per-cent carbolic lotion for a few hours will generally consummate a cure. Its cause is a specific coccus described by Rosenbach.

# PART III.

# TUBERCULOSIS:

## ITS ASEPTIC AND ANTISEPTIC TREATMENT.

# CHAPTER VIII.

## *NATURAL HISTORY AND TREATMENT OF TUBERCULOSIS.*

### I. ETIOLOGY OF TUBERCULOSIS.

Koch's discovery of the specific bacillus of tuberculosis has brought about a reconstruction of pathological classification and nomenclature that commends itself by clearness and simplicity. Miliary tuberculosis of the lungs and other internal organs, scrofulous affections of the lymphatic glands, the various forms of surgical tuberculosis, as, for instance, white swelling and caries, finally the several forms of lupus, are manifestations of one and the same morbid process—namely, of cellular decay caused by the deleterious influence of a vegetable parasite, Koch's tubercle bacillus.

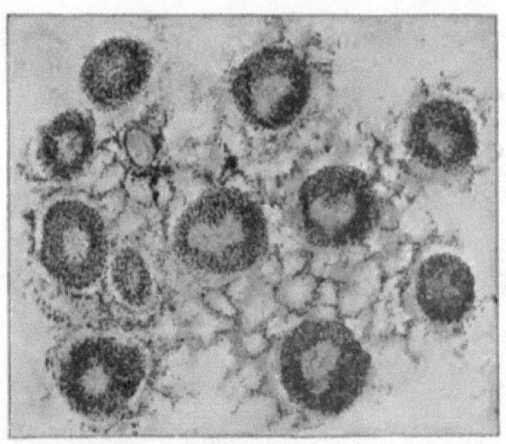

Fig. 191.—Miliary tubercles of lung, with central caseation (50 diameters). (Koch.)

The identity of this bacillus can be indubitably established by certain modes of staining. No other known microorganism will be affected by Koch's or Ehrlich's mode of staining like the tubercle bacillus. It appears under the microscope as a blue, elongated body of the length of half a red blood-corpuscle, and is found occupying alone or in company with other individuals a giant cell generally located in the center of a fresh tubercle. (Figs. 191, 192, and 193.)

The distribution of the tubercle bacillus is very unequal. It is found in large numbers where the invasion of the disease is recent, or where it is rapidly extending. It is very scanty in chronic affections like glandular scrofulosis or lupus.

The peculiarity of the tubercle bacillus is to incorporate itself with a white blood-corpuscle, and to influence it in such a manner as to convert it into a lymphoid cell of somewhat large proportions. This cell becomes sessile in some part of the body. After a while new lymphoid cells appear in the vicinity of the first cell, which by this time will have grown to the proportions of a multinuclear giant cell, containing a number of bacilli (Fig. 195). As the infection spreads along the periphery, peculiar changes are seen to occur in the center of the nodule composed of lymphoid cells. The nuclei of the lymphoid and giant cells lose their staining capacity and coagulate into a granular mass. The bacilli contained within them disappear, leaving behind, however, a crop of invisible spores that, transferred to a suitable soil, will readily produce a new growth of bacilli.

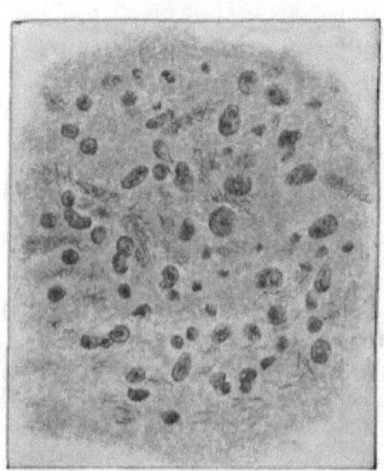

Fig. 192.—Part of one tubercle from foregoing illustration. Bacilli interspersed between nuclei (700 diameters). (Koch.)

With the formation of this coagulated mass of decayed cell-elements the *process of caseation* is established. The presence of this mass of necrosed tissue acts as an irritant upon the capillaries of the vicinity, and a wall of new-formed granulation tissue is thrown up around the focus. Should the infection of the neighboring tissues occur before the protecting wall of new-formed granulation tissue is completed, extensive *caseous infiltration* will be the result.

The barrier of new-formed granulations is also liable, here and there, to invasion by bacilli, and therefore caseation will generally extend in a rather irregular manner.

An increased exudation of blood-serum and white blood-corpuscles will finally bring about *emulsification of the cheesy focus*, which then represents the beginning of a cold abscess.

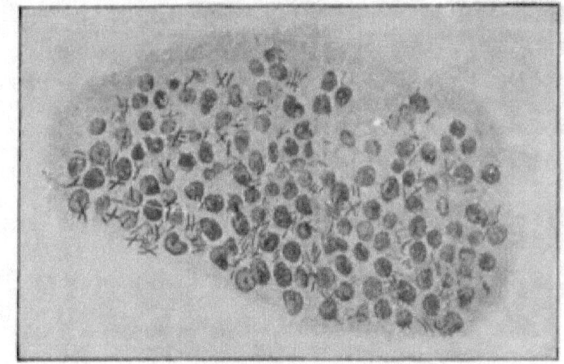

Fig. 193.—Part of miliary tubercle from a case of basilar meningitis (700 diameters). (Koch.)

# ETIOLOGY OF TUBERCULOSIS.

There is no organ of the human body that is exempt from the possibility of tuberculosis.

*The predisposition to infection* by the ubiquitous spores of the bacillus of tuberculosis is manifestly increased by any kind of deterioration of local or general bodily vigor. Malnutrition, whether due to an attack of measles or the whooping-cough, or to a chronic catarrh of the infantile gut caused by improper nursing, or to long-continued suppuration from an osteomyelitic sequestrum, is, as a matter of actual observation, very often followed by local and general tuberculosis.

*The most common way of infection is undoubtedly that by the lungs.* Catarrhal affections of the bronchial mucous membrane, regularly accompanied by superficial denudations of the epithelium, serve as portals for the entrance and implantation of the spores of the bacillus. And, *as the deterioration of the general state of health after measles is combined with a catarrhal condition of the bronchi, infantile tuberculosis is*

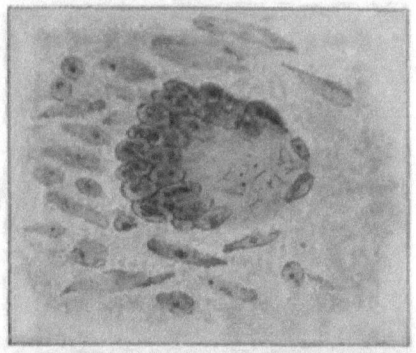

FIG. 194.—Giant cell containing bacilli taken from miliary tubercle (700 diameters). (Koch.)

*most commonly acquired after this eruptive disease.* For unknown reasons the pulmonary tissues of children do rarely become involved in serious tubercular trouble; but the virus is promptly transmitted to the *bronchial lymphatic glands* (Fig. 195), which undergo caseation, and, on account of their close vicinity to the thoracic duct and various vessels, serve as a depot for further distribution.

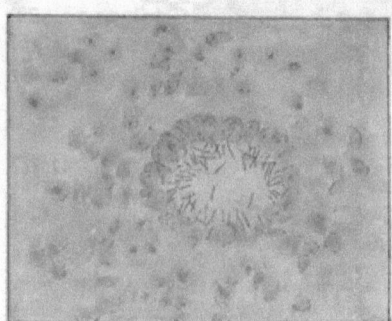

FIG. 195.—Giant cell, with radial arrangement of bacilli, from a caseous bronchial gland (700 diameters). (Koch.)

We owe to Ponfick proof of the fact that perforation of a caseous focus into the thoracic duct may cause a more or less general dissemination of tuberculosis. Koch himself has demonstrated another manner of distribution in the involvement and caseation of arterial walls. But the most common way of systemic tubercular infection was found by Weigert in the decay of the walls and perforation into the lumen of veins, which generally hold very intimate anatomical relations to caseous glandular tumors.

Entrance of small quantities of tubercular virus into the general circulation by the ways above indicated will lead to local tubercular affections of various organs, as, for instance, the bones, testicle, or joints. Massive invasion, on the other hand, will cause fatal general miliary tuberculosis.

*Tubercular matter carried along by the circulating blood is most apt to be arrested and to become sessile in the vicinity of the terminal arteries.* The views expressed in the chapter on the localization of acute infectious osteomyelitis seem to be applicable also to the localization of the tubercular process. (Page 209.)

Another rarer manner of tubercular infection is that by lesions of the skin. A Jewish circumciser suffering from pulmonary and faucial tuberculosis, communicated the disease to twelve infants by sucking their preputial wounds. This used to be the accepted manner of stanching hæmorrhage after ritual circumcision in former times.

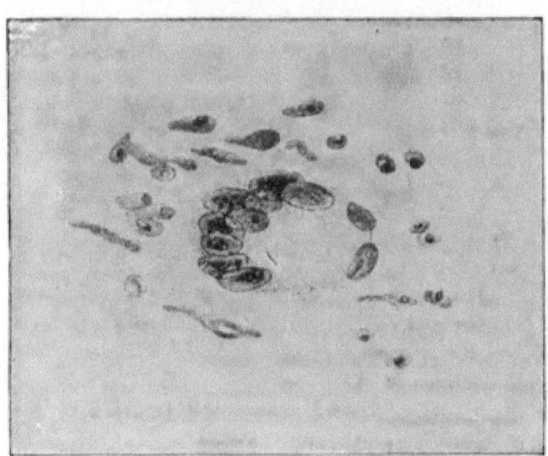

Fig. 196.—Giant cell containing one bacillus from Fig. 191 (700 diameters). (Koch.)

NOTE.—In 1879 the author was the victim of local tuberculosis of the pulp of the thumb, contracted by the infection of a small cut received during the amputation of a thigh for tuberculosis of the knee-joint, complicated with large tubercular abscesses of the thigh and of the medulla of the femur. A caseating elevated ulcer of the thumb developed and persisted for six weeks. The complaint healed after the final detachment and expulsion of two caseous plugs.

*The dissemination of tubercular matter during surgical operations,* done for the cure of the complaint, was first pointed out by *Koenig.*

It is well known that death by general tuberculosis is seen to follow exsection of the hip-joint with especial frequency. Upon this circumstance is based the statistically proved fact that the expectant or rather non-operative treatment of this complaint yields better results than an active operative therapy.

NOTE.—These facts find a ready explanation in the circumstances under which most early exsections of the hip-joint are carried out. The depth of the diseased joint; the difficulty of liberating the head of the femur, still held down firmly by undestroyed ligaments; the desire of operating subperiosteally, that is, with the employment of a good deal of blunt force; the forcible manipulations in distending the edges of the deep wound by retractors—all serve to propel any freed caseous matter into the cut orifices of veins and lymphatics. The result is that, by the time the local tuberculosis combated by the surgeon is healed, the patient succumbs to meningeal or pulmonary tuberculosis, probably chargeable to operative interference.

## II. COMPLICATION OF TUBERCULOSIS WITH PYOGENIC OR SUPPURATIVE INFECTION.

Tubercular decay of tissues by caseation is a generally slow process, as long as the affection remains subcutaneous—that is, occluded from access of air with its pyogenic organisms. But let a tubercular focus of the lung perforate into a bronchus, or let a group of caseous glands, or a cold abscess communicating with a distant focus of the spine or some joint, be opened without aseptic precautions, and the affection will have at once entered upon a new and more destructive phase. The formerly thin, flocculent discharge will assume a more purulent character, the production of pus will become prodigious, more or less fever will set in, and the symptoms of a rapidly progressive local destruction of tissue accompanied by hectic, will become more and more pronounced.

A new infection was thus implanted upon a soil already impoverished by ill-nutrition and preyed upon by a destructive parasite. To the slow decay of tuberculosis, the rapidly disorganizing forces of purulent infection were added. The seriousness of this contingency was justly comprehended by old-time surgeons, who abhorred meddling with a cold abscess or any covert strumous affection. Incision of a cold abscess then meant purulent infection of the cavity, extending to the often inaccessible primary focus of the disease, hectic fever, and rapid emaciation and decay of the patient.

Just appreciation of these remarks will at once impress upon the mind the great necessity of aseptic measures in our operative dealings with tubercular affections.

## III. TREATMENT OF TUBERCULOSIS.

### General Principles.

Considering the fact that about seventy per cent of all deaths are directly or indirectly caused by tuberculosis of various organs, principally consumption, and that the management of the infectious sputa of consumptives is careless in the extreme, it must be admitted that efforts at prevention offer no great hope of success. The sputa containing active bacilli or their spores are ejected on the ground or floor, dry there, and are converted into dust, which will penetrate everywhere and will cover everything with its deadly burden. The tent of the Indian and the palace of the millionaire are penetrated alike by dust containing dried and pulverized sputa of consumptives, and millions of spores of pyogenic cocci, derived from suppurating wounds, the discharges of which are carelessly thrown every day upon the ground, to be whirled up from there by draughts of air.

A more promising line of prevention can be cultivated in the proper nourishment and *régime* of the individual. The better the general condition of health, the fuller and more abundant the blood supply of this or that organ, the less the chance of its becoming the seat of tuberculosis. Or, if passing conditions of anæmia caused by illness or loss of blood have led to the establishment of a tubercular focus, raising of the general health by proper diet and exercise in the pure air of the sea or of high mountains, will check and often wholly eliminate the ravages of the disease. *A generous diet, with plenty of exercise in the open air, is the best preventive and systemic curative of tuberculosis. To the observance of scrupulous cleanliness in the household and in our personal habits must also be acceded a great protective, and in some measure a curative influence.*

### Local Treatment of Tuberculosis.

Knowledge of the true nature of the various forms of surgical tuberculosis has led to a clear understanding of the principles governing its successful treatment. Since we do not possess any therapeutic agent capable of destroying the bacillus of tuberculosis *in situ*, without interfering with the tissues that harbor it, chemical and mechanical influences must be brought to bear upon the tuberculous focus, with the object of destroying and removing all cell elements infested with the specific virus. In short, *the modern treatment of local tuberculosis is identical with that accepted for the cure of malignant new growths; it consists in a more or less complete removal of the affected tissues or organs by caustics, the knife, or the gouge, under aseptic precautions.*

1. **Cutaneous Tuberculosis.** Lupus (Fig. 197).—Various chemical caustics, the actual cautery, and excision are known to effect a cure of cutaneous tuberculosis. Internal medication has no effect upon it. The most destructive forms of lupus are those representing a complication of tuberculosis with pyogenic infection—as, for instance, *lupus exedens*. The miliary nodes nearest the surface caseate, break down, and perforate, and the way is open for the entrance of pus-generating cocci. Lupus of the face should be treated by caustics and scooping. The more radical treatment by excision is not to be commended in facial lupus on account of the disfigurement it is apt to cause. Relapses are frequent, and should be attacked over and over again as soon as they appear. Lupus of non-exposed parts of the skin should be exsected. The following case demonstrates the identity of lupus and tuberculosis:

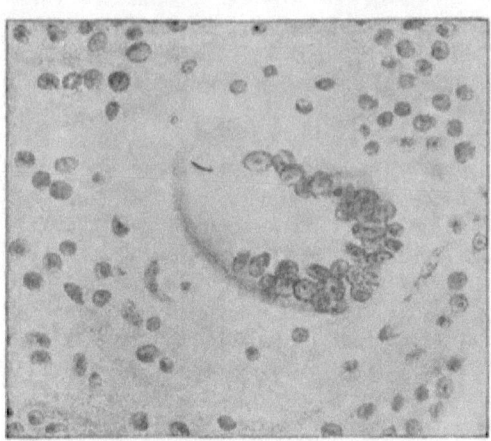

Fig. 197.—Section of lupous skin. Giant cell containing one bacillus (700 diameters). (Koch.)

CASE.—Otto Krim, aged five. Lupus exedens over the left external malleolus of the size of a silver dollar. The affection existed for nearly three years; about a year ago glandular swelling appeared in Scarpa's triangle of the left side and in the corresponding groin. Extensive scrofulous ulceration of the skin followed, and caseous glands lay exposed in the bottom of the inguinal wound. *February 4, 1887.*—Extirpation of the lupous patch and of the glandular masses from Scarpa's triangle and above Poupart's ligament. The peritonæum was exposed, and had to be stripped up to the external iliac vessels to permit complete removal of the glands. Primary union of the wounds about Poupart's ligament. The malleolar wound healed under a Schede dressing. *February 27th.*—Patient discharged cured.

2. **Tuberculosis of the Mucous Membranes.** — Scrofulous rhinitis, or coryza, is a very rebellious affection of the nasal mucous membrane. It is easily recognized by the chronic swelling of the mucous covering of the nasal cavity, the swollen upper lip, open mouth, hard hearing, and noisy breathing. Its surgical importance lies in its tendency to produce an early affection of the cervical lymphatic glands—scrofula. Ulcerative destruction of the mucous covering of the nasal bones opens the way for the ingress of pyogenic organisms, which bring about frequently more or less extensive necrosis. An intensely fetid odor makes the breath of these patients intolerable. Termination of this condition is best accomplished by removal of the necrosed bones in Rose's dependent position of the head (see Fig. 170). The sequestra are easily dislodged by the sharp spoon. The hæmorrhage is at first rather profuse, but soon subsides on irrigation with icewater. Daily instillation of the nasal cavity with a mild solution of corrosive sublimate (1 : 5,000) should be used until discharges cease to appear.

*Tuberculosis of the anal mucous membrane* is a most frequent cause of tuberculous *fistula in ano*. Simple slitting up of these fistulous tracks, lined with caseous granulations, and often dotted with miliary tubercle, will not accomplish their cure. Every nook and recess of the fistula must be carefully explored, and all caseous or granular matter must be removed by vigorous scooping and, if need be, excision. A thorough-going operation will always be followed by improvement, and in not too extensive cases by local cure.

*Tuberculosis of the urethra and bladder* is a most distressing complaint, and is hardly amenable to any form of treatment. Sedatives and, in cases where the affection of the neck of the bladder renders life intolerable on account of the unceasing painful strangury, median perineal cystotomy, followed by drainage, are indicated.

A common sequel of urethral tuberculosis is *caseous epididymitis* and *orchitis*. Testicular tuberculosis caused by urethral disease is generally bilateral. Single tuberculosis of the testicle, on the other hand, is generally of embolic origin. Its sovereign remedy is castration.

3. **Tuberculosis of Lymphatic Glands, or Scrofula** (Fig. 198).—Caseous chronic lymphadenitis is one of the most common affections of childhood and adolescence. Its foundations are generally laid by chronic affections of the oral, nasal, and aural mucous membranes, by tubercular affections of the cervical vertebræ, and by lupus and eczema of the face and scalp. The incipient stages

of the trouble can sometimes be controlled by timely attention to the causal disorders, an appropriate general treatment, and the local application of one or another preparation containing iodine in the shape of an ointment.

As soon as caseation has been well established, general and topical treatment of the milder sort will be of no avail.

The modern therapy of scrofulous lymphatic glands is dominated by the idea that they are not only the cause of present discomfort and suffering to the patient, but especially that within them is contained the seed for renewed infection, which by its dissemination through the circulation may cause other local affections or a fatal general malady. The close anatomical relation of most lymphatic glands to important venous trunks or their immediate affluents renders their early attachment by inflammatory deposit very easy. Cheesy degeneration will ultimately reach the wall of the vein itself, and dissemination of the tubercular virus through the circulation is the result.

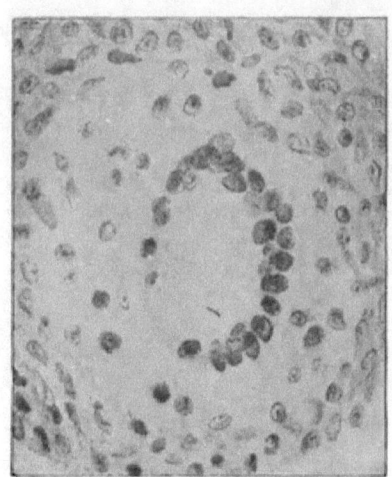

Fig. 198.—Giant cell containing one bacillus from a scrofulous gland of the neck (700 diameters). (Koch.)

The surgical therapy of cheesy lymphadenitis will have to be varied according to the stage of the disease, the chief object being always thorough removal or destruction of all infected tissues.

Where there is *central caseation only, and no fistula*, nor an appreciable abscess, *bodily excision of the glandular masses is most appropriate*. The neck being the most common seat of the trouble, a few words may be said regarding the detail of the operative treatment of scrofulous cervical glands.

The incision should be ample, and, if the tumors be very extensive, the formation of a flap is advisable. The capsule of the uppermost gland being split, the glandular body is shelled out of its nest. This is much facilitated by an assistant's holding aside the detached capsule with a small, sharp retractor while the surgeon suitably changes the position of the mass by turning it one way, then another, until all the looser attachments are divided. Great care must be exercised herein not to lacerate or crush the brittle substance of the gland.

Each gland has its afferent and efferent vessels, and these form a sort of pedicle, which must be tied off before it is cut.

In cases of very extensive involvement of the cervical glands situated both in the vascular and intermuscular interspaces (see page 208), it is very advisable *to cut the sterno-mastoid muscle across and in two*. The spinal accessory nerve will be found near its posterior margin, and should be saved.

The stumps of the divided sterno-mastoid muscle are raised from their proximal attachments, and one is turned up, the other is turned down. The otherwise difficult and even dangerous dissection of the glands from the vicinity of the large vessels is made much easier by the free exposure afforded by cutting the sterno-mastoid, which should be reunited by a number of catgut stitches after the completion of the exsection.

The manner of placing the drainage-tubes, the suture, and dressings, do not differ from the usual arrangement. Before closing the wound, a thorough mopping out with a strong solution (1 : 500) of corrosive sublimate is necessary, to make sure of destroying all spores of tubercle bacilli that may have escaped with cheesy matter from accidentally injured glands.

When dealing with *progressed central cheesy abscesses* of the cervical glands, a different course must be pursued. Incision of each abscess, followed by a thorough scooping away of all granulations and broken-down glandular tissue, is the proper treatment. *The sharp spoon can and should be used rather vigorously*, and no fear need be felt of injuring large vessels lying close by the walls of the abscesses, as there is a tough and thick wall of organized connective tissue interposed to protect them. A drainage-tube is to be inserted into each cavity.

*Caseous abscesses that have perforated spontaneously*, or have been opened inadequately, generally lead to tubercular infection of the subcutaneous tissue in the vicinity of the aperture. More or less extensive *undermining and bluish discoloration of the skin* are the consequence. The undermined, irregular edges show very little tendency to heal; they become inverted, and if healed, present an ill-shapen, uneven scar.

To aid and hasten the inadequate efforts of Nature, it is necessary to extirpate or gouge out the glandular bodies, to trim away all the undermined portions of skin with the curved scissors, *paying no regard to the extent of the resulting wound*. However large the denudation, it will heal rapidly and kindly under Schede's dressing, and, on account of the mobility and abundance of the cervical integument, the resulting cicatrix will be nearly linear in shape.

NOTE.—Glandular, cheesy abscesses on the necks of grown girls can be healed, without leaving a conspicuous scar, by repeated punctures with a stout aspirating-needle. The contents of the abscess being removed by aspiration, corrosive-sublimate lotion is injected through the cannula, and is again withdrawn. This is repeated until the lotion returns clear and limpid, when the cannula is taken out. The puncture-hole is protected by a drop of iodoformed collodion. The process is repeated whenever the abscess refills, until the cavity becomes closed. The author has cured two cases in this manner.

4. **Tuberculosis of Tendinous Sheaths.**—Weeping sinew or acute synovitis of the tendinous sheaths sometimes degenerates into a chronic affection of their synovial lining known under the name of *proliferating hygroma*. This rebellious affection is characterized by an elongated, fluctuating, irregular swelling of the carpal region. It is painless, but impedes the free use of the fingers. The swelling is due to a gelatinous thickening of the sheaths of the sinews. The tendons finally become adherent to the degen-

erated mass, thus losing their free mobility. The sacs frequently contain some more or less discolored synovia, and sometimes a large number of rice-kernel-shaped concretions of fibrin.

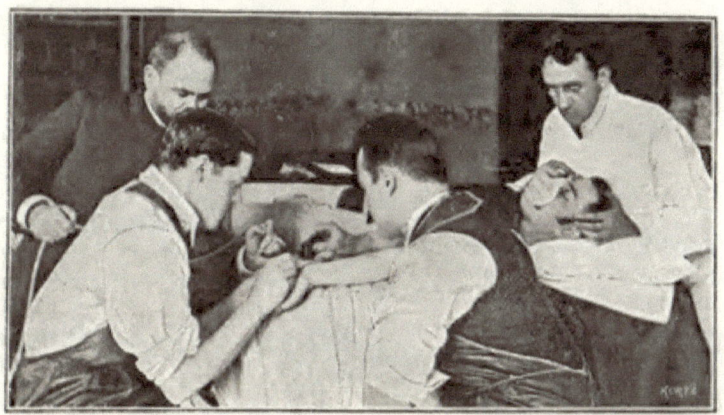

FIG. 199.—Group illustrating an exsection of tubercular tendinous sheaths of the palm.

Topical applications make no impression upon this disorder, which can be cured only by free incision and methodical removal of the fibrinous bodies and the gelatinous sheaths by careful dissection in artificial anæmia. If the new growth extend underneath the transverse carpal ligament, and can not be got at otherwise, the ligament must be divided to permit thorough removal. The carpal ligament, fascia, and skin are united by several tiers of catgut sutures, a slit is left open at each end of the incision, and a compressive Schede's dressing is applied to the arm and hand, which should be placed on a volar splint extending to the line of the metacarpo-phalangeal joints. *The patient is directed to actively move his fingers from the second day on*, and thus to fashion grooves in the blood-clot filling the interior of the wound, which are to become new tendinous sheaths after the substitution of the clot by new-formed connective tissue. (Figs. 199 and 200.)

FIG. 200.—Lines of incision on palmar and dorsal aspects of the hand for tendineal tuberculosis. (Case of Samuel H.)

CASE I.—Samuel H., medical student, aged twenty-five. Tubercular gelatinous synovitis of all extensors of right hand and of flexors of left hand. *December 30, 1886*.—Extirpation of diseased sheaths of extensor tendons of right hand under Esmarch at Mount Sinai Hospital. *January 12th*.—First change of dressings; primary union. By January 20, 1887, normal function re-established. *January 27th.*—Similar **treatment of flexor sheaths of left hand.** Double ligature and division of superficial palmar **arch; division of carpal ligament. Suture** of carpal ligament, fascia, and skin. *February 13th.*—**First change of dressings;** primary union. *March 15th.*—Function of **flexors normal.**

CASE II.—Mina Scheller, aged twenty-five. Tuberculous synovitis of **extensor tendons of both hands.** *March 26, 1886.*—Operation of right hand at **Mount Sinai Hospital. Primary union.** *April 6th.*—Operation of left hand; **primary union.** *January, 1887.*—**Function of** both hands perfect.

5. **Tuberculosis of Bone. Caries. Cold Abscess.**—Bone tuberculosis may **appear** in two ways: On one hand, it is either an independent affection of **the** shaft of a long bone, preferably in the vicinity of an epiphyseal line, or it is a deposit in the epiphysis itself, which by extension and perforation into the joint may cause tubercular arthritis; on the other hand, tubercular involvement of the bone may be caused in tubercular arthritis of the synovial type by **ulceration of** the cartilage **and** direct infection of the exposed bone. No bone **is wholly exempt** from tuberculosis. **The** skull, the spine, **the sternum, ribs** and scapula, **the pelvis, and the bones of** the extremities are all liable **to** infection.

The characteristic features **of idiopathic bone tuberculosis are thickening, the cheesy deposit, and, later on, ulcerative processes,** against which the exuberant production **of feeble and deciduous granulations conducts an** uneven and unsuccessful struggle. **In their turn the granulations also become infected and succumb to cheesy degeneration, and thus the process goes on interminably. Sequestra** of large size, as in acute osteomyelitis, **are never produced; but** the granulations contain **smaller or larger rudiments of dead bone, and a** good **deal** of **bony grit is to be felt in the** secretions.

*Cold abscesses* represent **the** accumulated result of cheesy degeneration and emulsification. They **travel** by well-known routes, and the surgeon is generally **able to** conclude from the place of their external appearance where their **source** is to be looked for.

Cold **abscesses contain** an enormous mass of infectious matter. **They** are a drain **upon the patient's** health, and should be therefore always evacuated. *Evacuation can be done in several ways, but it must under all circumstances be done with strict aseptic precautions.* The observance of asepticism **is of especial** importance **where the focus of** the disease **is** inaccessible, as **for instance** in Pott's disease.

NOTE.—Evacuation **by** puncture **with a** well-disinfected trocar, with subsequent injection of a solution of five parts of iodoform in one hundred parts of ether, was proposed by Verneuil, and has been found very effective by various surgeons, including the author. The injected ether evaporates in and distends the abscess cavity. Thus the iodoform enters every nook and corner of the irregular **hollow, where it exerts** the undeniably favorable influence of all iodides upon **the tuberculous process.** Undoubtedly, abscess cavities thus treated **fill** up much slower than after simple evacuation. Where the osteal process has reached its termination, they do not re**fill at all.** From one to two ounces of **the solution are to be** used, **and, after** thorough disten**tion** and gentle kneading for the sake of even distribution, the remnant should be permitted **to escape through** the cannula.

*Cold abscesses situated in the vicinity of accessible foci*, as, for instance, near the ribs, scapula, or about the extremities, can be treated much more radically. They should be incised to their full extent, and their pyogenic membrane and cheesy contents should be scraped away until bleeding, healthy tissue is reached. After this, the fistula leading from the abscess to the bone is searched, and the exact location of the diseased bone is ascertained.

The treatment of the affection of the bone consists in free exposure and thorough removal of all portions that are manifestly in a state of ulceration or cheesy degeneration. The foci are made accessible by a free use of the chisel and mallet. The sharp spoon and gouge must clean out the last vestige of granulating or cheesy tissue, until the bone presents a healthy and fresh surface. Finally, the external wound is closed by suture, due regard being paid to drainage, and the parts are dressed aseptically. Thus primary union of the entire wound may be accomplished.

The following example may serve as an illustration:

CASE.—Herman Mehle, barber, aged twenty-nine. Large cold abscess of interscapular space of dorsum, extending under the left scapula. *January 6, 1885.*—Incision, evacuation, and scraping of the cavity. A sinus leading toward the transverse processes of the second and third thoracic vertebræ was followed up by incision, and led to a number of small sequestra belonging to the heads of the second and third ribs. They were removed by gouging, and the abscess was closed by suture. Relapse of the cicatrices required renewed scrapings. *March 18th.*—Patient was discharged cured.

*Revision*—that is, exploration and supplementary removal of overlooked tuberculous masses by gouging and scraping—is a very necessary and perfectly harmless measure, that should be employed within three or four weeks after the primary operation, in case the remaining sinuses show no tendency to heal. *The appearance of exuberant ulcerating granulations about the orifices of the drainage-holes should be looked upon as an urgent indication for revision.* Anæsthesia can be rarely dispensed with on these occasions.

*Tuberculous foci in the vicinity of a joint are a great menace to its soundness.* Early detection and timely evacuation will have the character of a truly conservative step. The diagnosis of a single and central cheesy focus of a long bone is not easy to make; but the lymphatic habit of the patient, the local swelling of the bone, with elevation of the local temperature and distinct spontaneous and pressure pain, may serve as valuable guides to its correct ascertainment. Slight stiffness of the joint nearest to the focus in the morning, with a hardly noticeable limp, which becomes more marked toward night, are significant warnings portending the gradual breaking down of the remnant of bone-tissue serving as a barrier against the invasion of the joint.

Where cheesy foci are suspected in the vicinity of a joint, probatory incision and exploration are justified.

In cases where the increasing swelling of the bone, a cold abscess, or the presence of sinuses with fever admit no doubt regarding the nature of the trouble, free incision and exposure by chisel and mallet must be practiced, followed by a painstaking removal of all degenerated tissues, sequestra, and cheesy deposits. The subsequent treatment of these wounds is identical with that advised after necrotomy for osteomyelitic sequestra.

### 6. Tuberculosis of Joints. White Swelling:

#### *General Part.*

Typical tuberculous arthritis, caused by perforation of an epiphyseal cheesy focus into the joint, or by an independent infection of the synovial membrane from a distant focus (bronchial glands) by way of the general circulation, is popularly known as *white swelling*. Mild cases of children, treated by an invigorating regimen and proper orthopedic measures, will yield very good results without serious operative interference.

Even when "starting pains" indicate loss of the cartilaginous covering and caries of the joint surfaces, a cure by anchylosis or with the preservation of more or less mobility is possible. Small or great periarticular abscesses, incised and drained under aseptic cautelæ, will heal kindly, and the ingrafting of the more intense purulent infection upon tissues whose power of resistance has been lowered by tuberculosis and disuse, will be avoided. A careless incision, or a spontaneous perforation, on the other hand, is generally the starting-point of widespread destruction, caused by suppurative infection from without. Then, to conserve the limb or life of the patient, the diseased joint must often be sacrificed.

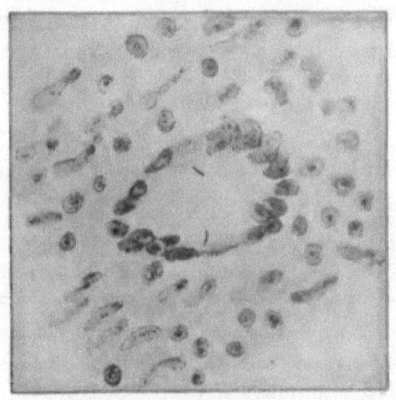

FIG. 201.—Giant cell containing two bacilli from fungoid granulations of the capsule of the hip-joint in morbus coxarius (700 diameters). (Koch.)

*a.* TECHNIQUE OF JOINT EXSECTION.—The technical rules to be observed in excising joints are governed by the following requirements:

(*a*) *Septic infection from without must be excluded* by strict adherence to the rules of asepticism. If a local septic condition, due to purulent infection by uncleanly management of a cold abscess or sinus, be present, this has to be first eliminated by free incision and drainage of burrowing phlegmonous collections and by frequent irrigation. Only after the return of the temperature to nearly the normal standard is exsection permissible.

NOTE.—Phlegmonous inflammation of a tuberculous joint is a much more serious trouble than that of a previous healthy joint. The cavities and sinuses preformed by the tuberculous process serve to disperse the new poison much more rapidly and widely than would otherwise be the case. Hence the formation of perforations and burrows up and downward between the muscles of the extremity occurs much sooner in tuberculosis than happens with a previously normal capsule. The typical mode of incision and drainage of the knee-joint, for instance, will be found insufficient in this contingency, and multiple perforation into the popliteal space will readily occur. Exsection of a knee-joint subject to the ravages of both tuberculosis and intense phlegmon will offer very slender chances of success, and amputation will have to be decided on.

The preservation of asepticism is greatly promoted by almost continuous irrigation of the wound during the time of operation. Corrosive sublimate (1 : 3,000) can be fearlessly used for any length of time *while Esmarch's constrictor is in situ,* as no absorption is thus possible (Woelfler). *In exsections done without artificial anæmia, very weak solutions of corrosive sublimate (1 : 5,000) or Thiersch's lotion should be employed.* At the conclusion of the operation, however, the wound should be well flushed with stronger (1 : 1,000) corrosive-sublimate solution.

(*b*) *Removal of all parts, soft or osseous, that are manifestly diseased,* whether carious, cheesy, gelatinous, or granulating, is a most important condition of success. On the other hand, no apparently healthy parts ought to be needlessly sacrificed.

NOTE.—Without antiseptics *partial excisions* of joints were much more dangerous than total ones. The reason of this was the fact that after total excision the conditions for effective drainage were much better than after partial exsections. Suppuration of resection wounds was the rule then, and is now the exception, hence partial excisions are just as safe at present as total ones.

*To prevent further dissemination of the tubercular virus from the site of the operation, ample incisions must be made.* They will enable the surgeon to reach every part of the diseased joint without the employment of undue force by retractors.

Diseased bones are removed by the saw in adults; in children, they can be pared off with a strong scalpel. Pockets filled with caseous matter are scooped out with the sharp spoon. *The entire capsule must be removed by dissection with curved scissors and a mouse-tooth forceps.*

(*c*) *To control hæmorrhage,* artificial anæmia should be used during the operation wherever possible. Where, as in the shoulder- and hip-joints, Esmarch's band can not be well applied, each vessel must be secured and tied as soon as it is exposed or cut.

*Artificial anæmia may be kept up till the dressings are completed;* but care must be taken to search out and tie every cut vessel before closing the wound. How to do this is described in the paragraph on artificial anæmia in amputations (page 69).

(*d*) *Preservation of the usefulness of the limb, or of the function of the exsected joint,* is the last, but not least, requirement to be fulfilled.

The knee- and occasionally the hip-joint will, as a rule, be more useful if firmly anchylosed than otherwise. Mobility of the other joints, however limited, is more desirable than anchylosis.

To favor anchylosis, the sawed surfaces of the bones to be united must be brought and kept in firm apposition by posture, suture or nails, and a contentive dressing.

*Where preservation of mobility is aimed at, the periosteal covering of the exsected bones must be preserved by subperiosteal dissection.* The periosteum can be stripped off easily with an elevator or Sayre's "oyster-knife," except at the site of the insertion of muscles, where the **aid of the scalpel** or a sharp raspatory must be accepted. The re-formation of the normal contour and function of the prospective **joint** depends **in a great measure** upon the preservation of the periosteum.

With drainage by rubber tubes, an exact suture of **the** external wound, and Schede's modification of the aseptic dry dressing, the operation is completed. Where Esmarch's constricting band was left *in situ* until the completion of the dressings, these must be made rather ample, and a good deal of elastic pressure by snug bandaging must be brought to bear upon the wound **to** control oozing and soiling of the dressings. The dressed limb must **be suspended** or otherwise elevated in a vertical position until **the** hyperæmia due to vascular paresis disappears. Care must be taken **to ascertain**, by **the** look of the tips of the toes or fingers, **that circulation is not** wholly cut off by strangulating compression of the bandage.

Should the oozings **penetrate the** dressing in the course of a few hours, the soiled surface of the **bandage must** be thickly dusted with iodoform powder to favor exsiccation. A **few** compresses of sublimated gauze are placed over the bloody spots, and are secured by a few turns of **a** roller bandage.

**In case** of continued oozing, **further** loss of blood **can** be checked by the temporary application of a Martin's elastic **bandage over** the dressings. If the soiling is too extensive **to admit the use of such** partial measures as those just indicated, the external **compresses composing** the dressing must be removed and replaced by clean **ones**. *The deepest part of the dressing, however, should not be disturbed.*

*b.* AFTER-TREATMENT.—Where, **as for instance, in the elbow,** mobility of the joint is aimed at, absolute fixation by splint should **continue only so** long as the drainage-tubes are withdrawn and **the** incisions **are firmly** healed. *Passive,* but especially *early passive motions,* so warmly recommended by older authors, *are harmful,* and not to be compared as regards their value with *active exercises.*

**The** disadvantages of early passive motions can **be** summed up in this: Before the re-establishment of the normal condition of **the** tissues pertaining to an exsected joint—that is, before **the** disappearance of the swelling and rigidity **of** the soft parts—all **motions,** active and passive, will be painful. Active motions will be **limited** to a harmless compass by the pain forbidding extensive movements; **but** passive motions, done without regard to the pain and struggles of the resisting patient, will be, **and as a** matter of fact often are, **carried far** beyond the limit of harmlessness. The forcible stretching and crushing together of the newly united parts and of the young connective tissue are inevitably followed by minute ruptures and lacerations. Renewed exudation **and a** diffuse state of **adhesive** inflammation are set up, which will cause the persistence **or even an increase** of the painful **swelling and** induration primarily found about **the exsected** joint. The greater the surgeon's energy the worse the **result,** and in many **cases anchylosis is** brought on by the very measures intended to prevent it.

If the surgeon, on the other hand, patiently awaits the time of spontaneous detumescence, which, with antiseptic measures and proper fixation, will occur at about the fourth or fifth week after the operation, gentle motions will cause no pain, and will encourage the patient to active exercise of the joint. The pain felt on excessive movement will serve as a wholesome check against undue zeal; the improvement of nutrition due to active exercise will hasten the definitive involution of the inflammatory products. Thus, day by day will the strength and amplitude of the active movements be increased, and by dint of painless attrition new articular surfaces will be ground and polished into shape. The psychological and moral part of the after-treatment is of the greatest importance here. *The conviction that active movements of the exsected joint are possible without pain will inspire the patient with courage.* Unceasing active exertion will work wonders, based upon the patient's confident expectation of a good final result.

The acute pain produced by frequent and merciless passive motion, and the subsequent tenderness engendered by it, will convert the after-treatment to a source of constant terror and moral depression to the patient. His courage will be shattered, and no amount of persuasion or coercion will induce him to inflict pain upon himself by active movements. And it will be a lucky circumstance if the physician's ill-conceived attempts at establishing a normal function are frustrated at an early date by the patient's resistance. Subsequently, rest and the disappearance of local pain will naturally elicit first timid, later bolder, attempts at active movement, and after all, an unexpectedly good function may thus result.

The aid afforded to Nature should be very discreet indeed, here as well as in other branches of surgery.

Aside from active movements, *massage and faradism* are powerful aids in re-establishing normal circulation and lost muscular power.

### Special Part.

*a.* SHOULDER - JOINT. —The application of artificial anæmia in exsection of the shoulder-joint is always difficult and sometimes entirely impracticable. After due cleansing and disinfection of the field of operation, the hand and forearm of the affected limb are enveloped in a clean towel wrung out of mercuric lotion (Fig. 202), and, the rest of the body being well protected by rubber sheets and clean towels, an ample anterior incision is carried from midway between the acromion and

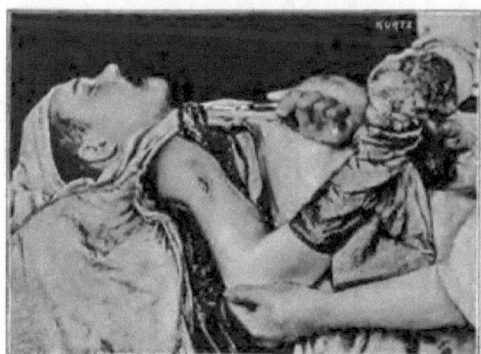

FIG. 202.—Exsection of shoulder-joint. Head of humerus turned out of glenoid cavity.

## TREATMENT OF TUBERCULOSIS.

the coracoid process down to the limit of the upper third of the humerus. The tendon of the long head of the biceps is held aside by a blunt hook. The capsular ligament and periosteum are raised from the bone by means of an elevator, or, where the insertions of the muscles offer greater resistance, by a sharp raspatory. This step will be very much facilitated by gradual inward and later by outward rotation of the humerus, to be done by an assistant holding the hand and bent elbow. *After decapitation of the humerus, the capsule is to be exsected by forceps and blunt scissors.* This, the most difficult part of the operation, will be very easy if the primary incision is ample. If found diseased, the glenoid fossa is thoroughly scraped, and,

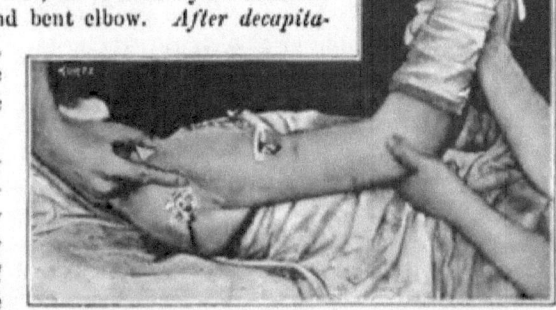

FIG. 203.—Exsection of shoulder-joint. Location of drainage on the posterior aspect of the shoulder.

a counter-incision having been made at the posterior aspect of the joint, a drainage-tube is inserted there. (Fig. 203.) The first incision is closed by several tiers of catgut sutures, and, the wound being dressed, the limb is bandaged to the thorax in a flexed position. Later on, an arm-sling will serve as an adequate support. (Figs. 204 and 205.)

The dressings are changed on the fourth day, when the drainage-tube can also be removed. In grown subjects the operation will generally result in a somewhat loose joint, lacking especially the power of active abduction.

CASE I.—Anna Haupt, aged sixty. Large subdeltoid cold abscess; no fistula. *May 25, 1879.*—Exsection of right shoulder-joint at the German Hospital. Head of humerus bare of cartilage and carious; caries of glenoid cavity. *August 3d.*—Discharged cured.

CASE II.—Willie Kunz, aged four. *January 25, 1882.*—Exsection of left shoulder-joint for cheesy osteitis of the head of humerus at the German Dispensary. *March 10th.*—Discharged cured.

CASE III.—August Arnold, aged three and a half years. *April 17, 1883.*—Exsection of left shoulder-joint for caseous foci in the head of the humerus at the German Hospital. *May 30th.*—Discharged cured.

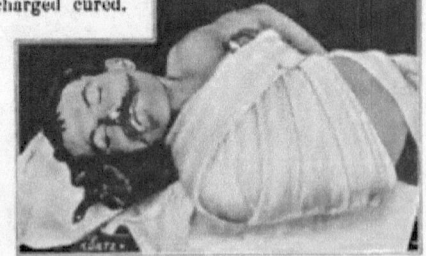

FIG. 204.—First dressing after exsection of shoulder-joint.

CASE IV.—Harry Gross, aged two. *September 30, 1884.*— Exsection of right shoulder-joint for caseous osteitis at Mount Sinai Hospital. Several relapses required renewed scraping of the fungous granulations. *January 15, 1885.*—Patient died of meningeal and peritoneal tuberculosis with ascites.

CASE V.—Carl Buchowsky, type-setter, aged twenty-eight. Synovial tuberculosis of right shoulder-joint of six years' standing; three fistulæ. *April 26, 1887.*—Exsection of the shoulder-joint at the German Hospital. In May, patient was discharged not cured, with two fistulæ, but with a very fair prospect of an ultimate cure, the cause of his discharge being a disciplinary breach of the rules of the hospital.

CASE VI.—Mr. Robert N., merchant, aged thirty-three. Had been suffering from chronic pulmonary trouble for many years; contracted tuberculous arthritis of right shoulder-joint in 1882, a fistula existing since November, 1889. Excision of shoulder-joint April 16, 1890, by a lateral incision, carried through the middle of deltoid muscle. Though dissection of attachments of the muscles to the tuberosities was difficult, the labor was repaid by the care with which the entire capsule could be excised. Patient was discharged cured, May 20, 1889.

FIG. 205.—Arm-sling. (Esmarch.)

*b.* ELBOW.—The patient's shoulder, hand, and part of his forearm are wrapped in clean towels soaked in corrosive-sublimate lotion. (Fig. 206.) The arm is vertically elevated for a few minutes, and elastic constriction is applied to the humerus below the shoulder. Langenbeck's posterior longitudinal incision will give most space. (Fig. 207.) In denuding the internal epicondyle, injury of the ulnar nerve should be guarded against by closely hugging the bone with the instrument. The diseased portions of the bones being removed, the entire capsular ligament is exsected, care being taken not to overlook any cheesy foci. One or more drainage-tubes are inserted, preferably through pre-existing sinuses, and the incision is closed by catgut sutures. The region of the elbow is enveloped in an ample Schede's dressing, held down by rather tight bandaging. The extended arm is fastened to a pair of lateral pasteboard splints, and is kept in the vertical position till the flushed appearance of the projecting tips of the fingers due to vascular paralysis has disappeared. (Fig. 208.)

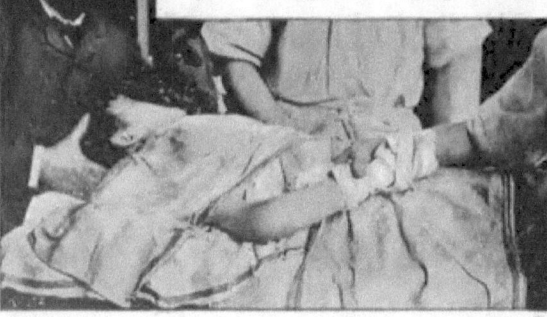

FIG. 206.—Exsection of elbow-joint. Patient ready for operation.

NOTE.—The simplest way of making suitable pasteboard splints is by *tearing* them out of a sheet of pasteboard. (Fig. 209.) The advantage of tearing over cutting is in the circumstance that the edges of the torn splint are not abrupt and hard, but become soft and then on account of the gradual thinning of the torn edge. Snug adaptation and a good fit result therefrom. Care must be taken to ascertain first the trend of the fiber of the pasteboard, as the edge of the splint torn across the direction of the fiber will turn out uneven, and a splint thus made is apt to break.

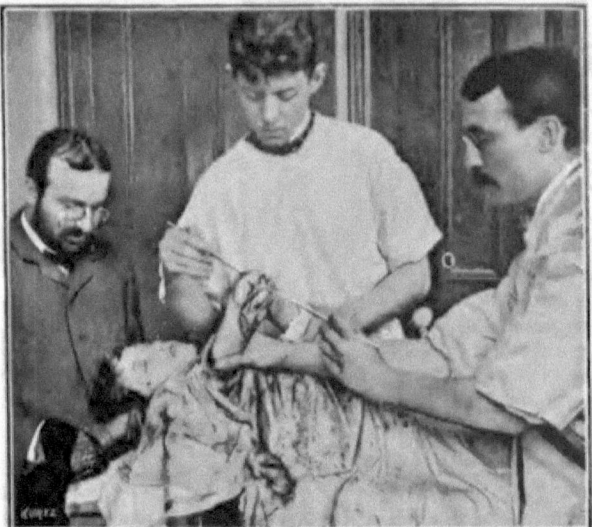

Fig. 207.—Posterior longitudinal incision of elbow-joint.

The dressings should be changed, and the drainage-tubes removed, a fortnight after the exsection. The elbow is to be redressed and put up at the same angle. As soon as the drainage-holes are healed, passive, but *especially active, exercises* should commence, aided by massage and faradism applied to the muscles. After partial exsection of the joint, little lateral mobility will be observed. In these cases no special apparatus will be required. But where much lateral

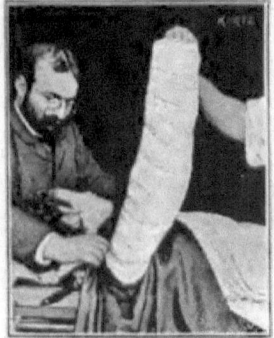

Fig. 208.—Finished dressing and elevation after exsection of elbow-joint.

Fig. 209.—Tearing into shape of pasteboard splint.

mobility, due to extensive removal of bones, is present, the use of an apparatus confining the movements of the joint to flexion and extension will be required. (Figs. 212, 213.)

NOTE.—The apparatus can be made by the surgeon without the aid of the instrument-maker in the following manner: Two strips of very light hoop-iron or sheet-zinc, about one-inch wide and from four to six inches long, are loosely riveted to each other at their ends, so as to form a hinge. Two pairs of such hinges are necessary. The patient's arm being protected by a few turns of a flannel bandage, a light silicate-of-soda wristlet and arm-band (Fig. 212) are applied. To these are fitted the hinges, one externally, the other internally, by giving their middle a suitable bend to allow for the expansion of the soft tissues on flexion of the joint (see front view). By

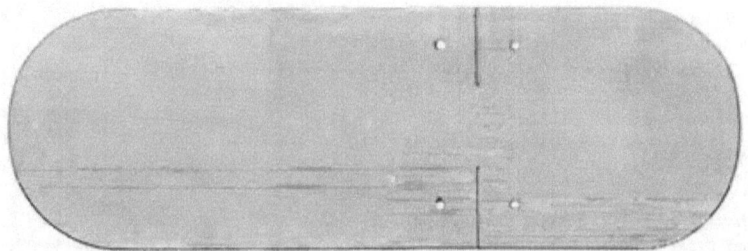

FIG. 210.—Pattern for angular pasteboard splint. (Esmarch.)

a few more turns of the silicate bandage, the hinges will become immured in the wristlet and arm-band. As soon as the splint is dry, it is split longitudinally on its anterior aspect, to permit its removal and further fitting. Shoe eyelets are put in along the edges of the longitudinal cuts for lacing. Two pairs of small-sized brass screw-eyes are let in on each side of the wristlet and arm-band, to serve for the attachment of solid rubber bands, which are to aid the efforts of the flexor muscles in bending the elbow. To prevent slipping down of the apparatus, a cap is made of a piece of sole-leather, softened in hot water, which is molded to the shoulder. It is left on till dry. A button is let into it to serve for suspending from it the apparatus by a short strap. Another strap slipped over this button is passed around the thorax of the patient, and is buckled in the opposite axilla. (Fig. 213.)

Flexion and extension are to be done by the patient at regular intervals from six to eight times a day, by raising first an empty pail from the ground twenty or thirty times. The elbow flexed by the rubber bands is extended by the weight of the pail. As the strength

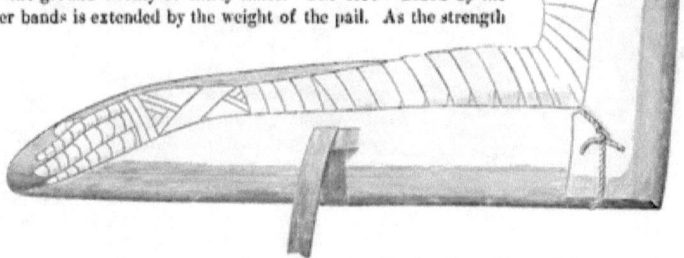

FIG. 211.—Angular pasteboard splint *in situ*. (Esmarch.)

of the flexors improves, active flexion is to be tried, and the weight of the pail is to be gradually increased by putting more and more sand or gravel into it. The apparatus is to be daily removed, for cleansing and the application of massage and faradism to the arm. The use of the apparatus can be abandoned with the disappearance of lateral mobility.

## TREATMENT OF TUBERCULOSIS.

The first of the nine cases of exsection of the elbow-joint performed by the author was done without aseptic precautions. Study of the history of this case and comparison with the other cases is earnestly recommended to the reader.

CASE I.—Joseph Keck, silk-weaver, aged thirty-nine. Synovial tuberculosis of right elbow, with cold abscess situated beneath the supinators; no fistula. *December 10, 1877.*—Total exsection of the joint at the rooms of the patient, without any aseptic precautions. Trochlea, ulna, and radius carious. Drainage, suture, and suspension in an interrupted wire splint. Wound was dressed with a compress, to be kept moist by immersion in tepid water. The thermometer indicated 103° Fahr. on the evening of the same day, and never descended below this figure until December 24th. Frequently the temperature rose to 105° Fahr. *December 13th.*—Wound fetid, inflamed, suppurating; stitches were removed, whereupon the wound gaped open, and was seen to be covered with a thick, adherent coating. *December 15th.*—Great swelling and dusky appearance of cubital region. Incision of abscess near triceps tendon. *December 17th.*—Rigor, elbow still more swollen. *December 18th.*—Rigor. *December 19th.*—Rigor and great debility. *December 22d.*—Rigor. *December 24th.*—Evacuation of another abscess from the upper angle of the wound, whereupon the temperature fell to 99° Fahr., and the dusky swelling of the limb moderated. Apparently the fever was due to osteomyelitis of the lower end of the humerus. *December 25th.*—Erysipelas set in, commencing from an abrasion caused by the splint. Temperature, 105° Fahr. *December 29th.*—Erysipelas extended to shoulder-joint, where it disappeared. *March 10th.*—Incised three abscesses of the forearm, wound granulating and contracting; removal of sequestrum of humerus. *June 14th.*—Removal of six small sequestra from humerus. Active and passive movements commenced. *July 12th.*—Flexion to 90°; extension normal. Sinuses were scraped in anæsthesia. Lateral mobility diminishing. *September 29th.*—Application of articulating apparatus. *October 30th.*—Patient was discharged cured, with normal flexion and extension, with limited pronation and supination, and slight lateral mobility. *May 1887.*—Arm sound and quite useful, in spite of slight lateral mobility.

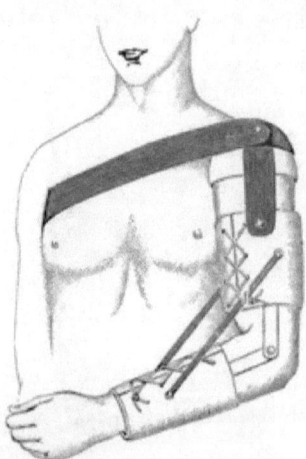

FIG. 213.—Elbow-joint apparatus in position.

FIG. 212.—Apparatus for after-treatment of exsection of elbow-joint.

CASE II.—Hermann Prieg, laborer, aged thirty-eight. *November 15, 1880.*—Total exsection of elbow-joint at the German Hospital for synovial fungous disease with fistula, under antiseptic precautions. Feverless course, primary union.

*February 27th.*—The patient was discharged cured, with limited motion and no lateral mobility.

CASE III.—Lena Bois, aged twelve. *March 14, 1882.*—Partial exsection of elbow-joint for caseous ostitis of the olecranon, from which a sequestrum was removed at the German Hospital. *April 30th.*—Discharged cured, with limited motion.

CASE IV.—Theodore Noirot, metal-worker, aged twenty-eight. *March 9, 1882.*—Total exsection of elbow-joint at the German Hospital for osseal tuberculosis of humerus, ulna, and radius. Primary union of the deep parts of the wound. *May 9th.*—Discharged cured with almost perfect function of the new joint.

CASE V.—Leonhard Fäth, aged seven. Cheesy tuberculosis of olecranon. *October 21st.*—Partial excision at Mount Sinai Hospital. *November 10th.*—Discharged cured with limited motion, which improved somewhat in the course of the following six months.

CASE VI.—Luigi Martini. *May 27, 1886.*—Total exsection for osseal tuberculosis of humerus, ulna, and radius at the German Hospital. Primary union. *June 6th.*—Discharged cured with limited motion. Owing to neglect of the parents, who failed to present the boy for after-treatment, the joint became almost entirely stiff.

CASE VII.—Charles Dunninger, aged two and a half. *April 22, 1886.*—Total exsection for extensive osseal tuberculosis at the German Hospital. Primary union and ultimately excellent function. Discharged cured August 1st. The discharge was delayed by the inability of the parents to take care of the child.

CASE VIII.—Nathan Blumenbach, aged seven. Extensive osseal tuberculosis with several abscesses. *February 9, 1886.*—Incision and drainage of the abscesses, followed by severe chill and fever, very likely due to septic infection at the time of the incision. *February 11th.*—Total exsection at the German Hospital, followed by prompt lowering of the temperature from 105° Fahr. to 99° Fahr. Primary union. *March 14th.*—Discharged cured, with good function.

CASE IX.—Rudolph Boenke, aged twelve. Cheesy osteitis of olecranon with abscess. *March 30th.*—Partial excision. A shell of the olecranon adhering to the triceps tendon was preserved. Suture; no drainage-tubes. *April 12th.*—Change of dressings; primary union. Elbow put up at a right angle. *April 14th.*—*Passive motion;* fixation at an acute angle. *Every few days passive motions were done,* and the arm was put up at a different angle. *This led to considerable irritation* and dense œdema of the elbow, compelling cessation of the passive movements. The mistake made in the after-treatment was further emphasized by the detachment and expulsion of the necrosed remnant of the olecranon. Two fistulæ discharging bloody serum remained open. *May 30th.*—The fistulæ were scooped out with the sharp spoon. No improvement following, June 10th, *the wound was reopened in ether anæsthesia.* Gelatinous infiltration of the soft parts surrounding the joint, tuberculosis of the radio-ulnar junction and caries of the resected bone-surfaces were found. Total exsection being performed, the arm was dressed and put up in a splint as usual, *and remained undisturbed for five weeks, after which active exercises were commenced. No passive movements were done at all. By August 1st, active flexion and extension were normal,* and the arm had regained its power almost completely.

*c.* WRIST AND HAND.—Langenbeck's dorsal incision affords the most favorable approach to the radio-carpal as well as especially to the intercarpal and metacarpo-carpal joints. (Fig. 214.) With artificial anæmia a very thorough removal of the diseased bones and capsular ligaments can be done. The wound is drained and closed by catgut sutures, and, being inclosed in an aseptic Schede's dressing, the hand is fastened to a short volar splint of wood, *which should not extend beyond the metacarpo-phalangeal joints.* The patient is directed from the second day on to practice active motions of the fingers. This will achieve two good purposes: First, extreme atrophy of the muscles will be prevented; and secondly, adhesions of the tendons and tendineal anchylosis will be avoided. The active movements, feeble and hardly perceptible at first, will become visibly stronger as the healing progresses, and thus a very acceptable degree of usefulness of the hand may be regained.

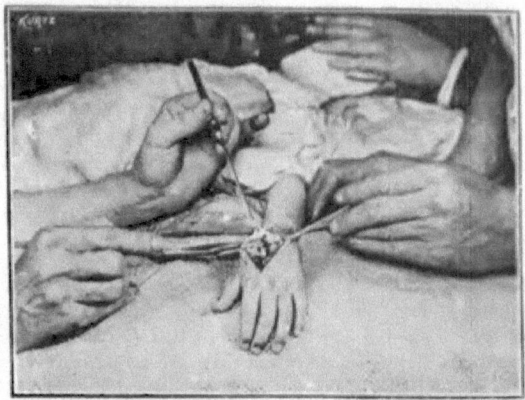

Fig. 214.—Langenbeck's dorsal incision for exsection of wrist.

CASE I.—Herman Rosengarden, clerk, aged thirty-four. *June 7, 1882.*—Total exsection of wrist at Mount Sinai Hospital for synovial tuberculosis with several fistulæ. Primary union. *August 7th.*—Discharged cured. When leaving, he played on an accordion.

CASE II.—A woman, aged thirty-eight. *August 25, 1885.*—Total exsection of left wrist at the German Hospital. Primary union. *September 30th.*—Discharged cured, with moderate function.

CASE III.—Matthew Dempsey, laborer, aged twenty. *June 22, 1885.*—Total exsection of wrist for osseal tuberculosis of carpal bones at Mount Sinai Hospital. Primary union and very fair function were secured. The discharge of the patient was delayed till the end of the year by several pulmonary hæmorrhages.

CASE IV.—Paul Klein, laborer, aged forty-one. *February 25, 1886.*—Total exsection of wrist for osseal tuberculosis with several fistulæ at the German Hospital. The patient was suffering from far-gone pulmonary phthisis. Primary union, but speedy relapse of tuberculosis in the interior of the wound and the cicatrix. *April 11th.*—Discharged not cured.

CASE V.—Max Friedmann, aged ten. *April 4th.*—Partial excision of wrist-joint on account of caseous osteitis of styloid process of ulna, with involvement of the radio-ulnar and radio-carpal joints. Primary union. *April 20th.*—Discharged cured, with good function.

CASE VI.—Ferdinand Ohle, aged five and a half. *March 22d.*—Total exsection of left wrist at the German Hospital for osseal tuberculosis. Wound healed by primary union. Patient remained in hospital for treatment of simultaneous tubercular disease of the knee-joint.

*d.* HIP-JOINT.—When in the presence of profusely discharging sinuses far-gone destruction of the hip-joint, especially with complication of the pelvis, is causing much suffering and a steady deterioration of the general health, excision of the joint is clearly indicated. The *modus procedendi* is as follows: The anæsthetized patient is laid upon his side, with the affected hip uppermost, the hip- and knee-joints lightly bent, and a solid cushion interposed between the knees. The body of the patient is carefully protected against wetting and exposure by rubber sheets, and the hip and buttock are shaved, scrubbed, and disinfected. A longitudinal incision, commencing two or two and a half inches above the tip of the trochanter, is carried down to the neck of the femur, until the margin of the acetabulum, the neck, and the trochanter are well exposed. The cartilaginous margin of the socket is nicked, and the soft tissues attached to the neck of the femur are cut away, the knife closely hugging the bone, an assistant aiding this procedure by inward and outward rotation of the thigh. The head of the femur being dislodged by flexion, adduction, and inward rotation, the head, neck, and trochanter are removed by the saw or a strong knife, whereupon the capsule is carefully excised with the aid of scissors and forceps, and the acetabulum and sinuses are thoroughly scraped. Cut vessels are immediately seized and deligated. The wound is well irrigated, then, after insertion of a drainage-tube, is packed with strips of iodoform gauze. A number of silkworm-gut sutures are insert-

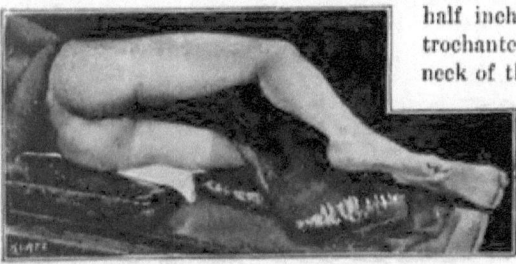

Fig. 215.—Exsection of hip-joint. Position of patient.

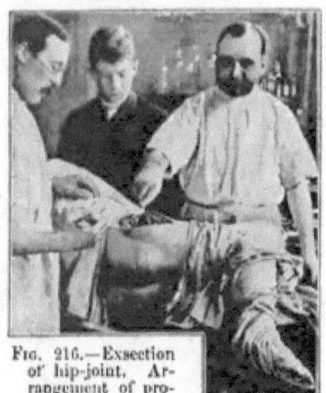

Fig. 216.—Exsection of hip-joint. Arrangement of protective cloths.

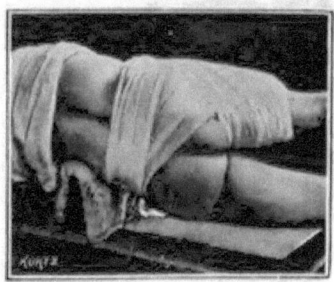

Fig. 217.—Completed dressing after hip-joint exsection.

ed, but remain untied for future use. An ample dressing is applied, and the limb is put in a weight extension apparatus. On the third or fourth day the gauze packing is withdrawn, and the sutures are closed. During the after-treatment the limb is to be kept fully extended and somewhat abducted. Should exuberant granulations appear, they have to be scraped away with the sharp spoon. As soon as the wound is nearly or completely healed, the patient should be permitted to exercise on crutches, and, when his strength permits it, should commence to walk in a Sayre's or Taylor's apparatus. This will prevent stretching of the young cicatrix and displacement of the trochanter upward. The use of the supporting apparatus should not be abandoned too soon.

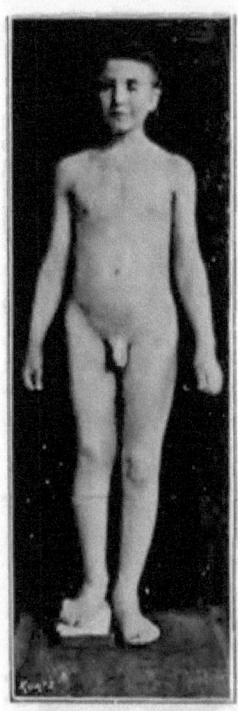

FIG. 218.—Exsection of hip-joint. Final result, Anterior view. (Dr. F. Lange's case.)

CASE I.—Nathan Spiegel, aged six. Coxitis of two years' standing. Under orthopædic treatment, four fistulæ developed. *August 9, 1887.*—Excision of hip-joint at Mount Sinai Hospital. The head of the femur was found necrosed and detached. Acetabulum much eroded. Discharged cured, with movable joint, December 22, 1887.

CASE II.—Sarah Friedman, aged nine. Coxitis with unopened femoral abscess, fever, and acute starting pains, which do not yield to treatment by weight extension. *September 1, 1887.*—Excision of hip-joint at Mount Sinai Hospital. The disease was mainly acetabular, a large cheesy focus containing two sequestra occupying the posterior aspect of the socket. Ligamentum teres detached from acetabulum. Discharged cured, with almost perfect motion and good power, January 15, 1888

CASE III.—Minnie Daly, aged sixteen. Very extensive hip-joint disease, which had been badly neglected while the child was under purely orthopædic treatment in a city institution devoted to that branch of surgery. The entire buttock and the pelvis presented a lamentable picture of a system of ill-drained cavities, the chronic retention having maintained constant fever and produced amyloid intumescence of the liver. The urine was loaded with albumen. In spite of these unpromising facts, the author deemed it his duty to make an attempt at saving life, and accordingly excision of the hip was done January 3, 1888, at Mount Sinai Hospital. The necrosed head of the femur was found floating in a cavity freely communicating with a pelvic abscess. The operation was very rapidly accomplished, and with the loss of very little blood, but the wretched patient did not have enough vitality to overcome the shock, and died, January 4, 1888.

CASE IV.—John Cohn, clerk, aged nineteen. Ununited fracture of the neck of the femur, with incipient tuberculosis of hip-joint. Intracapular fracture was sustained in March, 1888, and was treated with weight extension for two months. On failure of union, walking was attempted, but the pain caused by it was unbearable. *August 6, 1888.*—Exposure and removal of upper fragment. Two bony facets were seen on the surface of contact of the fragments. The acetabulum was filled with tuberculous brittle granulations, round ligament ulcerated at base. Patient was discharged cured, December 18, 1888.

FIG. 219.—Hip-joint exsection. Lateral view. (Case of Dr. F. Lange.)

CASE V.—Jacob Weber, aged eight. Acetabular coxitis, with profusely discharging sinus. *October 24, 1888.*—Excision of hip-joint at German Hospital. Perforation of acetabulum with small pelvic abscess. Discharged cured, April 7, 1889.

CASE VI.—Albert Gaupp, aged thirteen. Anchylosed hip-joint; caseous ostitis of os ilium, with complicated sinuses and pelvic abscess. *August 12, 1882.*—Incision and drainage of various sinuses and of the pelvic abscess; removal of a considerable portion of the ilium and os pubis with mallet and chisel at the German Hospital. *January 21, 1883.*—Discharged much improved.

CASE VII.—Samuel Amster, aged ten. Tubercular coxitis, with sinus, of two years' duration. *December 3, 1885.*—Exsection of hip-joint above the trochanters at Mount Sinai Hospital. Removal of the acetabulum, which was found perforated. After-treatment with weight extension. *January 18 and 26, 1886.*—Revisions of wound, on account of the presence of exuberant granulations in the drainage-tracks. *May 10th.*—Discharged cured. In November the patient was readmitted on account of pelvic disease. A fistula had been established below the anterior-superior spine, leading to the inner aspect of the ilium. *December 15th.*—Three sequestra were removed by an incision made along the crest of the ilium. In June, 1887, the patient was discharged cured.

CASE VIII.—John Renk, aged thirty-nine. Anchylosis of right hip-joint, with shortening of limb, the result of hip disease contracted in childhood, which was treated orthopedically. No fistula. Tuberculous ostitis of ilium and adjoining part of os pubis. *March 17, 1887.*—At the German Hospital, exsection of great trochanter and remnant of neck of thigh as a means to gain access to the diseased focus. An abscess was opened in front of the joint, and, being followed up, led to a number of sequestra located at the juncture of ilium and os pubis, which were removed. The softened and broken-down walls of the cavity containing the sequestra were scraped and gouged. Drainage and suture of the wound. Uneventful course of healing. In August the patient was still under treatment. A sinus persisted at the site of the operation. The discharge was very scanty and serous, however, promising early closure. Anchylosis firm again. Patient walking without support. Cured, October 1st.

The frequency of miliary tuberculosis following excision of the hip-joint was pointed out by Koenig, of Göttingen, who attributed it to an "*operative dissemination*" of the virus. If a small incision is made, and much blunt force is employed by the use of retractors and the elevator, tuberculous material deposited in the lymphatics in the neighborhood of the joint is projected into the general circulation, and may usher in meningeal or general miliary tuberculosis. The impossibility of here applying Esmarch's band offers another plausible explanation. To avoid the dissemination of the tuberculous virus during excision of the hip- (and shoulder-) joint, the following rules should be observed:

A large incision deserves the preference, as by it the diseased parts are freely exposed, and can be attended to without violent manipulation. The removal of the capsule can be rapidly and thoroughly accomplished. The knife, and not the elevator, should be used for stripping off the soft parts from the neck of the femur. Likewise, in removing the capsule the scissors deserve preference over the scoop. In evacuating cheesy pockets located within the os ilium, very sharp-edged scoops are to be used, or preferably a gouge and the mallet. During the chiseling, and especially while scraping, the irrigating stream should continually play over the field of operation, so that all the detached material may at once be washed away. In short, we may say. that, in a limited sense, the same rules apply to the removal of tuberculous as to the eradication of cancerous foci. The surgeon should rely upon the edge of his cutting instruments, which should be carried through apparently healthy tissues; he should, as much as possible, avoid the employment of blunt force as exercised by retractors and the elevator. The fulfillment of these conditions will greatly diminish the danger of operative dissemination as well as of local relapse. And these conditions can be fulfilled only if the site of the disease is laid bare by an ample incision.

*c.* KNEE-JOINT. — White swelling of the knee-joint in adults of the laboring class can, for various external reasons, rarely be treated by orthopedic measures. In children, a rational mechanical and general treatment will often reward the patience and skill of the physician by excellent results. Exsection of the infantile knee-joint is to be avoided as long as possible, on account of the great shortening that is caused by the removal of the epiphyses adjoining the knee, on which depends the growth of the thigh and tibia. In adults exsection is the shortest and safest way of eliminating the tedious morbid process, and substituting firm anchylosis for a useless joint. *Arthrectomy*, or exsection of the capsular ligament alone, as suggested by Volkmann, has not been attended with good success in the experience of the author. Two cases—one in an adult, the other in a child—resulted in relapse of the tubercular affection, although great care was taken in removing the entire capsule. A third case was permanently cured.

CASE I.—S. Lindholm, metal-worker, aged twenty-seven. *February 28, 1882.*—Arthrectomy and removal of the patella were done for fungous arthritis of the knee-joint. Primary union of wound followed. *March 22d.*—A relapse occurred in the cicatrix, which gradually involved the articular aspects of the femur and tibia. Amputation of the thigh was performed by Dr. I. Adler.

CASE II.—Fred. Ohle, aged five and a half. Tubercular arthritis of the knee-joint. *January 26, 1887.*—Arthrectomy was performed at the German Hospital. *March 22d.*—Revision and scraping of the entire cavity on account of tubercular relapse. In May the boy was still under treatment.

CASE III.—George Kuhn, butcher, aged twenty-six. *July 6, 1882.*—Arthrectomy and removal of carious patella was performed at the German Hospital. *November 5th.*—Discharged cured with slight mobility of joint.

In children, exsection should be strictly limited to the removal of actually diseased parts of the bones. By Schede's plan of dressing the wound, the hollow space remaining between the incongruent joint-surfaces will be filled up by an organizing blood-clot, and firm union may be attained.

FIG. 220. Hahn's suprapatellar incision for exsection of knee-joint.

CASE IV.—Eva Greenburg, aged eight. Osseal tuberculosis of the knee-joint with sequestrum in the external condyle; granular ostitis of the internal condyle; multiple cheesy deposits in the thickened capsule; subluxation backward of the tibia with rectangular contraction. *August 12, 1886.*—Partial exsection of knee-joint at Mount Sinai Hospital. After the removal of the sequestrum, a deep recess was left behind in the intercondylar notch. Patella and entire capsule were removed; the ham-string tendons were divided to prevent recontraction. The tibia was superficially pared, and the bones were held in apposition by a nail driven diagonally through femur and tibia. Plaster-of-Paris splint over a Schede's dressing. Several relapses in the popliteal space required repeated scrapings. The patient had one attack of erysipelas. By reason of these complications, cure was delayed. *February 27, 1887.*—Patient was discharged cured with firm anchylosis.

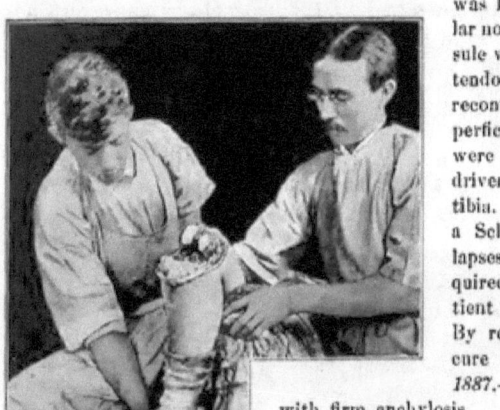

Fig. 221.—Exsection of knee-joint. Exposure of articular planes.

Total exsection of the knee-joint is usually done by the author in the following manner After careful shaving, scrubbing, and disinfection of the region of the knee, the foot and leg and the thigh of the diseased limb are wrapped in clean towels wrung out of corrosive-sublimate lotion. The limb is held elevated in the vertical position for five minutes to deplete its vessels, and the constricting elastic band is applied well up near the root of the thigh. The knee is flexed, and an incision, commencing at the middle of one condyle of the femur, and extending in a semicircular line *above the patella* to the middle of the other condyle, is carried into the joint. (Fig. 220.)

NOTE.—The transverse incision above the patella, proposed by Eugene Hahn, of Berlin, has many advantages over the incision made below the knee-pan. The chief one is the free access it affords to the bursa of the quadriceps, which must be carefully exsected along with the capsule.

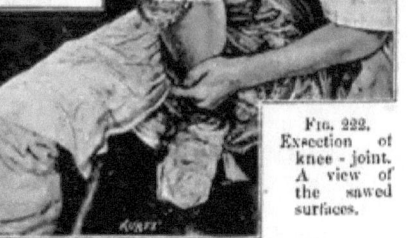

Fig. 222. Exsection of knee-joint. A view of the sawed surfaces.

The crucial ligaments are cut close to their attachment to the femur, and the patella, semilunar cartilages, and entire capsule, together with the bursa of the quadriceps, are exsected with mouse-tooth forceps and curved scissors. Care must be taken not to overlook some small bursæ situated behind the head of the tibia, which regularly communicate with the interior of the joint.

The condyles of the femur are sawed off, the plane of section corresponding to the transverse diameter of the epiphysis of the femur. (Fig. 222.)

NOTE.—Disregard of this rule will lead to anchylosis in the bow-leg position.

FIG. 223.—Steel nail.

The articular aspect of the tibia is sawed off at a right angle to the long axis of this bone. All visible orifices of vessels are secured by ligature. They can be made visible by compressing the vicinity of the wound with both hands.

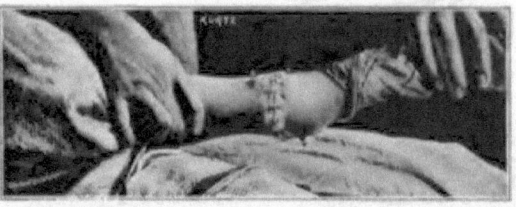

FIG. 224.—Exsection of knee-joint. Sutured wound. Anterior view.

If the transverse incision was not made long enough to permit of an easy arrangement of the drainage-tubes in the angles of the wound, it should be sufficiently lengthened. The inner ends of the tubes should reach into the popliteal space just behind the sawed surfaces, and the tubes must not be compressed and occluded by the tension of the soft parts surrounding them.

The limb is placed upon a long cushion covered with a clean towel wrung out of corrosive-sublimate lotion, and, while the sawed surfaces are held in exact apposition, two or four long steel nails, previously well disinfected by heating in an alcohol flame, are driven diagonally through femur and tibia, so as to firmly lock the bones in the desired position. (See Fig. 79, page 87.) The cutaneous incision is united by a sufficient number of catgut stitches. The limb is raised by the foot from the cushion, which is

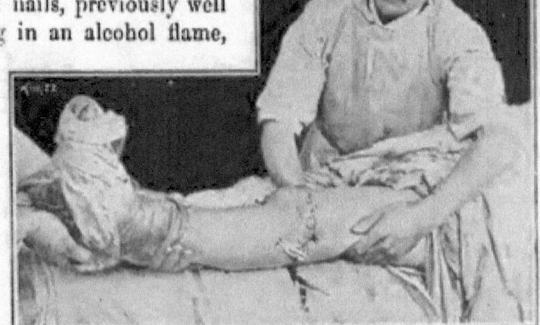

FIG. 225.—Exsection of knee-joint. Sutured wound. Lateral view. Heads of steel nails projecting from skin.

then removed. Strips of disinfected rubber tissue are slipped under the safety-pins, securing the ends of the trimmed drainage-tubes, and an oblong compress of iodoformed gauze is laid over the entire line of union. A suitable number of sublimated gauze compresses are arranged around the knee-joint, and two short lateral splints of veneer or thin board are firmly bandaged on to serve as a deep support. (Figs. 226 and 227.) Over these

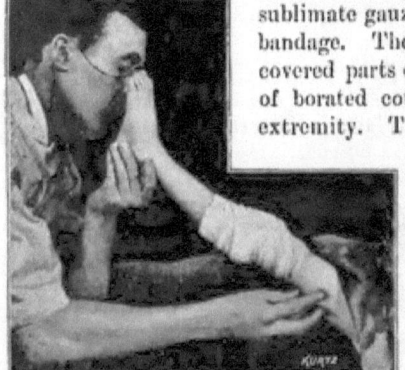

Fig. 226.—Immediate dressing of wound after exsection of knee-joint.

comes an ample external dressing of corrosive-sublimate gauze, also firmly held down by a gauze bandage. The towels are removed, and the uncovered parts of the limb are enveloped in a layer of borated cotton to equalize the outline of the extremity. Two long, lateral, pasteboard splints, held down by a muslin or crinoline bandage, complete the dressing for children or adolescents. (Fig. 228.) The more voluminous limbs of adults are better secured by a solid circular plaster-of-Paris splint.

The limb is vertically elevated, and the constricting rubber band is removed. Return of circulation is attested by the pink color of the toes. As soon as these turn pale, the extremity can be brought into the horizontal position.

If asepticism was well maintained, little aseptic fever and no severe pain will follow the operation. The dressings should remain undisturbed for thirty days, to afford a good chance for bony union. After thirty days the splints and dressings can be removed, and the nails and drainage-tubes can be withdrawn. The remaining sinuses are to be dressed lightly, the limb is incased in a silicate-of-soda splint, and the patient is ordered to walk about on crutches, whether osseous union be present or not. Gradually the use of crutches is dispensed with, and the patients generally learn to walk very well on an elevated sole, compensating the shortening.

Of twenty-one cases of total exsection done by the author for tuberculosis, twenty recovered. One died of meningeal tuberculosis.

CASE I.—Fred. Fuchs, aged seven. Osseal relapsing tuberculosis after arthrectomy, done by Dr. F. Lange in June, 1885. *March 4, 1884.*—Total exsection, done at the German Hospital, reveals two periarticular abscesses and five cheesy foci in tibia and femur. Suppuration of wound. *March 10th.*—Incision of abscess on outer aspect of

Fig. 227.—Deep support of exsected knee-joint by short lateral board splints.

knee. *April 23d.*—Separation of epiphysis of tibia. Separated epiphysis firmly united to femur. In April symptoms of meningeal tuberculosis developed, to which patient succumbed May 31st.

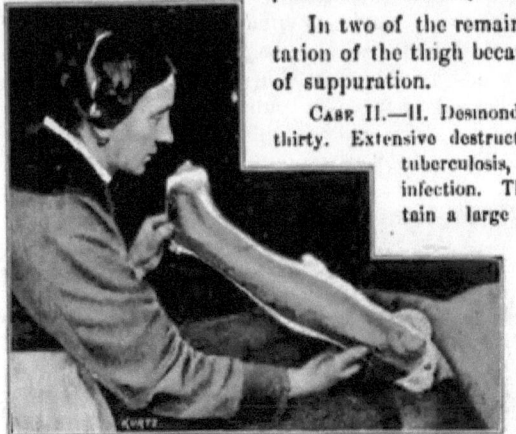

Fig. 228.—External long lateral pasteboard splints after exsection of knee-joint, applied over complete dressing.

In two of the remaining twenty cases amputation of the thigh became necessary on account of suppuration.

CASE II.—H. Desmond, professional athlete, aged thirty. Extensive destruction of right knee-joint by tuberculosis, complicated with pyogenic infection. The knee, leg, and thigh contain a large number of abscesses. Profuse secretion from seven fistulæ. The case was not suitable for exsection, and amputation was advised. But, at the patient's urgent request to make an attempt to save his limb, February 14, 1884, total exsection was done at the German Hospital. As suppuration was expected, the extremity was fixed to an interrupted dorsal suspension splint made of hoop-iron and plaster bandages. Profuse suppuration followed with evident prostration, and, April 19th, amputation of the thigh was performed. The wound healed by granulation, and in June patient was discharged cured.

CASE III.—Johanna Rose, aged thirty-nine. Far-gone destruction of knee-joint, five fistulæ, continued fever. As the patient would not consent to amputation, excision was done October 16, 1888, at the German Hospital, with little hope of success. The fever subsided, however, on account of the better drainage, but reappearance of the tubercular process prevented union. An amputation was done by Dr. J. Adler in February, 1889. The patient was cured.

Eighteen patients were cured, with preservation of the limb. In seventeen of these, firm bony anchylosis was secured. One case terminated in the formation of ligamentous union.

CASE I.—Niclas Gies, carpenter, aged fifty-four. Synovial tuberculosis with high temperatures and emaciation following a slight traumatism. Contraction of knee at an acute angle, with constant violent pain. *February 19, 1886.*—At the German Hospital, puncture yielded a small quantity of turbid bloody serum. In anæsthesia the limb was straightened, and the joint was incised, irrigated, and drained. The fever at once disappeared, but flocculent pus commenced to exude from the tubes, confirming the assumption of tuberculosis. In view of the patient's age, his wretched general condition, due partly to disease and to chronic alcoholism, amputation was thought to be advisable. The plan of operation was changed at the operating-table, and total exsection of the knee-joint was done. Hæmorrhagic synovitis and a large cheesy deposit in the bursa of the quadriceps were found. Five nails were employed, with an aseptic dressing and pasteboard splints. Temporary compression by Martin's elastic bandage was applied to control secondary oozing. Esmarch's constrictor was removed after the completion of the bandage. A feverless course of healing followed. Change of dressings was done on the twenty-second day. Four nails were found loose, and were withdrawn. *May 8th.*—Scraping of drainage-tracks and removal of fifth nail. Ligamentous union was found and a plaster splint applied. *June 12th.*—The sinuses were healed, and the patient was walking, without the aid of stick or crutches, in a light silicate-of-soda splint, though union of the bones was not perfect.

The other sixteen cases were in brief as follows:

CASE II.—Willie Bohn, aged three and a half. Osseal tuberculosis with fistulæ. *February 2, 1879.*—Total exsection. *April 2d.*—Patient discharged cured.

CASE III.—Charles Harris, aged twelve. Osseal tuberculosis with fistulæ; contracture and subluxation backward. *June 13, 1884.*—Total exsection at the German Hospital. Hahn's incision; two nails; plaster-of-Plaster splint. Some fever and deep-seated œdema of the region of the knee followed. Sawed surfaces and flesh-wound united by primary union. The nails being withdrawn on the twelfth day, some pus exuded from their tracks, showing that the nails had not been well disinfected. Several revisions were required on account of unhealthy granulations in the drainage-holes. *February 4, 1885.*—Patient discharged, with firm anchylosis and no fistula.

CASE IV.—Sussel Baerenknopf, aged nine. Osseal tuberculosis; several fistulæ; subluxation. *August 26, 1885.*—Total exsection at Mount Sinai Hospital. Nails; plaster splint. *September 25th.*—Change of dressing. Drainage-tubes and nails were withdrawn; firm anchylosis. *October 10th.*—Patient discharged cured.

CASE V.—Leonard Peters, waiter, aged nineteen. Synovial tuberculosis; no fistula. *August 27, 1885.*—Total exsection at the German Hospital. *September 27th.*—Plaster splint, dressings, drainage-tubes, and nails removed. *October 9th.*—Sinuses healed. *October 19th.*—Discharged cured, with firm anchylosis.

CASE VI.—Bertha Deutsch, aged twelve. Synovial tuberculosis of five weeks' standing. Continuous high fever, with rapid emaciation. Probatory puncture yielded scanty bloody serum. *January 21, 1886.*—Total exsection at Mount Sinai Hospital. The capsule was found studded with innumerable miliary tubercles. The fever disappeared immediately after the operation. *February 20th.*—Plaster splint removed; wound healed by first intention. *March 10th.*—Patient discharged cured, with firm anchylosis.

CASE VII.—Lizzie Boettger, aged twenty. Osseal tuberculosis of eighteen years' standing; rectangular contraction with subluxation backward. No fistula. *February 12, 1886.*—Total exsection at the German Hospital. *March 10th.*—Change of dressings; primary union; three nails and drainage-tubes were removed. *April 4th.*—Patient complained of a good deal of pain in walking. A hard body could be felt under the skin on the outer aspect of the tibia. An incision exposed the head of the fourth nail, which had not been found at the first change of dressings. It was withdrawn with some force, a little blood exuding from its track. *May 9th.*—Patient was discharged cured.

CASE VIII.—Anna Sauer, aged twenty-two. Synovial tuberculosis with osseal ulceration of articular surfaces of both femur and tibia. No fistula. *May 10, 1886.*—Total exsection at the German Hospital. *June 12th.*—First change of dressings; primary union of soft parts; delayed union of the bones. *August 1st.*—Discharged cured, with firm anchylosis.

CASE IX.—Katie Walter, aged eighteen. Synovial tuberculosis with caseous deposits in several recesses of the capsule, notably around and behind the crucial ligaments. Caries of articular surfaces. No fistula. *May 18, 1886.*—Total exsection at the German Hospital. Slight fever following the operation, the dressings were removed May 26th. Marginal slough of the upper edge of the skin wound. *June 17th.*—Nails were removed; firm anchylosis. *July 26th.*—Patient discharged cured.

CASE X.—Emma Friedmann, aged twenty-seven. Synovial tuberculosis with caries of articular surfaces. No fistula. *April 18, 1887.*—Total exsection. *April 22d.*—Considerable secondary oozing necessitated a change of external dressings and plaster splint. Feverless course. *May 23d.*—Change of dressings; primary union; firm anchylosis. Tubes and three nails were removed; a fourth nail could not be found, but was removed by incision on June 2d. Patient was discharged cured, with firm anchylosis, July 1st.

CASE XI.—Hilda Mildenbach, aged thirty-two. *October 25, 1887.*—Total excision at Mount Sinai Hospital. Discharged cured, January 19, 1888.

CASE XII.—Ernst Marquandt, musician, aged thirty-three. Tuberculosis of left testicle and right knee-joint, with fistula and closed abscess of calf. *March 20, 1888.*—Castration at Mount Sinai Hospital. *April 27th.*—Excision of knee. Discharged cured. *June 24, 1888.*—The patient, who also suffered from a chronic lung affection, presented himself, in April, 1890, to the author, in a most flourishing state of health.

CASE XIII.—Solomon Weil, butcher, aged fifty. Excision of knee for tuberculosis and subluxation of forty-eight years' standing, November 9, 1888, at Mount Sinai Hospital. Discharged cured, December 23, 1888.

CASE XIV.—Herman Guentner, engraver, aged twenty-two. Excision of knee, February 8, 1889, at the German Hospital. Discharged cured, April 28, 1889.

CASE XV.—Nicolas Straub, stableman, aged fifty-six. Excision of knee, February 19, 1889, at the German Hospital. Discharged cured, April 13, 1889.

CASE XVI.—William Weinert, cigarmaker, aged forty. Excision of knee, February 22, 1889, at Mount Sinai Hospital. Discharged cured, April 5, 1889.

CASE XVII.—Solomon Chabelsky, tailor, aged twenty-seven. Excision of knee, April 25, 1889, at Mount Sinai Hospital. Discharged cured, June 16, 1889.

CASE XVIII.—Mamie Simon, school-girl, aged thirteen. Anchylosis of tuberculous knee-joint in subluxation and flexion at right angle. Excision, November 8, 1889.—Discharged cured, December 16, 1889.

NOTE.—To prevent the disagreeable necessity of cutting down for searching out a nail buried in the tissues, Dr. F. Lange's suggestion of fastening a silk ligature to the head of each nail before driving it in seems to be very appropriate.

*f.* ANKLE AND FOOT.—Tuberculous affections of the ankle-joint, or of the joints formed by the tarsal and metatarsal bones, require, in case of the presence of one or more sinuses, exsection of the diseased parts. The long-continued discharges and lack of active exercise are very apt to reduce the general condition of the patient to serious anæmia and marasm, and, the disease extending to most of the complicated structures of the foot, may finally require amputation.

Early operations, especially in children, yield good functional results, as the extent of the removal can be limited to the parts actually involved.

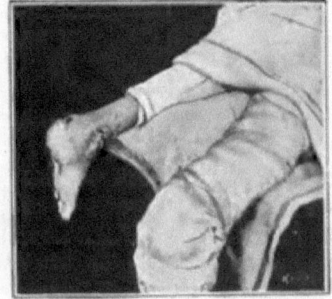

FIG. 229.—Arrangement of patient for Mikulicz's operation.

Exsections of the ankle or of other joints of the foot are not followed by good results in grown subjects, on account of the technical difficulty of a complete removal of the synovial membrane. Relapse of the tubercular process often supervenes, making amputation a necessity.

In tuberculosis of the calcaneum or the astragalo-calcaneal joint, *Mikulicz's osteoplastic exsection of the tarsus deserves employment.* The lower ends of the tibia and fibula are sawed off as in Syme's amputation, and the articular surfaces of the cuboid and scaphoid bones are also sawed off, so as to fit the section of the tibia and fibula. (Fig. 230.) Nutrition of the anterior part of the foot is maintained by the *dorsalis pedis artery*, and the patient soon learns to walk on the balls of the toes, as in *pes equinus*. (Fig. 231.)

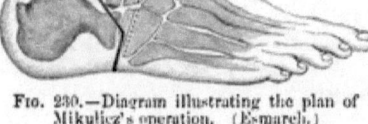

Fig. 230.—Diagram illustrating the plan of Mikulicz's operation. (Esmarch.)

CASE.—Hermann Mehle, barber, aged thirty-four. Synovial tuberculosis of the astragalo-calcaneal joint, with several fistulæ situated to the right and left of the tendo Achillis. *August 20, 1885.*—Osteoplastic exsection of tarsus at the German Hospital. Primary union of the deep parts of the wound and of the bones. Marginal sloughing of limited extent of the upper edge of the wound delayed the cure somewhat. *October 10th.*—Patient was discharged cured.

NOTE.—This operation was employed by the author successfully in two more cases. In one, an epithelioma of the calcaneal region; in the other, extensive chronic ulceration, due to frost-bite of the heel, was the indication to its performance.

The preparation of the foot to be operated on is of very great importance, and thorough removal of effete epidermis and dirt is a necessary condition of asepticism (see page 64). In exsection of the ankle, the bilateral incision gives very good access to the ankle-joint, though excision of the capsule will be found, at best, difficult to accomplish.

It being desirable to produce a movable joint, subperiosteal dissection is to be aimed at, as in exsection of the elbow. As soon as the sinuses are healed, active use of the foot on crutches, aided by a shoe and brace, or a silicate-of-soda splint, should be encouraged. The tendency to posterior or lateral deviation of the foot will be best met by the long-continued use of a supporting apparatus of one kind or another.

CASE I.—Caecilia Raab, aged twenty-two. Synovial tuberculosis of ankle-joint with several sinuses. *November 9, 1882.*—Exsection of ankle-joint at the German Hospital. Healing

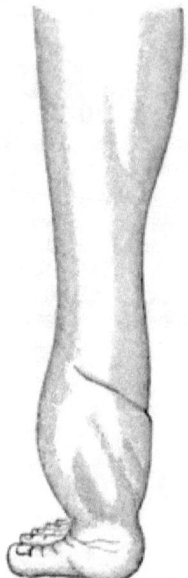

Fig. 231.—Shape of foot after Mikulicz's operation. (Esmarch.)

of the wound progressed favorably, when, November 30th, the patient contracted acute lobar pneumonia, in consequence of which she died December 2, 1882.

CASE II.—George Eitt, aged six. Tuberculosis of ankle-joint caused by a cheesy focus in the astragalus. *January 11, 1883.*—Partial exsection of ankle-joint, part of the astragalus and the malleoli being removed. *March 13th.*—Scraping of the sinuses on account of relapsing tuberculosis. Sinuses persisted until the summer of 1884, when Dr. F. Lange, then on duty at the German Hospital, performed total exsection, which resulted in a cure of the tuberculosis, but with pseudarthrosis. *July 20, 1885.* —The author exsected the ligamentous mass interposed between the lower aspect of the tibia and fibula and the calcaneum, and fixed the latter to the tibia by a steel nail driven through from the planta pedis. Primary adhesion followed, with the formation of a slightly movable union of the tibia and calcaneum. *September 5th.*—The boy was discharged cured. In January, 1886, the brace worn until then was dispensed with.

CASE III.—Henry Holzfaller, aged four. Osseal tuberculosis of ankle-joint. *March 20, 1883.*—Total exsection at the German Hospital. *May 25th.*—Patient discharged cured with serviceable joint.

CASE IV.—Frida Schmoltz, aged three and a half. Osseal tuberculosis of ankle-joint with fistula. *September 19, 1883.*—Removal of external malleolus and part of astragalus, which contained a caseous deposit. *October 15th.*—Wound completely healed. Plaster-of-Paris splint applied. *October 31st.*—Silicate-of-soda splint applied, and patient directed to use the foot. *August 4, 1885.*—Normal position of foot; function perfectly re-established.

CASE V.—I. S., aged eight. Osseal tuberculosis of ankle-joint with three sinuses. *September 26, 1883.*—Partial exsection of ankle-joint; astragalus and inner malleolus were removed. *November 15th.*—Patient discharged cured, with improving function and normal position of the foot.

CASE VI.—Jacob Deibel, farmer, aged twenty-three. Synovial tuberculosis of ankle and of astragalo-calcaneal joints. *March 12, 1886.*—Removal of both malleoli and of entire astragalus at the German Hospital. *April 20th.*—Patient discharged cured, with fair function of the foot, walking with the aid of a stick.

CASE VII.—Abraham Moses Goldenberg, aged four. Osseal tuberculosis of ankle-joint and sinuses. *November 8, 1886.*—Total exsection. Several relapses required repeated scraping with a sharp spoon. *June 3, 1887.*—The patient was discharged cured.

CASE VIII.—Lizzie Holzhauer, aged fourteen. Osseal tuberculosis of ankle-joint with sinus. Excision of astragalus, March 16, 1889, at the German Hospital. Discharged cured, June 5, 1889.

CASE IX.—Mollie Marks, aged two. Very far-gone osseal tuberculosis of ankle-joint. Total excision at the German Hospital, October 27, 1887. Discharged cured, December 18, 1887.

PART IV.

# GONORRHŒA:

## ITS ANTISEPTIC TREATMENT.

## CHAPTER IX.

### NATURAL HISTORY AND TREATMENT OF GONORRHŒA.

#### I. ETIOLOGY OF GONORRHŒA. GONOCOCCUS.

In examining the purulent secretion produced by a virulent case of urethral gonorrhœa, the observer will detect with the microscope a number of dark, round objects resembling grains of fine gunpowder, that are vividly oscillating, and can be clearly distinguished from the adjacent pus-corpuscles. The use of a stronger lens will reveal the fact that each individual coccus is divided in two unequal halves. If staining is employed, the body of the coccus will appear colored, and the dividing-line will become very conspicuous in the shape of a light, colorless streak. (Fig. 233.)

Fig. 232. Pure culture of gonococcus (700 diameters). (From Bumm.)

Frequently an indication of incipient secondary division of each half of the coccus can be seen. Thus four cocci will be united to a seemingly single body, which can be aptly compared with four coherent biscuits, divided into equal quarters by two cross-shaped grooves.

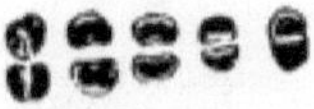

Fig. 233. Development and fission of gonococcus. (From Bumm.)

*The favorite location* of the gonococci found in the urethral secretions *is within the pus-corpuscles.* This peculiarity belongs exclusively to the coccus of gonorrhœa detected by Neisser in 1879, and represents its most important characteristic. (Fig. 234.)

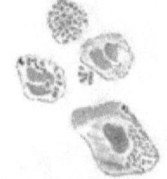

Fig. 234. — Epithelial cell studded with gonococci; pus cell, its protoplasm filled with gonococci; another pus cell gorged with gonococci; a group of free cocci alongside of a normal pus-cell (700 diameters). (From Bumm.)

*Gonococci are to be found in the secretion of every case of gonorrhœa,* provided that no germicidal injections were used.

Infection of the urethra with pus containing gonococci *always produces gonorrhœa,* and secretions that do not contain gonococci are invariably non-infectious if brought upon the urethral mucous membrane.

Gonococci have a peculiarly *invasive faculty,* by which they penetrate first the superficial layers of the epithelial membrane, and gradually by further proliferation the submucous layer. (Fig. 236.) The route of their inroads is along the intercellular substance. An intense hyperæmia of the capillaries and other blood-vessels adjoining the seat of the primary infection leads to a massive emigration of white blood-corpuscles into the affected epithelium. This and the growth of the gonococcal colonies lead to a rapid

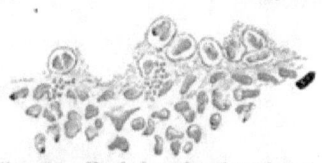

Fig. 235.—Vertical section through mucous membrane, showing first colonization of gonococci (700 diameters). (From Bumm).

disintegration of the epithelium, which is washed away by the lymph-serum in the shape of single cells or in coherent epithelial flakes. Loss of the epithelial investment is often followed by the exudation of a croupous membrane, beneath which clumps of gonococci are to be seen in process of active proliferation. Gonococci can be found occupying at this stage the interstices of the subepithelial tissues, their columns extending inward along the lymphatics, whence, according to various authors (Kammerer), they may be transported to the endocardium, the joints, and the synovial sheaths of tendons.

With the deeper invasion by the gonococci goes *pari passu* the dense infiltration of the infected tissues with leucocytes, the extent of which serves as a gauge of the intensity of the infectious process.

At the acme of the process, generally reached about the

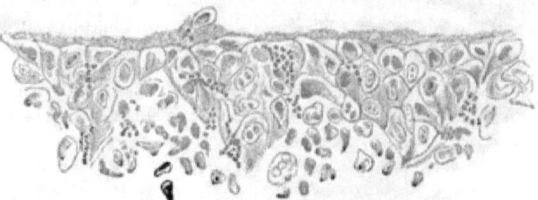

Fig. 236.—Invasion of epithelium by gonococci (700 diameters). (From Bumm.)

end of the second or third week, a regeneration of the lost epithelial layer commences. Complete restitution of the epithelium signalizes the termination of the malady, which, however, is attained only in favorable cases under favorable conditions. Generally primarily unaffected parts of the mucous membrane become involved by spontaneous extension of the infective process, or by the improper use of instruments; or portions which have recovered succumb anew to gonococcal destruction.

The regeneration of the epithelium is always accompanied by hyperplasia, which somewhat resembles by its

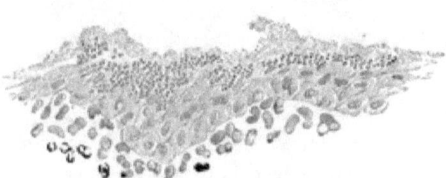

Fig. 237.—Proliferation of gonococci in the epithelium (700 diameters). (From Bumm.)

tubular formations epitheliomatous mucous membrane (Bumm). These foci of epithelial hyperplasia are often coincident with the seat of the most intense primary affection. They also correspond with those parts of the submucous layer at which the most intense inflammatory infiltration was present.

As regeneration progresses, the hyperplasia of the mucous membrane and the infiltration of the submucous connective tissue disappear by absorption. In some cases, however, *cicatricial transformation of the new-formed connective tissue of the submucous layer takes place instead of absorption, and organic stricture develops.*

*The transient hyperplastic conditions existing immediately after the termination of the gonorrhœal process*, and which generally give rise to a scanty secretion called **gleet**, *are mistakenly called strictures* by various authors.

In contradistinction to stricture, which is a *permanent condition*, they must be declared to be *transient stenoses* of the urethral caliber, which in most cases do disappear without or with the methodical introduction of a full-sized bougie or sound. The salutary effect of dilatation upon these coarctations of the epithelial and submucous layers is explained by the hastening of the absorption of the cellular infiltration by pressure.

It is true that, if neglected, some of these coarctations will not be absorbed, but will become veritable cicatricial strictures. Nevertheless, *it is an error to declare each and every narrowing of the urethral caliber observed shortly after a gonorrhœal attack a " stricture of wide caliber."* The term of "incipient stricture" is less objectionable, though often incorrect, as many of these "strictures" disappear spontaneously.

Note.—The presence of various micro-organisms, aside from the gonococcus, in recent and chronic urethral discharges, seems to point to the fact that *most cases of urethritis represent a mixed form of bacterial infection*. There is no doubt that the inoculation of *pyogenic microbes* into a gonorrhœally affected mucous membrane forms an important element determining the intensity and perniciousness of some very bad cases. This assumption is also more in accordance with the theory of the development of metastases, notably of gonorrhœal rheumatism. Bumm is very reserved in regard to the acceptance of Kammerer's investigations, who found gonococci in recent effusions produced during an attack of gonorrhœal rheumatism. On the other hand, we know that rheumatic attacks are occasionally provoked by an instrumental examination of the urethra of a patient afflicted with "simple" or "catarrhal" or "traumatic" urethritis, in which the absence of gonococci is indisputable. Finally, the frequent presence of simple pyogenic organisms in rheumatic effusions is generally accepted. It seems, then, that pus-generating organisms play an important part in cases of gonorrhœic and non gonorrhœic urethritis, and that the metastatic processes complicating urethral inflammations are mostly chargeable *to their and not to the presence of gonococci*. Hence the name "urethral rheumatism" would be preferable to "gonorrhœal rheumatism."

## II. TREATMENT OF GONORRHŒA.

1. **Acute Gonorrhœa. Clap.**—For practical reasons it will be found most convenient to divide the male urethra into two easily distinguished parts.

*The first part comprises the anterior portion* of the urethra, extending from the meatus to the "cut-off muscle," or *compressor urethræ*, which is situated in the membranous portion. All secretions originating in this anterior portion of the urethra will readily escape by the meatus into the linen of the patient.

The *second* or *deep portion* of the urethra consists of a fraction of the membranous part, together with the prostatic portion—in short, of all that is situated *behind* the "cut-off muscle."

This posterior portion of the urethra is correctly called the *neck of the bladder*, as it forms one cavity with the bladder whenever this becomes distended with urine. The internal sphincter alone, unable to resist long, yields readily to the pressure of the urine. The voluntary contraction of the compressor urethræ becomes, then, the only barrier to the escape of the urine, and water is voided immediately after the relaxation of this muscle.

Discharges secreted in the posterior part of the urethra can not escape outward past the compressor muscle, and do not appear at the meatus in the shape of an external discharge, as those of the anterior urethra. They accumulate in the neck of the bladder, and are voided only with the urine, which is rendered somewhat turbid by this admixture.

*A very useful practical test* for determining the seat of urethral inflammation *is that suggested by Ultzmann.*

The patient is made to pass his water consecutively into two tumblers, so that the amount voided should be about evenly distributed in the two vessels. *Whenever the anterior urethra alone is the seat of inflammation, only the first half of the urine will be turbid,* or at least will be found containing flakes and threads ; *the second portion will appear perfectly clear.*

*In cases of deep-seated urethritis—that is, when the neck of the bladder is affected—the first tumbler will receive flaky and turbid urine, and the water held by the second glass will appear also turbid, but somewhat less so than the first portion.*

An additional and most important symptom of the affection of the neck of the bladder is *frequent micturition,* in acute cases accompanied by severe spasm and the escape of a small quantity of blood at the end of the act. Simultaneously with the severe contraction of the vesical muscles, anal tenesmus is observed.

In every case of recent gonorrhœa the infectious process is confined to the anterior urethra, and first to its foremost portion alone. It extends from the meatus backward to the compressor urethræ, where it generally stops. In exceptional cases only does it penetrate to the deep urethra, as the "cut-off muscle" seems to serve as an effective barrier to its extension backward.

NOTE.—Forcible urethral injections made from a syringe containing too large a quantity of fluid, or the premature introduction of a sound, are frequent causes of the infection of the neck of the bladder.

The seat of the most intense inflammation of the urethra is in its naturally widest parts—that is, in the fossa navicularis and the sinus bulbi. Here we find located the majority of all strictures.

*a.* ANTERIOR GONORRHŒAL URETHRITIS.—The treatment of anterior gonorrhœal urethritis should be very discreet in the first invasive stage of the disease. It should consist of rest and appropriate general sedative management. Locally, cold applications will be found very grateful and effective.

## TREATMENT OF GONORRHŒA.

As soon as the turbulent first onset has abated, local treatment by disinfectants should commence. Since the œdematous swelling of the parts is still prominent, introduction of any instrument for the purpose of irrigation will have to be done with some force. It will cause abrasions of the tumid epithelium, and thus will open new portals to gonococcal and pyogenic invasion. Hence irrigation at this period is to be condemned.

Urethral injections, on the other hand, done with a properly shaped syringe of moderate capacity, are very useful. Sigmund's syringe, having a blunt conical nozzle, is an appropriate instrument. It holds three eighths of an ounce of fluid, which quantity is sufficient. (Fig. 238.)

Fig. 238.—Sigmund's urethral syringe.

The strength of the solutions employed should also be determined by the intensity of the local symptoms. Strong solutions will cause intense smarting, and on that account the injections will not be made frequently enough by the patient. In very sensitive cases an entirely unirritant tepid solution of salt water (6:1,000, or a teaspoonful to a quart) can be employed with much benefit. As the symptoms abate, sulphocarbolate of zinc (fifteen grains to six ounces), or permanganate of potash (one grain to six ounces), can be substituted for the saline solution.

The main object of these first injections is the cleansing of the urethra ; *hence the injections must be made frequently,* at least six times in a day, or oftener. Each injection should be preceded by urination, and should be a double one—the first syringeful to wash out the pus ; the second syringeful to act upon the mucous membrane. This second injection should be retained in the urethra for two minutes. The strength of the injections should be increased *pari passu* with the abatement in the acuity of the local symptoms, but the solutions should never be made corrosive.

*Every patient should receive practical instruction from the physician regarding the proper manner of injecting.*

NOTE.—The author saw a case of chronic gonorrhœa that had successively passed through the hands of three colleagues, none of whom convinced himself whether the patient was making the injections properly or not. Phimosis was present, and the patient was in the belief that the injections had to be made under the prepuce. No wonder his clap had remained uninfluenced by this treatment.

In the later stages of acute gonorrhœa *irrigation of the anterior urethra* will be found a very satisfactory and effective mode of treatment. It should be done by the physician himself at least once daily, or as often as possible, in the following manner :

A pint bowl is filled with tepid water. To this is added enough concentrated solution of permanganate of potash to color the water to the hue of light claret. A straight or slightly beaked female catheter of metal (Fig. 239), five inches in length (No. 8 English caliber), is lubricated *with glycerin,* and is introduced as far as the compressor-urethræ muscle. Whenever the beak of the instrument comes in contact with the muscle this will contract, and will resist further introduction. The patient stands in front of the sitting physician, and is made to hold a pus-basin or tin pan under his scrotum and penis. The physician fills with the solution a hand-syringe holding four or five ounces, and injects the fluid through the catheter into the urethra, whence it will readily escape by the meatus into the pus-basin. This is repeated until the solution is exhausted. Irrigation should be preceded by micturition.

FIG. 239.—Short metallic catheter for irrigation of anterior urethra.

With proper diet and *régime*, ordinary cases of gonorrhœa will be cured by this treatment in from three to six weeks.

NOTE.—To prevent soiling of the patient's linen by profuse urethral discharges, the following simple arrangement will be found effective and convenient. A child's sock is fastened with a safety-pin to the interior of the skirt of the patient's undershirt. In the toe of the sock is thrust a small ball of cotton, which is then drawn over the penis, and is held there by the sock. Whenever occasion permits, the soiled cotton is replaced by clean material, and thus no tell-tale blotches will be made on shirt and drawers.

*b.* DEEP-SEATED GONORRHŒAL URETHRITIS.—Spontaneous extension of gonorrhœal infection beyond the cut-off muscle to the posterior part of the urethra is a comparatively rare occurrence. More frequently infection is carried to the deep urethra by too large injections or the premature insertion of sounds. *As long as in a case of anterior gonorrhœa the discharges are profuse and creamy, and the mouth of the urethra œdematous and red, no sound should ever be passed.*

Infection of the deep urethra invariably provokes an unmistakable complex of symptoms—namely, frequent urination, which is followed at its termination by a violent spasmodic pain and the escape of some bloody urine or a few drops of pure blood.

Ordinary injections, or even irrigations of the urethra as above described, are utterly unable to reach and to influence the course of deep-seated gonorrhœa. To cleanse and disinfect the diseased part, an efficient germicidal solution must be brought exactly in contact with the morbid mucous membrane of the *posterior urethra*. If we inject a solution into the bladder, its chemical properties will be at once destroyed by the admixture of urine, hence means must be found by which we can make the unchanged solution come in contact with the seat of the disease. For this purpose *Ultzmann's method of irrigating the neck of the bladder* will be found very effective.

As soon as the most acute invasive stage of the affection shall have become mitigated by rest, sedatives, balsamics, and proper diet—that is, in about the third or fourth week—a quart of a mild, tepid solution of permanganate of potash (1 : 5,000) is prepared. A not too small-sized soft gum (Nélaton's) catheter (Fig. 240) is lubricated with glycerin, and is introduced as far as the compressor-urethræ muscle. A hand-syringe holding about four ounces of fluid is filled with the solution, which is then injected into the catheter, and will be seen escaping from the meatus alongside of the instrument. After this preliminary washing of the anterior urethra, the patient is directed to assume the recumbent posture. The soft catheter is again lubricated, and is passed gently into the bladder. This process will be very much facilitated by the injection of a small quantity of glycerin through the catheter when it is about to pass the cut-off muscle. A small amount of pressure will overcome the tension of the compressor, and the arrival of the point of the instrument in the desired locality can be tested by injecting an ounce or two of the prepared lotion. Should it escape from the urethra, this would be a sign that the eye of the catheter has not passed the com-

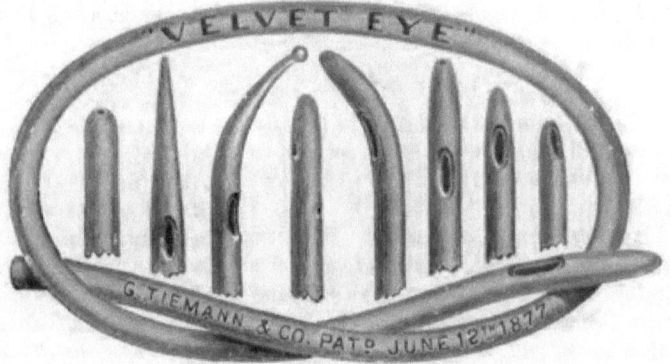

FIG. 240.—Nélaton's soft gum catheter.

pressor muscle. If, on removal of the syringe, the lotion is seen to escape at once from the bladder through the catheter, then it may be concluded that the eye of the catheter is in the cavity of the bladder, and that it has been introduced too far, and needs to be withdrawn an inch or a little more or less. *Should, on renewed injection, the lotion all enter the bladder, but fail to escape through the catheter, this is a positive sign that the beak of the instrument is just beyond the cut-off muscle—that is, in the posterior part of the membranous portion.* Fluids injected into this place will readily enter the bladder, as their pressure can easily overcome the internal sphincter; but recontraction of this muscle will prevent their escape until the beak of the instrument is pushed into the vesical cavity. According to the irritability of the patient, from one to four ounces of the lotion are slowly injected while the point of the catheter is located in the space between the cut-off and internal sphincter muscles. As soon as the patient complains of pressure, injection should cease, and the catheter should be gently pushed within the vesical cavity, whence it will at once conduct the injected fluid into a vessel placed between the thighs of the patient. It is better not to inject too large a quantity at the beginning, as this is liable to bring on vesical spasm, resulting in a violent and irresistible expulsion both of lotion and catheter.

The injections are to be repeated in this manner until the lotion is seen to return clear from the bladder. The final injection is voluntarily passed by the patient. This is to satisfy him that his bladder is empty, and that the sensation of the desire to urinate is not caused by retained fluid.

The improvement following this procedure is very apparent, though not lasting, and daily repetition will be necessary until the frequency of micturition will have been very materially reduced.

The author has never seen any untoward consequences following this gentle and very efficient mode of treating deep-seated urethral gonorrhœa. The danger of cystitis or inflammation of the testicle will be rather abated than increased by this treatment if it be carried out properly and without violence. The possibility of performing the entire procedure without any abrasion, undue pressure, or injury of the inflamed parts, ranks it high above all measures in which unyielding sounds, catheters, or caustic holders are placed in the neck of the bladder for purposes of cauterization. Their use is often followed by epididymitis, and is deservedly held in bad repute.

Where the affection extends over the whole urethra, treatment of the neck of the bladder and of the anterior urethra can and ought to be carried out simultaneously until the secretion escaping from the meatus be reduced to a minimum, and until the frequent urgency to urinate and the turbidity of the water give way to a marked extent.

*Gonorrhœal catarrh of the neck of the bladder should not be mistaken for acute cystitis.* Pus will be found in the urine in

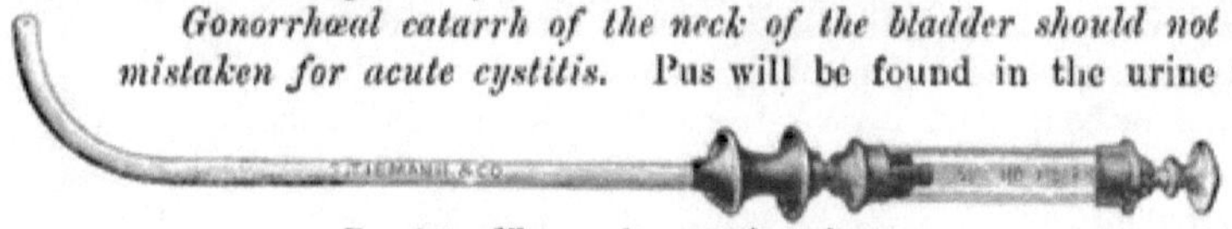

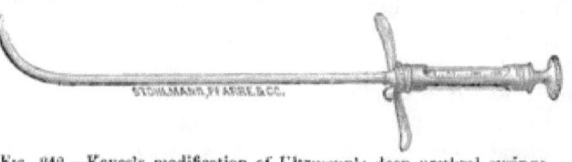

FIG. 241.—Ultzmann's prostatic syringe.

both cases, but in cystitis febrile disturbances accompanied by alteration of the general health will be observed, and pressure pain above the symphysis pubis will be noted aside from the periodical pain located in the perineal region, which follows urination, and which is the diagnostic sign of the affection of the deep urethra only.

Should irrigation of the deep urethra not effect rapid or complete cessation of the affection, *instillation of a few drops of a five-per-cent solution of nitrate of silver* will be found very beneficial. This is done by Nélaton's catheter or Ultzmann's deep urethral syringe. (Figs. 241 and 242.) The point of the filled instrument is dipped in glycerin, and is gently introduced just within the compressor-urethræ muscle. When the barrel of the syringe is at an angle of forty-five degrees with the body of the recumbent patient, its beak is just within the neck of the bladder. Three, four, or five drops of the nitrate-of-silver solution are expelled from the syringe, and enter the deep

FIG. 242.—Keyes's modification of Ultzmann's deep urethral syringe.

urethra. Intense smarting and spasm of the neck of the bladder follow the injection, but soon disappear if the patient retain the reclining posture for a short while.

These deep injections of nitrate of silver are a very effective though painful means of checking a gonorrhœal inflammation of the deep urethra, and deserve more frequent employment than they receive at present. The procedure does not entail any danger, and is rather a preventive than a cause of epididymitis or cystitis.

### 2. Chronic Gonorrhœa. Gleet:

*a.* INFLAMMATORY STENOSIS (INCIPIENT STRICTURE) AND PERMANENT OR CICATRICIAL STRICTURE OF THE URETHRA:

(*a*) *Anterior Urethra.*—The termination of acute gonorrhœa is never abrupt. It is always inaugurated by a period characterized by the escape of a scanty amount of purulent discharge. During this period subacute attacks or relapses of the affection may be precipitated by any cause inducing hyperæmia of the urethral mucous membrane. Sexual irritation, alcoholic indulgence, severe bodily exercise, offer mainly occasions for this occurrence.

When an acute gonorrhœa has reached this stage, the progress of the recovery often seems to suffer a halt, due principally to secondary hyperplastic changes of the mucous and submucous tissues. The daily introduction of a full-sized sound or bougie for a week or two is generally sufficient to produce rapid absorption of the interstitial exudation and a permanent cure.

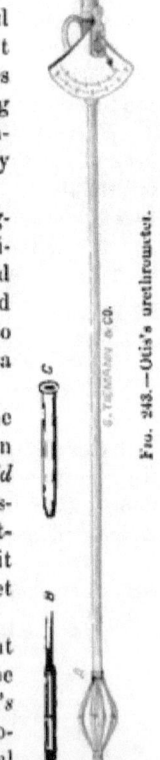

FIG. 243.—Otis's urethrometer.

A contracted meatus is an effective impediment to the application of the sound, and requires an adequate division of the narrow urethral orifice. *Meatotomy,* however, *should never be carried too far,* its only object being the easy admission of a full-sized steel sound. It is made with a blunt-pointed tenotomy knife, and the hæmorrhage caused by it can be easily checked by the introduction of a small pledget of iodoformed gauze into the slit.

Should the patient positively decline meatotomy, blunt dilatation of the part of the urethra, which is the seat of the inflammatory swelling and contraction, can be done by *Otis's urethrometer.* (Fig. 243.) The closed instrument is introduced beyond the coarctation, then it is opened until the dial indicates that the bulb has been dilated to full caliber, and then it is drawn with some force through the narrowed portion of the urethra. The author has seen very good results follow this use of Otis's instrument, though the procedure does not deserve preference over meatotomy and dilatation by the steel sound.

The absorption and disappearance of these "incipient strictures" is very much hastened by the local application of a strong (five-per-cent) solution

# 340   RULES OF ASEPTIC AND ANTISEPTIC SURGERY.

of nitrate of silver.  *To enable an exact application of the caustic under the guidance of the eye, the endoscope must be used.*

The endoscope is a cylindrical silver tube of from four to six inches in length, and of various calibers. (Fig. 244.) An obturator facilitates its painless introduction, and a flange or shield made of hard rubber, having a "dead finish," permits an easy handling of the instrument. Strong artificial light or sunlight is needed for endoscopy. The patient reclines on a tall chair, or sits on the edge of a table, his back supported by a suitable rest, the examiner occupying the space between the patient's legs. To protect the patient's clothing against soiling with blood or chemicals, a piece of rubber cloth (eighteen inches square), provided with a small central slit just long enough to permit the slipping through of the penis, is spread on the pubic region. Thus the only object exposed to view will be the patient's

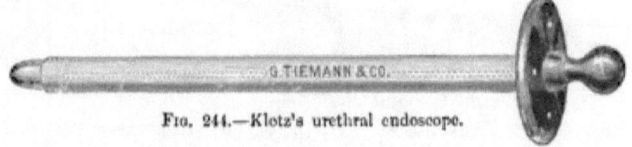

Fig. 244.—Klotz's urethral endoscope.

penis. Over the rubber cloth a clean towel is laid for wiping off fingers, etc. A basin containing a number of slender match-sticks, their ends armed with tufts of absorbent cotton, is at hand, and a pus-basin is next to it, to receive the soiled sticks. On a little table adjoining the operating-chair are a small, wide-mouthed bottle of glycerin and a few glass salt-cellars or hour-glasses for the reception of such solutions as may be required. Of these the author uses two—a five-per-cent solution of nitrate of silver and a ten-per-cent solution of the same substance, both in dark bottles.

An endoscopic tube of suitable size being selected, it is lubricated with a little glycerin, and is introduced well into the bulbous portion of the urethra. The obturator is withdrawn, and the surgeon by his head-mirror directs a ray of sun- or lamp-light into the bottom of the tube, where the mucous membrane of the urethra is visible in the shape of a typical image, consisting of several concentric folds uniting to a central, funnel-shaped depression.

In sunlight *the normal mucous membrane* is pale, of about the same hue as the normal buccal lining, and on it are visible a number of delicate tracings, produced by minute vessels. It is very smooth and glossy, and the folds of the image are flexible and rather delicate, and present *no change of color on deeper introduction or withdrawal of the tube.*

*Inflamed urethræ* show an entirely different aspect. The most delicate manner of introducing the instrument is apt to cause slight hæmorrhage, which sometimes is very troublesome, as the blood fills up the tube faster than it can be mopped away, frustrating for the time being all further manipulation. When the mucous membrane, exposed in the bottom of the endoscope, is dried off with a pledget of cotton, it has a dull, dead gloss, or velvety appearance; it shows a more or less intense, *uniform* shade of red, scarlet, or purple. The folds of the endoscopic image are few and coarse, and not so flexible as those of the normal urethra.

Gradually withdrawing the tube with short stops, the entire length of the urethra can be thus inspected.

In chronic gonorrhœal urethritis the inflammation will be found limited to more or less well-circumscribed portions of the urethra. These parts, examined by urethrometer or bulbous bougie, quite frequently show a well-marked though moderate contraction, which can also be demonstrated to the eye through the endoscope.

In withdrawing the tube, new parts of either normal or uniformly red, inflamed mucous membrane will present themselves to the examiner's eye. Suddenly, however, the field of vision will become *pale, perfectly anæmic, and ivory-colored.* This change of color is due to depletion of blood and the anæmia of the constricted part of the urethra, caused by the distention produced by the dilating instrument. As soon as the end of the tube is withdrawn from the stenosed part, the formerly bloodless tissues are seen to *suddenly flush up and become of exactly the same color as the rest of the inflamed mucous membrane.* Examination by the bulbous bougie (Fig. 245) will show that the seat of this phenomenon corresponds exactly with the locality of the narrowing of the urethral caliber.

FIG. 245.—Metallic bulbous bougie.

In cases where gleet has persisted for several months, these constricted places appear in the endoscope of a pearly color, which is due to the considerable thickening of the epithelial layer.

The application of the nitrate-of-silver solution to these "incipient strictures" will be found to materially hasten their absorption, if it be supplemented by the introduction of a full-sized sound. The applications are made through the endoscope every other day with a camel's-hair brush or a wad of absorbent cotton fastened to the end of a long match-stick. They cause a slight smarting, which does not persist very long. Occasionally they are followed by slight hæmorrhage on the day subsequent to the application, which, however, is without any significance.

Most of these "incipient strictures" get well under the treatment just described, *and do not require urethrotomy.*

But, when the embryonic connective tissue of these stenoses of inflammatory character becomes definitely transformed into fibrillar connective tissue—that is, a fully developed cicatrix—it represents *a permanent—that is, organic—stricture* that can not be cured by simple dilatation and topical applications. True, it may be gradually dilated to the normal caliber, but the dilatation will be evanescent, and speedy recontraction will follow the cessation of the treatment.

The appearance of a cicatricial or permanent stricture in the endoscopic field of vision differs in many ways from that of an inflammatory stenosis. This diagnostic distinction is all the more valuable, as an examination by the bulbous bougie, although capable of demonstrating the presence of a narrowing of the urethral caliber, does not divulge anything regarding the nature of the stenosis.

The most characteristic feature of permanent strictures is the unchanging anæmic, pale condition of the mucous membrane about the stricture in the endoscopic field of vision. *The sudden flushing up* on withdrawal of the endoscopic tube, seen in the contractions of recent date, *is absent*. The second characteristic is the *peculiar rigidity of the urethral wall* at the site of the stricture. On withdrawing the endoscope, the rigid walls of the urethra show a tendency to remain patulous, so that, instead of a small and rapidly changing image of soft, pliable mucous membrane, a comparatively long stretch of the urethra can be looked over at a glance, resembling somewhat the walls of a short tunnel.

Absorption and disappearance of a cicatricial stricture are a very exceptional occurrence, whether it be subjected to treatment or not. *To sufficiently widen a strictured urethra, urethrotomy, followed by methodical dilatation, is required.*

Such a cure as is not infrequently observed to come from treatment of an inflammatory stenosis—that is, a *perfect restitution* of the normal state of affairs—*is never to be expected* after the treatment of a cicatricial stricture, be this treatment dilatation alone, or cutting combined with subsequent dilatation. The cicatricial ring will become wider than before, but its rigidity and unnatural appearance will remain unchanged.

The cases in which the cicatricial bands can be divided in their entirety yield the comparatively best results. But the worst strictures involve the entire thickness of the spongy part of the urethra, and to effect complete division in these cases the entire thickness of the urethra would have to be cut through, which is an impracticable and sometimes dangerous procedure.

CASE.—M. F., aged forty-two, had a series of old cicatricial strictures involving the entire anterior portion of the urethra. One seated in the fossa navicularis was very tight, another one at the bulbo-membranous junction was very massive, so that it could be felt through the perinæum. Blunt dilatation with steel sounds, up to No. 34 of the French scale, always produced cessation of the profuse discharge, but, recontruction to the old condition always following within forty-eight hours, internal urethrotomy was decided on. *August 20, 1885.*—The operation was performed with Otis's urethrotome. The urethra was dilated to No. 30, and then two parallel incisions were made along the entire length of the roof of the pendulous portion. Some hesitation of the bulbous bougie was noted at the bulbo-membranous junction, therefore Otis's instrument was reintroduced, dilated to No. 32, and the still narrow part of the urethra once more cut. Smart hæmorrhage was observed, but not more than the length of the incision justified, and after some compression it ceased. On returning to the patient after the lapse of two hours, the writer found him lying on his blood-soaked mattress in a pool of blood, in a most deplorable state of prostration and anxiety. The scrotum and penis were swollen out of proportion, and had assumed a blue-black color, and blood was issuing from the meatus at varying intervals. A large English web-catheter was introduced and tied into the bladder, and only persistent digital pressure exerted over the bulbous portion for more than two hours succeeded in arresting the loss of blood, and checked further bloody infiltration of the penile and scrotal tissues. Fortunately, infection of the wound was avoided by careful asepsis, and thus, no fever and inflammation following, the entire enormous extravasation was readily absorbed. Introduction of large sounds was commenced on the twelfth day, and after a somewhat prolonged convalescence the patient recovered. With the regular use of the full-sized steel sound, and an occasional irrigation of the neck of the bladder, the patient succeeds in maintaining a very comfortable state of health.

In the case just related, complete division of the posterior stricture situated at the bulbo-membranous junction, led to the injury of the bulbar artery, imbedded in the cicatricial mass constituting the stricture. Had the wound been infected by the use of uncleanly instruments, suppuration and decomposition of the large bloody infiltration might have brought the patient into very great danger.

A serious objection to Otis's otherwise excellent urethrotome is the great difficulty of thoroughly cleansing the complicated instrument.

The Author's Aseptic Urethrotome (Fig. 246, A) is simple in construction, thoroughly reliable and firm; it is precise in action, and, being self-registering, obviates the necessity for a urethrometer, the functions of which are performed by the cutting instrument itself. It is composed of five easily detachable parts, three steel rods and two screws. One of the rods is provided with a laterally grooved bulb of small size (B), acting as a wedge, which, by the aid of a stout thumb-screw, serves to spring apart a pair of congruent steel blades (C). The amount of separation of these steel blades (somewhat resembling a pair of old-fashioned draper's shears), reduced to millimetres, corresponding to the urethral caliber, is indicated by a dial placed above the ring that serves for the fixation of the instrument. The correct adjustment of the thumb-screw is secured by a small check-screw which represents the proximal end of the urethrotome. The third rod, a small knife, hidden in the slightly curved beak of the instrument, can be withdrawn so as to correspond to the place of widest separation of the shear-blades. The caliber of the closed instrument is exactly fifteen millimetres. It permits of a distention to forty-five millimetres, and in these particulars coincides with the minimal and maximal dimensions of Otis's urethrotome. It can be taken apart in fifteen seconds, and can be put together in about double that time.

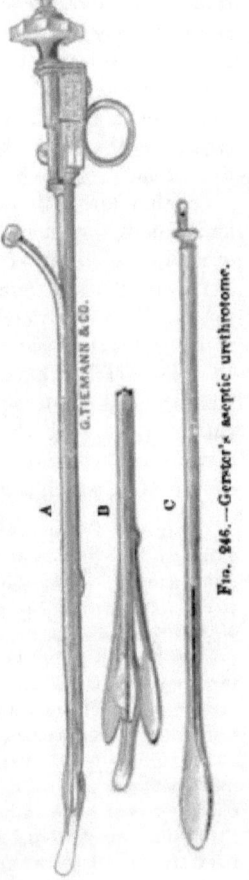

Fig. 246.—Gerster's aseptic urethrotome.

The *modus operandi* is as follows: The urethra being properly prepared, the closed instrument is lubricated with glycerin, and passed in well beyond the stricture or strictures. After this the shear-blades are separated by means of the thumb-screw to the desired caliber, and the instrument is drawn forward until it becomes arrested by the resistance of the stricture. Now the hidden knife is drawn into position, and the whole instrument, being firmly grasped, is steadily pulled forward. Thus the stricture is gradually dilated so as to offer a favorable degree of tension for the effective use of the knife, which will readily cut all the resistent tissues composing the stricture. The moment that the stricture is cut along its entire linear extent, and to the proper depth, the distended part of the instrument, which serves the purpose of a bulb, will slide through the site of the stricture, thus indicating that an adequate amount of division has been accomplished. A series of strictures can thus be divided one after the other without the necessity of removing the instrument from the urethra.

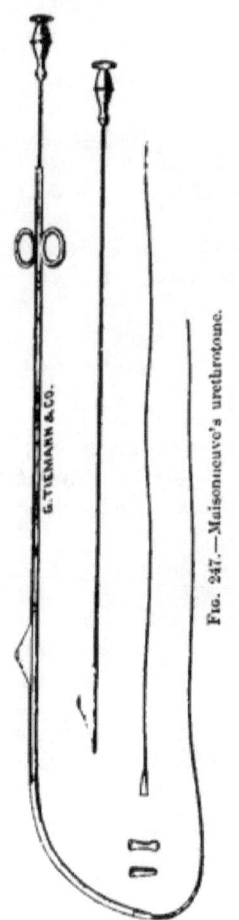

Fig. 247.—Maisonneuve's urethrotome.

It is advisable in cases of comparatively tight and very dense stricture, where a great disproportion exists between the normal caliber of the urethra and the caliber of the undistended stricture, not to attempt a complete division of the stricturing bands at one stroke, as the great amount of traction required to accomplish a full distention of the stricture would threaten a circular rupture of the urethra on a line just beyond the proximal limit of the stricture. It has been found much safer and also easier to cut such strictures *gradatim*. What is meant is this: that, the instrument being introduced, the first cut is to be made at a moderate degree of dilatation. The knife being slipped back, and the instrument somewhat closed, it is again passed behind the stricture, when a second cut is made, the dial indicating this time five or ten millimetres more of dilatation than was accomplished by the first cut. And this should be repeated until thus gradually full division is accomplished.

For very tight strictures Maisonneuve's instrument is most proper. (Fig. 247.)

Careful disinfection of the surgeon's hands and instruments, and irrigation of the urethra with a watery tepid solution of permanganate of potash (1 : 2,000), should precede every step or operation that may lead to wounding of the urethral mucous membrane. As a lubricant, iodoformized vaseline (1 : 30) should be used. The operation should terminate with a renewed irrigation of the urethra.

Whenever strictures are cut that have their seat near the bulbo-membranous junction, a new, large-sized, English elastic catheter should be tied into the bladder for twelve hours, and the patient should be kept in bed for a day or two. These precautions are rarely necessary in cutting strictures located in the pendulous portion, as it is not difficult to prevent hæmorrhage by the application of a compressory bandage to the penis. A gutter of light pasteboard is applied to the under side of the penis, which is first enveloped in a layer of cotton, and the splint is firmly secured by a few turns of a roller bandage. The penis and scrotum are held up to the belly by a snugly-fitting T-bandage. This preventive appliance can be abandoned on the second day after the operation.

If ammoniacal urine be present, its condition should be influenced before operation by the internal administration of boric acid, benzoate of soda, lactic acid, or turpentine, so as to become at least of neutral, or what is still better of acid, reaction.

A full-sized steel sound is to be introduced twice weekly, *the first application not to commence before the fifth day after the operation.* Much pain to the patient will be avoided by first introducing a copiously anointed smaller-sized sound, which will carry a good deal of the lubricant into the urethra, and will render the subsequent use of a full-sized instrument comparatively painless and easy.

With the precautions above described, the author has not observed a case of urethral fever following either internal urethrotomy or the use of dilating instruments in the urethra. His experience extends over seventy-one cases, in which strictures were cut successfully from within. No febrile or inflammatory complications were ever observed.

(b) *Deep Urethral Strictures.*—Strictures of the deep urethra are located in the membranous portion. Their development is preceded by a stage of epithelial and submucous hyperplasia, identical with the process observed in the anterior urethra. This hyperplastic condition is amenable to successful treatment by dilatation and caustics, but unheeded, will develop into permanent stricture.

Internal urethrotomy of a deep-seated stricture is a much more grave undertaking than the cutting of a stricture of the anterior urethra. Both the danger of hæmorrhage and the difficulty of controlling it, should it occur, render the operation serious. Hæmorrhage from the posterior part of the urethra, lying behind the "cut-off" muscle, may long remain unrecognized on account of the absence of free bleeding from the meatus, as the escaping blood will flow back into the bladder, and can be expelled only with the urine. For these reasons treatment by gradual dilatation should be carried on whenever possible, and urethrotomy should be reserved for cases only that do not yield to dilatation after patient trial, or will not brook delay. When an operation is decided on as necessary, *external urethrotomy deserves the preference over the internal operation, especially in cases complicated by ammoniacal cystitis.* Hæmorrhage will be easy to control. The good drainage resulting from the external incision will prevent urine infiltration, and ready access to the bladder will facilitate antiseptic irrigations of the organ.

*External Urethrotomy.*—The anæsthetized patient is brought in the lithotomy position, his hands being bandaged to the feet, which are then wrapped in clean towels, wrung out of corrosive-sublimate lotion. The perinæum and anal region being shaved and rubbed off with the same lotion, the operation begins. Irrigation of the wound by Thiersch's solution is carried on during the entire operation. When a staff or even a filiform bougie can be carried into the bladder to serve as a guide, the operation will offer no difficulty whatever. As soon as the urethra is opened and the stricture exposed, its division can be accomplished by the use of a blunt-pointed tenotomy knife. External urethrotomy without a guide is not as easy, but its difficulties can be overcome by patience and circumspection.

While an assistant exerts gentle pressure over the distended bladder, the bottom of the urethral wound being well exposed by small, sharp retractors or fillets of silk drawn through the lips of the urethral incision, one or two drops of urine will be seen exuding from one or another point of the stricture. A fine probe is inserted into the point in question, and will often penetrate the stricture. A narrow, grooved director is insinuated along the probe, and serves to guide a sharp-pointed tenotomy knife through the contraction, which then can be divided without difficulty.

Should this expedient fail, on account of inflammatory swelling of the tight part of the urethra, suprapubic aspiration of the bladder may serve to tide over the difficulty. Relief of the distention of the bladder is often followed by decrease of the swelling, and a few hours after the operation urine will be found escaping through the urethra, when the true channel can be searched out and dilated.

CASE.—N. S., laborer, aged 42, impermeable stricture of the membranous portion of the urethra. *March 11, 1883.*—External urethrotomy without guide. The stricture being exposed, most diligent search failed to ascertain the direction of the channel, which was obscured by the intumescence and great vascularity of the parts. The distended bladder was finally emptied by suprabubic aspiration, and the patient was brought to bed. Six hours later the bladder had refilled, and urine was seen to trickle from the wound whenever the patient strained. Renewed search was rewarded by the finding of the right track, which was divided on the grooved director without much trouble or pain to the patient. *May 20th.*—Patient was discharged cured.

A modification of another expedient, proposed by the venerable Petit, was also successfully employed by the writer.

CASE I.—John Smith, negro hostler, aged thirty-one, suffered from impermeable stricture of the deep urethra with dangerous distention of the bladder. The usual expedients for entering the bladder having failed, external urethrotomy was determined upon, and was carried out December 2, 1876. The distal part of the stricture being exposed, no entrance could be effected. As there was no aspirating needle on hand, a slender trocar was inserted into the middle of the strictural mass, and was pushed forward in the direction of the urethra, toward the center of the prostate, under the guidance of the left index-finger placed in the rectum. The point of the **instrument** was several times caught in the mass of the prostatic gland, **but finally entered the** median canal and the bladder, this being attested by the escape **of urine. A grooved director** was pushed in along the cannula, which was withdrawn, and the **stricture was divided** with a tenotomy knife. A sharp attack of fever and cystitis followed, **but the patient** fully recovered, and was discharged cured, March 5, 1877.

CASE II.—George G., saloonkeeper, aged forty, acute retention due to impassable stricture of the membranous portion—the pendulous portion also the seat of a number of strictures; in fact, the entire pendulous and membranous portions forming one stricture. *July 26, 1888.*—External urethrotomy *without guide*, at the German Hospital. The urethra could be felt throughout as a cord-like, hard mass. This **was** incised in the region of the urethral bulb, and hence a filiform bougie could be introduced in a retrograde direction toward the end of the penis. Division of strictures by Maisonneuve's, then by Otis's urethrotome. The proximal part of the urethra could not be recognized; hence a fine, long trocar was thrust through the cicatrix and **the** middle of the prostate **into** the bladder, then this track **was** dilated with **the knife,** and a large soft **catheter was** left *in situ* traversing the **whole urethra. No fever or reaction** followed, and the urine became acid. *August 25th.*—**Patient was discharged cured,** with directions to use a sound.

Strictures located in the anterior urethra can be simultaneously divided by Gerster's urethrotome or the tenotomy knife before the patient recovers from the anæsthetic. The bladder is then washed out with Thiersch's solution, and the **wound is dressed** with a pad of iodoformed **and** a compress of sublimated **gauze, held in place by** a T-bandage. In the **presence of** fetid **urine, the use of a** drainage-tube **is** advisable. Before **applying the dressings the wound** should be rubbed out with a small **sponge dipped in iodoform powder.** Anointing **of the** perinæum and **buttocks with vaseline is necessary to** prevent eczema. **The** external dressings ought to be **changed** whenever soaked ; **the** iodoform pads, however, should **not be disturbed without necessity as** long as they are adherent. Daily **sitz-baths in a weak** (1 : 10,000) corrosive-sublimate solution will tend **to increase the comfort** of the patient, and will aid the healing of the **wound.**

The daily introduction of a full-sized steel **sound need not be commenced** before **the** fifth day, and should be continued **at increasing** intervals for at least a year after the operation.

Altogether, the author **performed external** urethrotomy twenty-seven **times. Twenty-four patients recovered, three died.** The fatal **cases were as follows:**

CASE I.—Mr. S. O., **tailor,** fifty-four **years** old, suffering from tight, deep-seated **stricture of the** urethra, complicated **with purulent and fetid** pyelo-nephritis. The **urine** remained ammoniacal, and the **fistula** never closed. He died, August 5, 1886, of uræmia, five months after the operation, done March 25, **1886.**

CASE II.—Abraham Goldfish, aged seventy-seven, **suffering from deep-seated urethral stricture, fetid cystitis,** and extensive urine **infiltration of the perinæum, due to a false** passage made by a physician. External urethrotomy **was performed, November 1, 1886,** at Mount Sinai Hospital, with much relief **of the** subjective symptoms, **but** the patient succumbed to septicæmia and septic nephritis on November 18, 1886.

CASE III.—Christian Schlenker, engineer, aged twenty-seven. Tight stricture of membranous portion, with chronic fetid cystitis and pyelo-nephritis. External urethrotomy, June 3, 1887, at the German Hospital. The urgent symptoms were abated, but the patient continued to have fever, lapsed into hectic, and died August 24, 1887. The autopsy revealed far-gone double tuberculous nephritis.

One case deserves special mention on account of its rarity :

CASE.—S. E., shopkeeper, aged sixty-three, sustained, in 1875, a *compound fracture of the left horizontal ramus of the os pubis*, from which he recovered after a long term of illness. In the spring of 1882 increasing difficulty of micturition became noticeable, and finally led to retention of urine. *June 25, 1882.*—The author saw the case in consultation with Dr. I. Schnetter. A metallic sound could be passed easily as far as the membranous portion, but was there arrested by a grating, hard body, thought to be a sequestrum or a stone. External urethrotomy was done June 27th, and an irregularly shaped sequestrum, one inch long and one sixth of an inch thick, was withdrawn with some difficulty. Patient recovered without fistula, and was cured in about six weeks.

Three times external urethrotomy was successfully done for deep-seated stricture and the relief of coexistent prostatic abscess. Four times internal and external urethrotomy were done simultaneously. In four cases permanent lip-shaped fistulæ remained behind and required closure by Szymanovsky's plastic, which succeeded in each instance. In one of these cases internal urethrotomy of two strictures of the pendulous portion was done simultaneously with Szymanovsky's plastic successfully.

*b.* VEGETATIONS OF THE URETHRA.—Venereal vegetations, such as are frequently observed under the prepuce of men suffering from gleet, occasionally occur in the urethra, principally in the fossa navicularis and in the sinus bulbi. They maintain a rebellious urethral discharge that can be stopped only by their removal. Their diagnosis can be made by the aid of the endoscope, which also affords the best means of access for their treatment. The use of the curette, or a small wire snare, or of chromic acid in crystals, will readily destroy them, and will terminate the urethral discharge depending on their presence.

*c.* GRANULAR URETHRITIS.—One of the most tedious affections of the urethra is a chronic inflammation of the mucous membrane following an attack of acute gonorrhœa, characterized by an irregularly distributed hyperæmia and scanty discharge. The velvety mucous membrane bleeds at the slightest touch, and the condition resists every form of local treatment for a disproportionately long time. It seems that the intractability of this affection depends in a great measure upon constitutional disorders ; at least the author observed it most frequently in anæmic individuals of a scrofulous habit. Measures directed to the improvement of the general condition, and supplemented by the local application of a five-per-cent solution of nitrate of silver by the endoscope, seem to have been more efficient than anything else, though it must be admitted that a few cases resisted every kind of treatment, and had to be given up as entirely unmanageable.

*d.* CHRONIC CATARRH OF THE POSTERIOR PART OF THE URETHRA, AND CHRONIC CYSTITIS.—Chronic catarrh of the membranous and prostatic part of the urethra is frequently observed following an acute attack of gonorrhœa, in subjects formerly addicted to masturbation, or those indulging in general, and especially in sexual, excesses. In these cases no external urethral discharge is visible, but frequent micturition is present, and both portions of the urine, passed into two tumblers, show turbidity, the first portion, however, being more turbid than the last.

Treatment by gradual dilatation with full-sized sounds is perfectly useless in this affection, and may even lead to epididymitis in some cases. *Methodical irrigation of the neck of the bladder*, on the other hand, by means of a soft gum catheter and hand syringe, as described in a preceding paragraph, will be very often found beneficial. Of all substances, a 1 : 2,000 tepid solution of permanganate of potash has been found most generally applicable. A quart china bowl is filled with warm water, and enough of a concentrated solution of the salt is added to tinge the water a light-claret color. This test, by observing the depth of the tinction, is very sensitive if applied to weak solutions, and commends itself by its simplicity. Next to permanganate of potash, one-per-cent solutions of sulpho-carbolate of zinc or of acetate of lead deserve mention. But *nitrate of silver is the most efficient of all known remedies in obstinate cases of chronic deep-seated urethritis or prostatic catarrh*. A few drops of a five-per-cent solution are instilled, twice or three times a week, by Ultzmann's or Keyes's deep urethral syringe, as formerly described.

*Acute cystitis*, whether gonorrhœal or pyogenic, *is not amenable to instrumental treatment, which should only commence after the cessation of the invasive stage*. The object of medicinal irrigation is the disinfection and removal of fermenting urine and its decomposed contents, such as ropy mucus, blood, and pus.

If stone or a stricture be the causative agents, they must be removed; if imperfect evacuation of the bladder, on account of paresis, or enlargement of the prostate, is at the bottom of the trouble, regulated evacuation of the organ by catheterism must be employed. Aside from fulfilling these causal indications, recovery can be materially hastened by methodical irrigation.

*Irrigation with a metallic "double current" catheter, as recommended by various authors, is unsatisfactory.* Introduction of the rigid catheter is painful, and may be the source of various complications. The advantages of the double current are illusory, as much of the ropy mucus and other sediment found in the cul-de-sac of the bladder is not brought out by its use. A more gentle and much more efficient way of thoroughly emptying the deleterious contents of the inflamed bladder is as follows:

The patient is made to stand before the seated physician. This position is more favorable than any other, as in it the sedimental matter contained in the urine is made to gravitate toward the neck of the bladder, where it is readily stirred up and evenly distributed in the urine by the injections. Thus it will pass the catheter much easier than when it forms a sticky mass. A soft rubber catheter is introduced into the bladder, and a hand-syringeful of a tepid, weak solution of cooking-salt (one teaspoonful to a quart, about

6 : 1,000) is thrown in gently, and is allowed to escape at once. This is repeated until the returning saline solution is clear and limpid. After this, two or three ounces of a tepid 1 : 5,000 solution of permanganate of potash are injected and retained for one or two minutes, and the process is repeated until the returning fluid ceases to be discolored. By and by, as the bladder becomes more tolerant, the injection should be made more forcible, as a thorough stirring up and dislodgment of the ropy sediment by the jet of lotion is very essential to its complete evacuation. The strength of the medicinal lotion should also be gradually increased (to 1 : 1,000).

In cases of paresis, or when a tendency to vesical hæmorrhages be present, *cold*, instead of tepid, injections will be appropriate.

In obstinate catarrh the strength of the permanganate-of-potash lotion can be increased to 3 : 1,000. Alum (from 1 : 100 to 5 : 100), sulphate of zinc (from 1 : 100 to 2 : 100), and nitrate of silver (from $\frac{1}{2}$ : 100 to 2 : 100), will also be found very effective. Deodorization of fetid urine is readily effected by injections of a 3 : 100 solution of resorcine, which should be followed up by the employment of one or another of the medicinal solutions above mentioned (Ultzmann).

If the capacity of the bladder be very much diminished by long-continued spastic contraction accompanying gonorrhœal or calculous cystitis, gentle and gradual distention of the organ by salt water or medicinal injections of increasing volume will be followed by increasing tolerance. Thus micturition will gradually become less frequent, and the normal condition of things may be re-established.

NOTE.—Gradual distention of the shrunken bladder of elderly persons is dangerous, as it may lead to rupture of diverticula.

PART V.

## SYPHILIS:

### ASEPTIC AND ANTISEPTIC TREATMENT OF ITS EXTERNAL LESIONS.

# CHAPTER X.

## *ASEPTICS AND ANTISEPTICS APPLIED TO EXTERNAL SYPHILITIC LESIONS.*

1. Aseptic Treatment of Primary Induration.—The nature of the specific virus of syphilis is not known. In most cases its local and general manifestations are amenable to appropriate systemic and topical remedies.

It is not intended here to dwell upon the nature and treatment of syphilis as a general disease; only inasmuch as some of its more common local phenomena require surgical treatment will their consideration be deemed within the limits of this chapter.

The anatomical structure of the primary induration, of tuberous syphilides, and of gummy swellings, resembles closely that of recent tuberculous deposits; and their course of development and termination in central coagulation necrosis, fatty changes, or caseation, also bears much general resemblance to the affections caused by the bacillus of tuberculosis. But there is a third point of parallelism.

As long as softened tuberculous or syphilitic foci remain subcutaneous, and are not exposed to the influence of the air and its pus-generating germs, their course is bland and slow, and their tendency is to fatty degeneration, encapsulation, and final absorption. But, as soon as such a softening deposit comes under the influence of the pyogenic elements contained in the atmospheric air, its slow and bland character is changed to a most destructive one. Thus syphilitic nodes of the internal organs, being protected from contact with the outer air, rarely, if ever, terminate in ulcerative destruction: they generally tend to fatty involution, absorption, and cicatrization. Specific deposits of the outer skin, the mucous membranes—as, for example, of the nasal and oral bones—on the other hand, are all noted for their pronounced tendency to rapid ulceration or gangrenous destruction.

As an illustration of a parallel behavior of tuberculous foci, cold abscesses and articular tuberculosis may be mentioned. Before perforation, their course is mild and slow; but after the establishment of one or more sinuses they become the source of profuse secretion, and their course is characterized by rapid local destruction with general emaciation.

The explanation of this peculiar difference in the behavior of syphilitic indurations or tumors, essentially identical in morbid character, is to be found in the fact that the poor nutrition and low vitality of the cellular elements composing a primary or secondary syphilitic node, exposed to pyogenic infection by contact with the outer air, offer very favorable conditions for the rapid development and destructive multiplication of germs, that are notoriously deleterious even to healthy tissues. Pus-generating cocci deposited on the excoriated surface of a syphilitic focus, as, for instance, a primary induration of the prepuce, or a gummy swelling of the nasal bones, will, by their multiplication, lead to massive invasion and rapid ulcerative destruction of the densely infiltrated and poorly nourished node.

*Syphilitic ulcers of every kind present a combination of syphilitic and of pyogenic infection.*

If we succeed by appropriate systemic treatment in preventing the extension of the central softening of a syphilitic node to the surface, ulcerative changes also will thus be prevented. For example, the timely administration of large doses of iodide of potash may prevent necrosis of the nasal bones, which are the seat of a growing gummy swelling. Their dense infiltration pertains to syphilis; their necrosis, however, is caused by the invasion of pyogenic germs. But we possess another means for preventing ulcerative destruction of syphilitic deposits located in the outer skin. They are more exposed to pyogenic infection, but they are also more accessible to local remedies.

*The aseptic protection of the surface of the primary induration offers an easy remedy for preventing the formation of the primary ulcer or chancre.*

True, that the prevention of the ulcerative destruction of a primary induration of the prepuce will not prevent the systemic development of syphilis; but it will, nevertheless, constitute a valuable service rendered to the patient, who will be spared all the suffering, annoyance, and danger connected with the development of the primary ulcer.

If a patient, exhibiting a recent primary induration of the penis, presents himself for treatment before the appearance of the pustular excoriation, or before the epidermal film of the formed pustule is broken, and if the surgeon thoroughly cleanses and disinfects the affected parts, afterward carefully enveloping the penis in an aseptic dry dressing, ulceration of the indurated node—that is, the development of a primary ulcer—can be effectually prevented.

The node will lose its epidermidal covering, but the aseptic dressing will exclude pyogenic infection, and the course of development and involution of the syphilitic deposit will be as though it were subcutaneous. A small quantity of lymph will exude from the excoriated surface, will be imbibed by the aseptic dressing, and will exsiccate, thus forming a hermetic seal and protection to the diseased tissues.

Fatty disintegration of the infiltrated tissues will be followed by the formation of new epidermis, and when, after three or four weeks, the dressings come off, a cicatrized though still somewhat indurated portion of skin will be exposed to view.

Specific rash, and other manifestations of systemic infection, will appear in due course of time; but the incalculable extension of the ulceration to adjoining non-infiltrated parts of the skin, and the **formation of suppurative buboes and other complications, will be obviated.** The following case may serve as an illustration:

CASE.—H. B., aged twenty-five, presented himself January 2, 1887, **with a hard, elevated node, the size of** a nickel, occupying the dorsum penis, **and** another **smaller** induration near the frenulum. Suspicious cohabitation had been indulged in for some time until within a few days of the visit. Bilateral indolent inguinal lymphadenitis was noted, and the presence of specific infection was assumed. The patient was kept under daily observation, and was directed not to meddle with any blister that might appear on the indurated spots. *January 8th.*—A yellowish discoloration was observed occupying the apex of the larger node, and was looked upon as an indication that a pustule was forming. The entire penis was carefully cleansed with green soap and warm water, and was disinfected with a 1:1,000 solution of corrosive sublimate, good care being taken not to break the transparent layer of epidermis covering the discolored spot. A thick layer of iodoform powder was sprinkled over both indurated nodes, and a small patch of iodoformized gauze was placed over them—this being held down by a narrow, oblong compress of corrosive-sublimate gauze, snugly bandaged on **with a muslin roller. The meatus was left** exposed **for** micturition, and **the patient was directed not to interfere with the** dressings and to report daily. The first **dressing remained undisturbed until** January 17th, when its external part, getting disarranged, was removed. The strip of iodoform gauze was found firmly attached to the underlying indurated nodes, and had **the** appearance of a hard, **flat** cake, that had been evidently soaked through by lymph or serum some time since its application. Evaporation of its aqueous contents had converted it to the shape **just** described. It was left *in situ*, and a fresh outer dressing was applied.

At the same date (January 17th) the girl with whom the patient had held commerce, presented herself for examination at the author's request, and was found to be covered with a small, papulous, specific rash. The appearance of her throat, the universal adenitis, and two freshly-cicatrized spots on the labia minora, left no doubt of her being subject to florid syphilis. She remained under **prolonged** specific treatment, and in July, 1887, still exhibited pharyngeal ulcerations.

*January 25th.*—The dressings applied to the patient's penis became again disarranged, and had to be renewed. The immediate covering of the nodes, consisting of iodoform gauze, was still firmly adherent, and was left unchanged.

*February 12th.*—A general maculous rash appeared on **the patient's body, and** systemic treatment by mercurial inunctions was commenced.

*February 20th.*—The entire dressings came off—the strip of iodoform gauze in the **shape of a** perfectly dry scab, **to** the inner side of which was found attached a patch **of shiny scales,** consisting of effete epidermis. The nodes, which were formerly prominent, **had receded to** the level of the surrounding skin, and the induration, which still **could be felt, was** marked by a coat **of** fresh-looking young epidermis. The patient **received fifty inunctions of** blue ointment, which freed him **from** all cutaneous symptoms **of the disease. In** May, pharyngeal ulcerations appearing, the inunctions were **resumed. Size** and hardness of the initial sclerosis were visibly diminished by this time.

It seems **in the foregoing case that** the ulcerative destruction of the primary induration **was forestalled by** disinfection **and** subsequent aseptic management. Without them the imminent formation of an initial sore would **have** inevitably occurred. The treatment of the fully-developed chancre would certainly have **been** a much more disagreeable, painful, and filthy experience than the simple manipulation of once cleansing and protecting the initial induration. The site of the morbid process thus protected against "external irritation"—that is, pyogenic infection—ran, as it were, a subcutaneous and bland course of slow involution, the aggregate of discharge during forty-three days not exceeding the small quantity required to permeate a strip of four layers of iodoformized gauze, covering an area of about two thirds of a square inch.

**2. Antiseptic Treatment of the Primary Syphilitic Ulcer.**—The results obtained by the various time-honored and well-established forms of local treatment of the primary syphilitic ulcer all bear out the assumption that the specific alteration of the affected tissues only serves as a predisposing condition to the subsequent ulcerative destruction of the initial sclerosis. The ulceration is directly produced by the ingrafting of purulent infection on a soil, devitalized by the dense cellular infiltration, characteristic of initial sclerosis. The rapid destruction observed in chancre is always signalized by the detachment of the epidermis raised in the shape of a pustule, under which we find a yellowish, brittle necrobiotic nucleus, which is the first to succumb to the onslaught of the pyogenic organisms, deposited on it by the manipulations of the patient or otherwise.

*The various forms of local treatment successfully employed for the cure of chancre are all antiseptic in character.*

Their aim is either the prompt removal of the infectious discharge by prolonged baths and frequent moist dressings, or disinfection by weak or concentrated caustics, or a combination of measures directed toward a rapid mechanical removal of the deleterious secretions, with chemical disinfection. As the most powerful and most effective arrester of the destructive course of phagedenic chancre, the actual cautery is to be mentioned—the sovereign destroyer of all microbial parasites.

*a.* Chemical Sterilization and Surface Drainage by Medicated Moist Dressings.—The energy to be applied to the local treatment of an ulcerating initial sclerosis should be proportionate to the virulence and destructiveness of the morbid process. In most cases the resistance of the vital forces combating the morbid process will be sufficient to check the damage. This is attested by the numerous cases of neglected chancre that end ultimately in spontaneous cure. Hence, in most instances, a mild treatment by local antiseptic baths, combined with moist antiseptic dressings, will answer the purpose.

*Frequent removal of the soiled dressings* forms the most essential part of this plan of therapy. The patient is directed to provide himself with a wide-mouthed, one-ounce vial, which is filled with suitably proportioned small, square pieces of lint or gauze, over which is poured a moderate quantity of a one-per-cent solution of carbolic acid, or a $1 : 5,000$ solution of corrosive sublimate. The cork-stoppered vial can be easily carried by the patient, who is enjoined to dress the sore or sores at least once every hour, and oftener if the discharge be very profuse. In the morning and evening a prolonged local bath in the same solution is advisable. In many cases this plan will be sufficient to check the extension of the ulcer, and to bring about cleansing of its bottom.

Another mild form of antiseptic treatment consists of the application of iodoform powder to the ulcerating surface. The objectionable odor of the drug can be excellently masked by the admixture of equal parts of freshly roasted and ground coffee. As soon as the appearance of a cicatricial border is apparent, these modes of treatment should be abandoned in favor of the application of strips of mecurial plaster, which should be renewed in proportion to the amount of discharge. Cicatrization will be very much hastened by this change.

*b.* CHEMICAL STERILIZATION BY STRONG CAUSTICS.—Cases of greater virulence which do not yield within a fortnight or so to the mild plan of treatment by scrupulous cleansing and disinfection, or in which rapid extension of the ulcer does not justify temporizing, require the application of escharotics. The author has found *a fifty-per-cent solution of chloride of zinc* the most convenient and most effective of all chemicals recommended for the cauterization of chancre. Its application is to be done as follows: The ulcer and its vicinity are subjected to a careful cleansing by a mop of cotton dipped in a 1 : 1,000 solution of corrosive sublimate. Crusts and scabs overlapping the edge of the sore must be gently removed. A small piece of clean blotting-paper is applied to the ulcer and its vicinity with gentle pressure to remove all moisture. A moderate quantity of the caustic solution is applied to the sore with a glass rod or match-stick, care being taken not to corrode unnecessarily the surrounding healthy skin. Previous thorough drying of the integument with blotting-paper will best prevent overflowing of the caustic. All the nooks and indentations of the margin of the ulcer must be carefully covered by the solution. As soon as the base of the sore assumes the color of parchment, which will occur in from three to five minutes, cauterization is completed, whereupon the surplus of caustic should be removed by the application of another piece of blotting-paper. The eschar is dusted with a little iodoform and coffee-powder, and is protected from injury by a strip of moist lint or gauze.

If the cauterization was sufficient, further extension of the ulcerative process will be arrested thereby. In from two to six days, according to the depth of the eschar, a narrow line of demarkation will appear, and, the eschar being detached, a healthy granulating surface will become visible. This should be dressed with strips of mercurial plaster until cicatrization is completed.

Insufficient chemical cauterization will not check the ulcerative decay of the tissues. In proportion to the incompleteness of the application, partial or total extension of the ulcer will be observed. In some cases only a tongue of renewed ulceration will be seen extending outward from the margin of the eschar. In others, the ulceration will spread all around the cauterized patch, thus demonstrating the entire inadequacy of the application. The surgeon's error should be in favor of too much rather than too little of the caustic.

When the process is found to be extending more or less in spite of a previous cauterization, the deficiency should be corrected without delay by a renewed application.

*c.* STERILIZATION BY THE ACTUAL CAUTERY.—Phagedenic forms of chancre, occurring on the penis, lips, or fingers, and characterized by dusky swelling and a rapidly-spreading, more or less gangrenous decay of the tissues, can be rarely arrested by anything short of the energetic application of the actual cautery. In some cases renewed searing will be required to check the trouble brought under control in one portion of the ulcer, but extending further in another direction from a limited part of the lesion. It is especially important to search out all recesses overlapped by the undermined margin of integument, as they are the chief nidus of active infection. The thermo-cautery, or red-hot iron, should be well inserted in all of these recesses and sinuses, otherwise the result will be incomplete or entirely unsatisfactory. The wound should be packed with very narrow strips of iodoform gauze while the patient is still under the influence of the indispensable anæsthetic, and care should be taken to line all nooks and crevices of the irregular wound with the gauze. The object of this is to prevent retention, and to secure prompt disinfection of the discharges which needs must be absorbed by the dressings. The penis is enveloped in an ample compress, moistened with warm carbolic lotion (one per cent), over which is placed a piece of rubber tissue to prevent evaporation. On the penis, daily change of dressings is to be done after a hip-bath, which will very much facilitate their painless removal. The febrile disturbance regularly noted with these most virulent forms of specific ulcer, and the general debility and anæmia, which is its main predisposing cause, require appropriate roborant and anti-febrile general treatment. As soon as cicatrization shall have commenced, the affection is to be treated like a simple ulcer.

The foregoing view of the relation of suppuration to syphilitic lesions is based exclusively upon clinical data, and needs corroboration at the hands of pathologists more expert in systematic and exact research than the author. One object of these remarks was to arrange the clinical facts pertaining to syphilitic ulcerations under a general principle, from which the therapeutic measures usually employed for their cure could be easily and logically deduced.

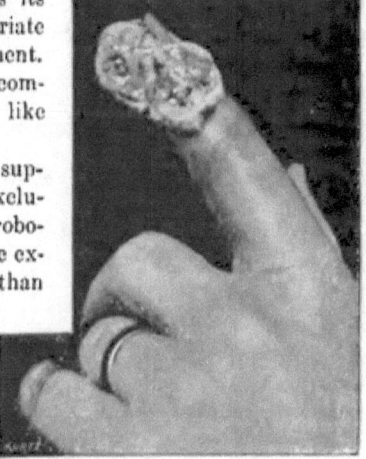

FIG. 246.—Specific ulcer of index finger.

# INDEX.

Abbe's catgut rings for enterorrhaphy, 162.
Abdominal drainage, 120, 145.
　operations, 119.
　suture, 146.
　toilet, 145.
　tumors, puncture of, 144.
Abscess, anal, 285.
　of bone, 219.
　cervical, 234.
　cold, 294.
　**formation of, 193.**
　glandular, **203, 234, 259.**
　of liver, 276.
　lumbar, 276.
　mammary, **237.**
　mastoid, 235.
　maturing of, **194.**
　metastatic, 195.
　pelvic, 260.
　perinephritic, 276.
　perityphlitic, 260.
　prevesical, 269.
　self-limitation **of, 194.**
　temporal, 235.
　tonsillar, 229.
Accidental **wounds, 29.**
　definitive **care of, 31.**
　infection of, by careless probing, **30.**
　temporary care of, 27.
Accidents in and after tracheotomy, 102–104.
Acetic acid, 11.
Active movements after joint exsection, 307.
Actual cautery for syphilitic ulcers, 358.
Adhesions, treatment of, in abdominal tumors, 143.
Æther pneumonia, 152, **156,** 157, 165.
　nephritis, 121.
Amputation, 61.
　for diabetic gangrene, 63.
Amputation of breast, female, 113.

Amputation of breast, **in males, 117.**
　statistics of, 118.
　technique of, 115.
Amputation of limbs, clean cases, 64.
　hyperæmia and oozing after, 73.
　intensely septic cases, 67.
　mildly septic cases, 66.
　management of stump after, 74.
　osteomyelitis of stump in, 65.
　statistics of, 61, 62.
Amputations, dressings after, 75.
Anæsthetics in herniotomy, dangerous depressing effect of, **129.**
Anal abscess, 285.
Anal fistula, 286.
　excision of, **286.**
　suture of, 287.
　tuberculous, 299.
Anatomy of connective-tissue **planes of neck,** 222.
Anchylosis, **bony, 87.**
Aneurism, **axillary, 52.**
　**carotid, 49.**
　**cyrsoid, 52.**
　femoral, **50, 51.**
　innominate, 50.
　palmar, 50.
　popliteal, 50, 51.
Ankle-joint, exsection of, 325.
Antisepsis, 27, 183.
Antiseptics applied to **primary syphilitic ulcers, 356.**
Apnœa after tracheotomy, 104.
**Apparatus for the after-treatment of the exsected elbow-joint, 312, 313.**
Appendicitis, 261, 262.
　**acute perforative, 264.**
　**acute, simple, 263.**
　**acute, with tumor, 265.**
　chronic, 272.

Appendicular stenosis, 262.
Aprons, 20.
Arm, suppuration of, 252.
Arteries, ligature of, 48.
Arterio-phlebectasia of foot, 51.
Artery, deligation of axillary, 52.
  carotid, common, 49, 186.
  carotid, external, 52.
  femoral, 50, 51.
  iliac, external, 50, 51.
  lingual, 97, 98, 99.
  operative injury of, 49.
  palmar, 50.
  ulnar, 50.
Artery forceps, 69.
Arthrotomy, 78.
  for dislocation, 82, 83.
  for elbow fracture, 83.
  for habitual dislocation, 8.
  for joint-hydrops, 78.
  for old irreducible dislocation of shoulder-joint, 83.
Artificial anæmia, 69.
  anus, 126.
Asepsis, 3.
  in peritoneal operations, 119.
Aseptic cap, 92.
  accidental wounds, 34.
  fever, 20.
  operating, manner of, 17.
  wounds, 5.
Aseptics of amputation, 64.
  no excuse for bad operating, 37.
  of the orifices, 96.
  of rectum, 167.
Autoinfection, operative, 52.
Axilla, evacuation of, 115.
Axillary glands, 252.
  vein, 59, 115.

Bacteria of putrescence, 185.
Bismuth, 11.
Bladder, aseptics of the, 173.
  treatment of, before ovariotomy, 145.
Bloodclot, healing under the, 6.
Boiled water for irrigation, 10.
Bone abscess, 219.
  tuberculosis, 303.
Boro-salicylic lotion, 10.
Bose's method of tracheotomy, 102.
Bottle-shaped wounds, 41.
Bow-leg, 86.
Bozeman's position, 168.
Breast, amputation, 113.

Breast, benign tumors of, 115.
Bursa, olecranic, 252.
  of quadriceps, 257.
  prepatellary, 256.
Button sutures, 46.

Cachexia strumipriva, 112.
Calculous kidney, 279.
Callus, deformed, 88.
Cancer of detached lobe of mammary gland, 23.
  of tongue, 97.
Cancerous lymph glands, 54.
Carbolic acid, 10.
Caries, 303.
Carpal exsection, 315.
Caseation, 294.
Caseous infiltration, 294.
Castration, 165.
Cataplasms, 200.
Catgut, 8.
  impure, 8.
  slipping of, 5, 57.
Catheterism, 159.
Catheters, cleansing of, 174.
Cervical abscess, 173, 234.
Change of dressings, 20.
Chisels, 212.
Cleanliness, surgical, 7.
Cleansing process of feet, 64.
Club-foot, 88.
Cold abscess, 294, 303.
Colotomy, inguinal, 156.
  lumbar, 156.
Compound fracture, 16, 24, 31, 33, 34.
  of cranium, 33.
Compressor urethræ, 333.
Contaminated accidental wounds, 31.
Continuous suture, 46.
Corrosive-sublimate lotion, 10.
Coryza, scrofulous, 299.
Cotton dressings, 15.
Creoline, 10.
Crinoline bandages, 15.
"Cut-off" muscle, 333.
Cynanche, parotid, 233.
  sublingual, 232.
Cyst of broad ligament, 149.
Cystitis, 319.
Cystotomy, perineal, 177.
  suprapubic, 177.
Czerny's suture for hernia, 136.

Definitive care of accidental wounds, 31.

Deformities, 86.
Diphtheria of fauces, 225.
  of intestine, 129.
Dislocation, habitual, 82.
  irreducible, 82.
Dissection, technique of, 36.
Dissemination, operative, of cancer, 53.
  of tuberculosis, 318.
Drainage, 47.
  abdominal, 120, 145.
  of peritonæum, 121.
Drainage-tubes, 9.
  removal of, 22.
  replacement of, 48.
  T-shaped, for cystotomy, 178.
Dressings, 11.
  change of, 20.
  first change of, in infected wounds, 27.
  for hand and forearm, 83.
  patterns for, 14.
Dry dressings, 12.
  spores, 192.
Dust, 5.
Dustless operating-room, 7.
Dyspnœa, expiratory, 105.

Elastic ligatures, 9, 151.
  in anal fistula, 288.
Elbow apparatus, 312, 313.
  fracture, 83.
  joint, exsection of, 310.
Embolism, septic, 193.
Emergencies, 23.
Emphysematous gangrene, 205.
Empyema, old, 242.
  recent, 241.
Endoscope, urethral, 340.
Enterectomy, 158.
Enterorrhaphy, Lembert-Czerny's method, 159.
  Senn's method, 162.
Epididymitis, tuberculous, 165, 299.
Erysipelas, 184, 289.
  phlegmonous, 290.
Esmarch's bandage, 69.
Estlander's operation, 242.
Excision of anal fistula, 286.
Exsection of ankle-joint, 325.
  of elbow-joint, 310.
  of hip-joint, 315.
  of joints for tuberculosis, 305.
  of knee-joint, 319.
  of shoulder-joint, 308.
  of wrist, 314.

External urethrotomy, 345.
Extirpation of axillary glands, 115, 253.
  of cervical glands, 52, 60.
  of inguinal glands, 57, 58, 260.
  of tumors, 52.

Face, carbuncle of, 224.
Fauces, diphtheria of, 225.
  Rose's position for operations in the, 227.
Faucial suppuration, 225.
Feet, cleansing process of, 64.
Femur, necrotomy of, 211.
Fever, aseptic, 20.
  septic, 193.
Fibrinous arthritis, 78.
Finger-joints, exsection of, 251.
  suppuration, 250.
Finger-nails, cleansing of, 7.
Fistula in ano, 286.
  excision of, 286.
  thoracic, 242.
  tubercular, 299.
Floating bodies, 79.
Follicular tonsillitis, 226.
Foreign bodies in larynx and trachea, 106, 107.
Fresh cadavers, infectiousness of, 191.
Funnel-shaped wounds, 41.

Gangrene, diabetic, amputation for, 63.
  of gut in herniotomy, 127.
Gastrostomy, 154.
Gauze, 14.
  corrosive-sublimate, 15.
  iodoformized, 15.
Gerster's urethrotome, 343.
Giant cell, in tuberculosis, 294.
Glandular tuberculosis, 299.
Gleet, 339.
Goitre, 111.
  tracheotomy for, 113.
Gonococcus, 331.
Gonorrhœa, 331.
  acute, 333.
  anterior, 333.
  chronic, 339.
  deep-seated, 336.
  posterior, 336.
Granular urethritis, 348.
Granulations, infection of, 198.
Green soap, 7.
Gross dirt, 192.
Gunshot fractures, 36.
Gunshot wounds, 35.

362  INDEX.

Habituation to septic influences, 197.
Hæmorrhage, in amputation of limb, 69.
 from nutrient artery of bone, 73.
 secondary, 8, 49, 72.
Hæmorrhoids, 167.
 Whitehead's operation for, 170.
Hæmostatic needle, 42.
Hahn's incision for exsection of knee-joint, 320.
Hair, aseptic management of, 92.
Hand, phlegmon of, 244.
Hernia, congenital, 135.
 radical operation for, 133.
 radical operation of, Czerny's, 136.
 radical operation of, Macewen's, 137.
 strangulated, 123.
Hernial sac, treatment of, 124.
Herniotomy, 121.
 castration in, 135.
 dressings after, 131.
 establishment of artificial anus in, 126.
 open section in, 125.
 radical, for congenital hernia, 136.
 undescended testicle in, 135.
Hilton-Roser's method of incising abscesses, 204, 208.
Hip-joint exsection, 315.
Hip-rest, Volkmann's, 131.
Hot applications, 201.
Hydrocele, 163.
 tapping of, 164.
Hygroma, proliferating, 301.
Hysterectomy, 151.

Immersion, continuous, 249.
Incision of knee-joint for suppuration, 256.
Incontinentia alvi, 288.
Infected wound, first change of dressings of, 27.
Infection by impure catgut, 8.
 of accidental wounds by careless probing, 30.
 portals of, 185.
Infectiousness of tonsillitis, 226.
Inflammation, 192.
Ingrown toe-nail, 253.
Inguinal glands, suppuration of, 259.
Injections, urethral, 335.
Injury, operative, to large arteries, 49.
Instrument-pouch, 26.
Internal urethrotomy, 342.
Interrupted suture, 44.
Intubation, 101.
Iodoform, 11.
 dusting box, 15.
 mania, 99.

Irrigation, 7.
 continuous, 249.
 of joints, 76.
 of the neck of the bladder, 336.
 of the urethra, 335.
Irritation, caloric, 190.
 by drainage-tubes, 47.
 chemical, 190.
 mechanical, 189.

Joint-exsection, 305.
Joints, after-treatment of excised, 307.
 hydrops of, 78.
 tuberculosis of, 305.

Kidney, surgical, 279.
Klotz's endoscope, 340.
Knee-joint, anchylosis of, in vicious position, 87, 323.
 exsection, technique of, 319.
 floating bodies of, 80.
 hydrops of, 78.
 suppuration of, 256.
 tuberculosis of, 312.
 vegetations of, 79.
Knock-knee, 86.

Langenbeck's rule for excision of tumors, 55.
Lange's position for nephrotomy, 277.
Laparotomy, exploratory, 139.
 treatment of navel in, 140.
 warm sponges and towels in, 145.
Laryngeal operations, 100.
Laryngofissure, 107.
Larynx, entrance of blood into, 96.
 extirpation of, 108.
 granuloma of, 106.
Laudable pus, 198.
Lawson Tait's aseptics, 120.
Lead-plate suture, Lister's, 46.
Leg, ulcer of, 253.
Leptothrix, 228.
Ligatures, 8.
Litholapaxy, Bigelow's, 175.
Little finger, suppuration of, 246.
Liver abscess, 276.
Lumbar abscess, 276.
 dressings, 278.
Lupus, 298.
Lymphadenitis, caseous, 299.
Lymphangitis, 199.
Lymph glands, cancerous, 54.
 suppurating, 52, 203, 234, 253, 259.
 tuberculous, 299.

Maas's operation for defects of integument, 94.
Malignant tumors, removal of capsule of, 54.
Mamma, amputation of, 113.
Mammary abscess, 237.
Mastitis, interstitial, 239.
Mastoid abscess, 235.
**Measles** and tuberculosis, **295**.
Meatotomy, 339.
Mechanical irritation, 189.
Mikulicz's operation, 326.
Moist dressings, 13.
Moss, 17.
Mouth, aseptics of, 96, **97**.
Mucous membranes, tuberculosis of, 299.
Multiple puncturing, Volkmann's, 200.
Myxœdema, 112.

**Nails, arrangement of, 87.**
 extraction of, after exsection of knee-joint, 325.
 for knee-joint exsection, 321.
Neck, caseous lymphadenitis of, 300.
 connective-tissue planes of, 222.
Neck of the bladder, cauterization of, 338.
 irrigation of, 336.
Necrosis of bone, 207.
 of bone by evaporation, 13
 of gut, 126.
Necrotomy, 208.
Needle hæmostatic, 42.
Needle-holder, 41
Nephrectomy, abdominal, 153.
 lumbar, 281.
Neuber's implantation, 214.

Œsophagus, retrograde catheterism **of, 154.**
 cancer of, **155.**
Olecranic bursa, 252.
Open exsiccative treatment **after plastic operations,** 92.
Open wound-treatment after amputation, 67, 69.
Operating bag, 25.
Oral cavity, 96.
Orchitis, tuberculous, 165, **299.**
Organization of blood-clot, **5, 6, 12.**
Osteomyelitis, acute **infectious, 205.**
Otis's urethrometer, **339.**
Ovarian tumors, 147.

Palliative excision of tumors, 53.
Palmar bursa, 246.
 suppuration, 247.
Passive movements, 75.

Passive movements after joint exsection, **307.**
Pasteboard splints, 311, 312.
**Patella,** suturing of fractured, 80.
Patterns for dressings, **14.**
Perineoplasty, 94.
Perinephritic abscess, **276.**
Peritonæum, denudation of, **120.**
 drainage of, 121.
 **great absorbing power of, 119.**
 **protection of, 141.**
 **stagnant blood-serum in, 120.**
Peritoneal irritation, saline purgatives in, 121.
 tuberculosis, **122.**
Peritonitis after abdominal section, 120.
Perityphlitic abscess, 260, 266.
 anterior parietal type, 268.
 mesocœliac type, 271
 posterior parietal type, 270.
 rectal type, **271.**
 Willard Parker's type, 267.
Pes valgus, **88.**
Phelps's operation, **88.**
Phlegmon, cause of, 183.
 **cutaneous, 199.**
 **development of, 191.**
 **retro-pharyngeal, 229.**
 **subcutaneous, 199.**
 **subfascial, 203**
 **treatment of, 198.**
Phlegmonous erysipelas, 204.
Plastic operations, 91.
 on extremities, Maas's method, 94.
Pleurisy, purulent, 240.
Pneumogastric nerve, cutting of, 60.
Pneumonia, from æther, 156, 157, 165.
Poulticing in phlegmon, consequences of, 194, 248.
Predisposition to tuberculosis, 295.
Prepatellary bursa, 256.
Prevesical abscess, 269.
Primary induration, syphilitic, 353.
 ulcer, syphilitic, 354.
Proctoplasty, 288.
Prostatic syringe, Ultzmann's, 338.
Protection of patient's body from wetting, 20.
 of surgeon's person, 20.
Pseudo-erysipelas, 290.
Ptomaïnes, 4, 192.
Puncture of abdominal tumors, 144.
Purse-string suture, 131.
Putrescence, bacilli of, 185.
Pyæmia, 196.
Pyonephrosis, 279.

Quadriceps, bursa of, 257, 320.
Quilled suture, 146.
Quinsy sore throat, 229.

Radical operation for hernia, 133.
　for hydrocele, 163.
　for varicocele, 164.
Rectal cancer, Kraske's operation for, 173.
　tampon-tube, 168, 170.
Rectum, aseptics of, 167.
　extirpation of, 171.
Retention of sweet serum in freshly healed wounds, 23.
Retractors, 40.
Retrograde catheterism of œsophagus, 154.
Retro-peritoneal abscess, 260.
Retro-pharyngeal abscess, 229.
Retro-visceral interspace, 222.
Revision for tuberculosis, 304.
Rose's position of head, 227.
Rubber tissue, 12, 13.

Sawdust, 16.
Saws, disinfection of, 65.
Scalp wounds, 32.
Scalpels, shape of, 37.
Schede's dressing, 12, 217.
Scrofula, 299.
Secondary hæmorrhage, 8, 49, 72.
　from perforation of popliteal vein and artery, 259.
Secondary suture, 46.
Sepsin, 4.
Sepsis, 3.
Septic embolism, 195.
Septic fever, 191, 192.
Sequestrum, diagnosis of, 210.
Shock after laparotomy, 153.
Shoulder-joint, curing habitual dislocation of, 8.
　curing irreducible dislocation of, 83.
　exsection of, 308.
Sigmund's urethral syringe, 355.
Silk, 9.
Silk-worm gut, 9.
Sinuses after operations, their true cause, 47.
Soiled accidental wounds, 31.
Solutions for disinfection, 10.
Small wounds the school for asepticism, 23.
Spanish windlass, 30.
Splints of pasteboard, 311.
Sponge packing, disadvantage of, 43.
Sponges, 8.
　in laparotomy, 141.

Spray-apparatus, 141.
Staphylococcus, 183, 184.
Starcke's irrigation-tube, 249, 250.
Sterilization, chemical, 7.
Strangulated hernia, 123.
Strangulating hernial band, 124.
Streptococcus, 184, 289.
Stricture, urethral, 339.
　incipient, 341.
　permanent or cicatricial, 341, 342.
Stump, management of, after amputation, 74.
Styptic solutions, abuse of, 244.
Submaxillary capsule, 222, 232.
Suction lead, 46.
Suppuration, cause of, 183.
　cutaneous and subcutaneous, 199.
　spread of, 193.
　superficial, 199.
Suppurations on the face, 223.
　of the fauces, 225.
Surgical kidney, 279.
Suspension, vertical, in phlegmon of hand or arm, 249.
Suture, abdominal, 146.
　of anal fistula, 286.
　secondary, 46.
Sutures, 8, 44.
　removal of, 22.
Suturing fractured patella, 80.
Syphilitic external lesions, 353.
Syphilitic ulcer, caustic treatment of, 357.
　primary, 356.
　moist treatment of, 356.
　treatment by the actual cautery of, 358.
Syringing of freshly healed wounds reprehensible, 23.

T-bandage, 169, 170.
T-splint, Volkmann's, 77.
T-tube, 178.
Tampon cannula, Gerster's, 97, 99.
Tampon-tube, rectal, 168, 170.
Temporary care of accidental wounds, 29.
Tendinous sheaths, tuberculosis of, 301.
Teratoma of occiput, 106.
Testis, necrosis of, 166.
　removal of, 165.
Thiersch's solution, 10.
　spindle-apparatus, 43.
Thomas's operation for mammary tumors, 114.
Thoracic fistula, 242.
Thrombosis and embolism after amputation of breast, 118.

## INDEX.

Thrombosis of pulmonary artery, 118, 141.
  of innominate and axillary veins after breast-amputation, 118.
Thrombosis, septic, 193.
Through-drainage, 47.
Thumb, suppuration of, 246.
Toilet, abdominal, 145.
Tongue, 97
Tonsillar abscess, 229.
Tonsillitis, 225.
Tonsils, cauterization of, 227.
Tracheotomy, accidents in and after, 102–104.
  apnœa after, 104.
  for removal of foreign bodies, 106, 107.
  for goitre, 113.
  avoidance of hæmorrhage in, 102.
  inferior, 103.
  for laryngeal tumors, 105.
  preliminary, 97, 100.
  statistics of, 104.
  superior, 102.
Trendelenburg's T-shaped drainage-tube, 178.
Trocars, disinfection of, 7, 76.
Tuberculosis, 293.
  of ankle-joint, 325.
  of bone, 303.
  cutaneous, 298.
  dissemination of, 295, 318.
  general treatment of, 297.
  of joints, 305.
  of knee-joint, 319.
  local treatment of, 298.
  of lymphatic glands, 299.
  of mucous membranes, 299.
  of peritonæum, 122.
  prevention of, 299.
  and pyogenic infection, combination of, 297.
  of tendinous sheaths, 301.
  of testicle, 165, 299.
Tuberculous infection, direct, 296.
  through the lungs, 295.
Tumors, extirpation of, 52, 55.
  treatment of pedicle of non-ovarian, 56.

Ulcer of leg, 255.
Ultzmann's method of irrigating the neck of the bladder, 336.
  prostatic syringe, 338.
  test, 334.
Uræmia from æther, 121.
Urethral endoscope, 340.
  injections, 335.
  irrigation, 335.
  stricture, 339, 341, 342.
  syringe, Sigmund's, 335.
  tuberculosis, 299.
  vegetations, 348.
Urethritis, 333.
  granular, 348.
Urethrometer, Otis's, 339.
Urethroplasty, 93.
Urethrotome, Gerster's, 343.
Urethrotomy, external, 346.
  internal, 342.
Uterine appendages, removal of, 150.
  stump, 152.

Van Lennep's rubber rings for enterorrhaphy, 162.
Varicocele, 164.
Vein, ligature of axillary, 59.
  femoral, 57–59.
  jugular, 60.
Veins, exsection of, 58.
  lateral closure of, 56.
  management of, in operative wounds, 43, 57, 69, 72.
Venereal vegetations, urethral, 348.
Vermiform appendix, 260.
Vertical suspension of limbs, 249.
Vesical tuberculosis, 299.
Vessels needed for operating, 18.
Volkmann's hip-rest, 131.
  multiple puncturing, 200.
  suspension splint, 249.
  T-splint, 77.

White swelling, 303.
Wounds, funnel-shaped, 41.
  bottle-shaped, 41.


THE END.

www.ingramcontent.com/pod-product-compliance
Lightning Source LLC
Chambersburg PA
CBHW032024220426
43664CB00006B/361